HIGH PERFORMANCE COMPUTING SYSTEMS AND APPLICATIONS

edited by

Jonathan Schaeffer
University of Alberta, Canada

SPRINGER SCIENCE+BUSINESS MEDIA, LLC

 Electronic Services <http://www.wkap.nl>

Library of Congress Cataloging-in-Publication Data

A C.I.P. Catalogue record for this book is available
from the Library of Congress.

ISBN 978-1-4613-7567-8   ISBN 978-1-4615-5611-4 (eBook)
DOI 10.1007/978-1-4615-5611-4

Printed on acid-free paper.

HIGH PERFORMANCE COMPUTING SYSTEMS AND APPLICATIONS

THE KLUWER INTERNATIONAL SERIES
IN ENGINEERING AND COMPUTER SCIENCE

Contents

Preface

This book contains the proceedings of HPCS'98, the 12th annual international symposium on High Performance Computing Systems and Applications. The HPCS series originally started in 1987 as a two-day conference (Supercomputing Symposium '87) at the University of Calgary. By bringing Canadian researchers with a common interest in supercomputing together, the intent was to create an academic, industrial and government environment for fostering the growth of computational research. In 1991 the symposium broadened its scope to include international papers, and changed its name to High Performance Computing Systems (HPCS).

There are a plethora of conferences that advertise using the keywords parallel, distributed and/or high-performance computing. With so much competition, it is imperative that an event be attractive to prospective authors and attendees. Towards this goal, several things were tried to strengthen the conference:

1. Changing the program emphasis to high performance computing *and applications*. This will allow the HPC system developers and the end users to get together and interact.

2. Offering an impressive lineup of international high-profile keynote and invited speakers.

3. Increasing the refereeing standards. Only 38% of submitted papers were accepted.

4. Offering an attractive publication venue. Kluwer Academic Press kindly agreed to publish the conference proceedings.

5. Adding a poster session, allowing students to showcase their work.

6. Locating the conference at a unique venue. I doubt many of the attendees will soon forget the novelty of West Edmonton Mall.

Feedback suggests that all these changes enhanced the conference. It is hoped that next year's conference can build on our success.

This book contains the keynote, invited and refereed papers of HPCS'98. For the keynote speakers, submitting a paper was optional. I would like to express my sincere thanks to John Gustafson and David Bailey for kindly turning their excellent presentations into papers for this book. The annual C3 meeting was help at HPCS'98; Brian Unger (University of Calgary) and Sid Karin (San Diego Supercomputer Center) gave overviews of the current state of HPC in Canada and the United States. Phil Tannenbaum (NEC) gave a light-hearted talk on the correlation of computer price, performance and weight. Sadly, Phil did not have time to turn his talk into a paper for this book.

HPCS'98 gratefully acknowledges the sponsorship of the following companies (in alphabetical order): C3, Canarie, Digital Equipment Corporation, HNSX/NEC, IBM, IEEE Canada, Silicon Graphics, Sun Microsystems, Super*Can, and Telus. The University of Alberta provided support, as well as the Faculty of Science, Computing and Networking Services, and the Departments of Chemistry, Computing Science and Physics.

I would also like to thank the program committee, which included: Hamid Arabnia (University of Georgia, U.S.A.), Henri Bal (Vrije Universiteit, The Netherlands), Luc Bauwens (University of Calgary, Canada), Virendra Bhavsar (University of New Brunswick, Canada), Brian d'Auriol (University of Akron, U.S.A.), Geoffrey Fox (Syracuse University, U.S.A.), Guang Gao (University of Delaware, U.S.A.), Peter Graham (University of Manitoba, Canada), Qian Ping Gu (University of Aizu, Japan), Laurie Hendren (McGill University, Canada), Robert Ito (University of British Columbia, Canada), Wayne Karpoff (Myrias Computer Technologies Corp., Canada), Mariusz Klobukowski (University of Alberta, Canada), Thomas Kunz (Carleton University, Canada), Christian Lengauer (University of Passau, Germany), Keqin Li (SUNY, New Paltz, U.S.A.), Michael Quinn (Oregon State University, U.S.A.), Alexander Reinefeld (Paderborn Center for Parallel Computing, Germany), John Samson (University of Alberta, Canada), Ken Sevcik (University of Toronto, Canada), Ajit Singh (University of Waterloo, Canada), Rick Sydora (University of Alberta, Canada), Duane Szafron (University of Alberta, Canada), Ron Unrau (University of Alberta, Canada), and David Wishart (University of Alberta, Canada).

Organizing a large conference like this is not possible without the help of many people. In particular, I would like to thank: Sylvia Boyetchko, Michael Byrne, George Carmichael, Sunrose Ko, Marc La France, Nadine Leenders, Wally Lysz, Steve MacDonald, Diego Novillo, Janet Service, Ron Senda, John Samson, Denise Thornton, Ron Unrau, and Norris Weimer. Steve MacDonald and Diego Novillo need to be singled out for their enormous commitment to the conference. I would especially like to thank the organizers of the applications streams: Mariusz Klobukowski (chemistry), Rick Sydora (physics) and David Wishart (biology).

Finally, we appreciate the support of Scott Delman from Kluwer Academic Press for giving us the opportunity to publish these proceedings.

JONATHAN SCHAEFFER
jonathan@cs.ualberta.ca

This book is dedicated to the C3 initiative. I hope that HPCS'98, in some small way, will help make this vision a reality.

I Keynote Presentations

1 MAKING COMPUTER DESIGN A SCIENCE INSTEAD OF AN ART

John L. Gustafson

Ames Laboratory, U.S.A.

gus@ameslab.gov

Abstract: Computer architecture has evolved much like the architecture of buildings: we try some things, they either work or do not work very well, and a few established experts develop a "feel" for the choices to be made. As Connection Machine architect Danny Hillis has complained, computer science does not teach us how to build better computers—at least, not the way physics teaches us how to build better engines or chemistry teaches us how to create better polymers.

A recently developed tool may be a step toward deterministic, automatic computer design. It creates two profiles that are functions of time: one for the application demands and one for the hardware speed. The matching of these profiles has proved a remarkably accurate predictor of application-architecture fit. Through analytical models for the hardware profile, you can obtain precise answers to questions about optimum sizes for installed memory, caches, and even the numerical precision (word size) for a given clock rate. Perhaps most relevant of all, it provides a quick way to detect which applications will work well on parallel computers and which should be left on serial computers. We may be returning to an era in which computer design is a science instead of an art.

Keywords: computer design, computer architecture, application map, computer performance, HINT.

1.1 WHY COMPUTER SCIENCE IS NO GOOD

The section title is from Danny Hillis' 1986 book *The Connection Machine* (Hillis, 1986). Computer science is no good, he explains, because it does not tell us how to build a better computer. "...Quantum chromodynamics is not much use in designing bridges. Computer science is not much use in designing computers... Our current models seem to emphasize the wrong details."

It is now twelve years later. There have been a few improvements in the art of computer design, but it is still an art. We cannot present an application, then write a program that will design optimal computer hardware for that application automatically. Our approach is to try many things, to let the marketplace make decisions, and to trust the experienced opinions of a few practitioners of computer design who have acquired a feel for what will or will not work well.

If you design an airplane, science tells us a great deal about the choices you must make. An understanding of the physics of airflow lets you automatically optimize the shape of wings for better lift or fuel efficiency. If you design a building, there are structural analysis programs that reveal behavior of the building during an earthquake. They can even optimize the structure automatically. Why is this sort of thing not possible for computer design?

Parts of computer design are amenable to simulation. Network load is a good example. You can simulate a proposed network on a computer and refine it to minimize "hot spots" and maximize throughput. But how do you select the cache size, or the width of the memory bus? What is the importance of short cycle counts for floating-point operations? From personal experience, I would say designers still make choices artistically. The people who make the choices often know what they are doing, but not well enough to be able to tell a computer how to do it. Hence, it is not a science.

My thesis is this: computer science does not tell us how to build better computers because it has so far failed to explain application performance as a function of hardware features. Moreover, it has failed because it seems hopelessly complicated to analyze how the performance of a range of computer applications depends on specific hardware choices. All you can do is make changes and see what happens.

A recently developed tool may be a step toward deterministic, automatic computer design. That tool, HINT, creates two profiles that are functions of time: one for the application demands and one for the hardware speed. The matching of these profiles has proved a remarkably accurate predictor of application-architecture fit. Through analytical models for the hardware profile, you can obtain precise answers to questions about optimum sizes for installed memory, caches, and even the numerical precision (word size) for a given clock rate. Perhaps most relevant of all, it provides a quick way to detect which applications will work well on parallel computers and which should be left on serial computers. We may be returning to an era in which computer design is a science instead of an art.

This paper presents the motivation for HINT and shows how it can be an accurate predictor of computer performance. More information on HINT can be found at http://www.scl.ameslab.gov/Projects/HINT/HINThomepage.html.

1.2 BENCHMARKS AND COMPUTER DESIGN

For many of us, "benchmark" is a dirty word. It reminds us of misleading information and pandering advertisements. It summarizes the fact that you must often use a simplistic estimate of what a computer can do, when what you really want to know is how it will run the real workload you have in mind.

I have seen people say they do not believe in benchmarks, and then in the same breath start talking about megahertz or peak megaflops. Those are benchmarks too! They are the most simplistic benchmarks of all: machine specifications. Every computer advertisement touts machine specifications in the hope that potential buyers will associate those specifications with higher performance on their applications. More-over, many potential buyers do, blindly.

If benchmarks really worked as performance predictors, we would indeed have a science. The procedure would work like this:

1. Measure some essential aspect of real applications.

2. Measure some essential feature of computers.

3. Show, by experiment and theory, correlation between the measurements in steps 1 and 2.

4. Model the cost of varying the computer features (or altering the application program).

5. Vary computer and software designs to optimize performance divided by cost.

In the first few decades of electronic computing, we thought we had the first two steps solved: use arithmetic operation count as the essential aspect of the application, and use arithmetic operations per second as the essential feature of the hardware. This seemed sufficient because the arithmetic was the hard part. It was the part that cost the money, and the part that took up all the execution time. This has changed.

1.3 MOORE'S LAW DOES NOT APPLY TO LATENCY

Gordon Moore's observation that chip density doubles every 18 months, commonly known as "Moore's law," is often misunderstood to mean that every aspect of computer performance doubles every 18 months. If this were true, we would not need computer architects. We would be able to sit back and watch technology carry our application programs to loftier levels of performance year after year. While it seems to be true that processor speed (by some measure) also doubles roughly every 18 months, you cannot say the same for memory speed. Bandwidth improves more slowly, and latency much more slowly. It can even happen that latency gets worse, because larger memories take more time to address. There was a crossover in the 1970s regarding which was more expensive: memory references or arithmetic. This is illustrated in Figure 1.1.

The 1970 ILLIAC IV took one microsecond either to fetch a word from memory or do a floating-point operation. By 1976, the CRAY-1 demonstrated scalar floating-point arithmetic taking six or seven clock cycles... but a memory reference took 11 to 13 clock cycles. This phenomenon was not restricted to supercomputers. Array processors of that era, which were about 1,000 times cheaper than the Cray computers, also took fewer clock cycles for a floating-point operation than for a memory fetch. On current generation microprocessors, most arithmetic operations execute in one or two clock cycles, but fetching data from a memory location (not in cache) can take 50 to 100 clock cycles.

Call this exchange of the relative importance of the arithmetic speed and memory speed "The Great Crossover" in computer design. At about the same time, the total cost of the wire connections began to overtake the cost of gate transistors, and the total time spent moving signals began to overtake the total switching time. All are symptoms of the same phenomenon.

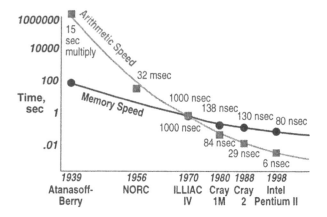

Figure 1.1   The crossover of memory and arithmetic performance.

When we experienced "The Great Crossover," we continued to use computer design principles that had served us well before the change. Since those methods now frequently lead us astray, computer design has become mysterious and artistic instead of scientific and deterministic. This is what prompted Hillis to write his astute complaint. If we can replace rules used before the crossover with ones that recognize the inversion in costs, we can make computer design a science again. By "computer design," I mean the process of answering the question, "Which computer designs are best for which applications?"

1.4   THE ORIGINAL COMPUTER DESIGN GUIDELINE

Here is, in short, the guideline that worked well for the first three decades of electronic computer design, or more precisely, for algorithm-architecture fit:

1. Measure the number of arithmetic operations in the application. The smaller the number, the faster it will run.

2. Measure how many arithmetic operations per second the computer design can produce. The higher the number, the faster it will run.

There is plenty of evidence that this guideline no longer works. Higher "Peak Advertised Performance" numbers derived from flops/sec correlate *inversely* with application performance. Figure 1.2 shows this for four 1990-vintage supercomputers.

One reason for this is that a high peak flops/sec rating sells computers. Therefore, in allocating component costs, a computer designer has a choice between making the memory faster or the arithmetic faster. It is now much easier to make the arithmetic faster, resulting in unbalanced designs with impressive specifications.

Many people are still having trouble with the paradigm shift, and refuse to believe the mounting evidence that neither the arithmetic speed nor the algorithmic arithmetic complexity matter very much any more. Hence we see government programs that place the achievement of teraflops or petaflops performance paramount. We see aca-

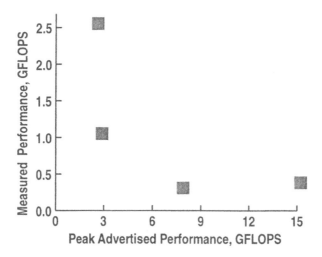

Figure 1.2 The fallacy of using "peak flops/sec".

demic papers based on the PRAM model, which assumes that memory access is zero cost. Large research laboratories express indignation when they measure the flops/sec ratings of an application and discover they are getting less than 10% of "peak." Millions of personal computer buyers now believe that the megahertz rating of the processor determines the performance of the computer (ignoring the speed of memory, cache, and disk storage).

1.5 THE DIFFICULTY OF FINDING SOMETHING BETTER

Part of the reason we cling to this outdated guideline is that we have not had a viable alternative. Computer science has not provided us with a way to substitute memory references for flops in algorithm analysis. Memory hardware has become so complicated that it is seldom a simple thing to state its speed in simple and usable terms. If we simply tried to translate the previous guideline into memory accesses, we arrive at:

1. Measure the number of fetches and stores in the application. The smaller the number, the faster it will run.

2. Measure the memory bandwidth of the computer design. The higher the number, the faster it will run.

The problem, of course, is that it is very difficult to "measure the number of fetches and stores in the application" because memory is hierarchical. The application description, even if done at the level of assembly language, does not reveal which data references will be in registers, primary cache, secondary cache, *etc.* It is also very difficult to "measure the memory bandwidth of the computer design." Besides the caches, questions of memory interleave, bank recovery time, and pipelining defy simple analysis. In measuring arithmetic, you can look at how many clock cycles are

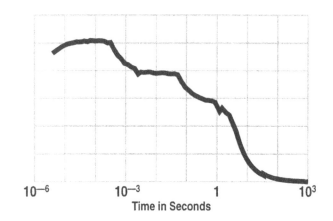

10⁻⁶   10⁻³   1   10³
Time in Seconds

Figure 1.3   General hardware behavior over a large dynamic range.

required for a square root or an integer add (using register variables so that memory is not an issue). In addition, bandwidth is not enough. We need to know about latency.

Walter Tichy writes that (Tichy, 1998): "... computer science cannot accurately predict the behavior of algorithms on typical problems or on computers with storage hierarchies. We need better algorithm theories, and we need to test them in the lab."

The (Hockney and Jesshope, 1988) model tells us to look for an asymptotic peak speed r_∞ and a measure of acceleration, "$N_{1/2}$" (the problem size that results in half the asymptotic speed). This assumes, like so many historical models, that memory is flat. If you actually try to measure one of these performance curves on a real computer, and vary the size over a large dynamic range, what you see is performance that *drops* in each memory regime, as shown in Figure 1.3.

The traditional model is too inaccurate to be useful. Because we have lost our simple guideline based on counting arithmetic work, we have had to retreat to a lower form of reasoning: trial and error. Workstations and personal computers are available with varying amounts of cache, for example, and you can do experiments by running applications to see if the benefit of increased cache is worth the expense. However, trial-and-error cannot provide *predictability*. If you ask a hardware engineer how fast a new computer design will run an application, the engineer will reply that the only way to know is to port the application and time it. If you ask a programmer how fast a new application will run on a computer, the programmer will reply that the only way to know is to port the application and time it. We have lost predictability.

1.6   MEASURING MEMORY PERFORMANCE: A CAR TRANSPORTATION ANALOGY

Since most of us have a feel for human transportation issues, perhaps we can take guidance from an analogy. In U.S. automobile advertisements, especially for high-performance cars, you sometimes see claims like "0 to 60 mph in 6.4 seconds." You also might see, in technical brochures, the amount of time to go a quarter mile or

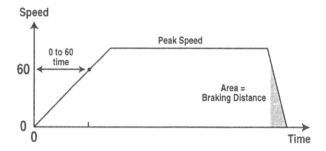

Figure 1.4 Idealized view of car performance model.

some other distance, and the peak speed of the car. These are crude measures of transportation hardware analogous to simple computer benchmarks.

When it comes to measures of the *application* of the car, the car industry makes distinctions between "highway" and "in-town" driving. Everyone knows that high-performance cars lose much of their nominal advantage when faced with many short travel segments instead of long stretches of road.

So to predict "architecture-application fit", how would you approach the problem of estimating how long a particular car would take to travel two miles on an idealized road? (Assume the car starts the trip at rest and ends the trip at rest.) You could divide the distance travelled by the peak speed. This would overestimate the performance because the car must accelerate to the peak speed. You could use the "0 to 60 mph" rating to figure the acceleration, and perhaps a braking distance for the deceleration at the end of the trip. The idealized mental picture, derived from the benchmarks, is something similar to that shown in Figure 1.4.

Now you can adjust the time spent at the plateau to make the distance traveled (the area under the curve) equal two miles, and this will estimate the total time. This is as simplistic as the Hockney and Jesshope model is for computer performance: a startup cost and a peak rate are the only parameters.

For a limited range of straight distances, you can use Figure 1.4 to estimate travel time. Simply integrate the speed curve until you reach the distance in question. Instead, suppose I pose the much more difficult question: "How long will it take to get from home to work in that car, if there is no traffic?" Perhaps the route to work looks something like Figure 1.5. Instead of having to have a map of the entire route, you only need to know the distribution of travel segments, as shown in Figure 1.6.

After a little reflection, you realize how inadequate the performance data is for answering the question, "How long will it take to drive this route with this car?" What you really want is something like what is shown in Figure 1.7. This accounts for the way the car changes gears (short time scale), and the need to refuel or rest (large time scale). These result in changes in performance for different time regimes. "Speed" here means total distance traveled divided by total time, not instantaneous speed. This has the effect of smoothing the curve as a moving average. You could use this graph to estimate how long the car will take to go 50 feet, 2 miles, or 100,000 miles in a single path. You can determine travel time for any trip by applying the histogram of

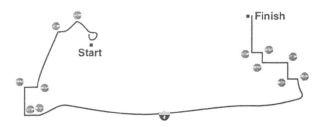

Figure 1.5   Transportation task schematic.

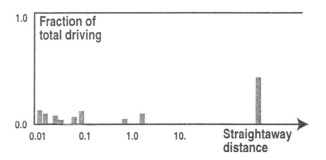

Figure 1.6   Histogram of transportation "subtasks."

the total trip (Figure 1.6) and summing the time taken on each. It would be tedious to do, and would ignore interfering factors like other users of the road and the random time spent waiting at stoplights, but it would answer design questions scientifically. An automotive engineer could ask, for example, whether it would be better to have three or four forward gears. A driver could ask at what point walking is faster than driving at the low end, or flying in a commercial airliner is faster than driving at the high end.

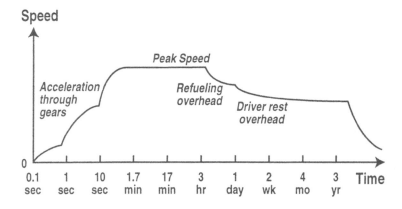

Figure 1.7   Realistic graph of car performance.

1.7 APPLYING THE TRANSPORTATION MODEL TO COMPUTING

The transportation model shows how performance questions can be factored into a graphical task description and a graphical hardware description. For computers, we need an "acceleration" curve for computer hardware that describes its speed as a function of time (or problem size), and a "map" of the target applications that shows the distribution of sizes of its component tasks. If we can measure those things reliably, we can approach computer design as scientifically as a problem in Newtonian physics.

Obtaining an acceleration curve is not as easy as it sounds. The naïve approach of timing a loop of increasing length is not very satisfactory. It tends to be easier to cache data in contrived examples than in real applications. It also is difficult to define the "speed" in absolute terms as opposed to hardware-dependent terms (like instructions per second or operations per second). Since 1993, Quinn Snell and I refined a tool called HINT (Hierarchical INTegration) that has a fractal memory access pattern and defines performance in terms of answer quality instead of computer activity (Gustafson and Snell, 1995). The hierarchical integration method yields answer quality that increases linearly with each iteration, and requires linearly more memory with each iteration. Quality improvement per second (QUIPS) serves as a measure of computing performance that works on any digital computer. HINT can use either integer or floating-point arithmetic of any precision in any number base. (It rewards higher precision since it improves the answer quality obtainable.)

HINT graphs can use either storage or time as the horizontal axis. Storage tends to be more informative when analyzing programs with known memory demands. Time is a better axis for scalability, since computers of widely disparate performance will still occupy roughly the same range of execution times. Figure 1.8 shows examples of HINT graphs based on storage. Figure 1.9 shows HINT graphs as a function of execution time, and uses a log scale for the performance to view a wide range of computer speeds.

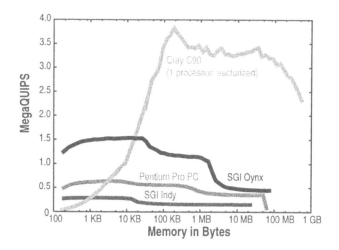

Figure 1.8 HINT graphs as a function of storage used.

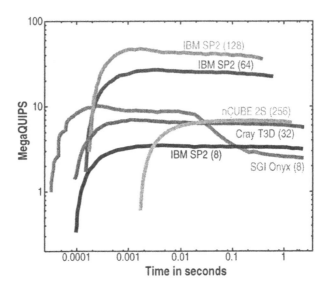

Figure 1.9   HINT graphs as a function of time.

1.8   THE APPLICATION MAP

How can you obtain the equivalent of the transportation histogram of road segments? There are dozens of papers on instrumenting code to discover memory behavior, and this is certainly one approach. These instrumented codes usually ignore the effect of the operating system. Operating systems interrupt the instruction stream and seize control of the memory bus for a variety of reasons, such as servicing memory paging or polling timeshared processes. These interrupts have a large effect on the state of the caches, and instrumented code does not bring this effect out for analysis.

You can also use hardware analyzers to find the *Application Map* (AppMap) for a given program. This is easiest to envision for a computer with a single memory bus. You would acquire several gigabytes of data for a few seconds of execution by recording every address that appears on the memory bus. If we treat this as a signal, a Fourier transform of the signal gives a "power spectrum" of the activity in the computer. It will show the fraction of time spent from small loops (that use registers or primary cache) to sweeping operations that traverse the main memory and might not loop at all. We, as well as other researchers, are engaged in this activity, but we see it as a confirmation experiment and not as a primary way to obtain AppMap information.

Instead, we have discovered a purely empirical approach that shows considerable promise (Gustafson and Todi, 1998). We find the statistical correlation between each point on the HINT graph and the performance of the application in question, over a large set of measurements from different computers. This yields correlation as a function of the memory hierarchy. Figure 1.10 shows a representative AppMap obtained in this manner.

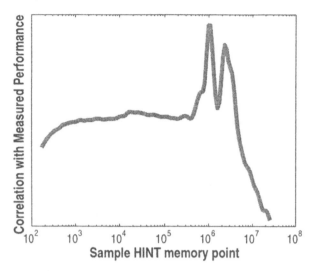

Figure 1.10   AppMap for Hydro2d application.

Currently we have about 30 different architectures in our database, ranging widely in type of parallelism, degree of parallelism, memory hierarchy, and integer versus floating-point emphasis. We have found AppMaps for dozens of programs ranging from small benchmarks to full applications. When we use the AppMap to predict performance, it generally ranks the machines from slowest to fastest with at most one or two interchanges. Since the HINT graphs cross one another, it is unlikely this would be possible unless the HINT-AppMap pairing was identifying the correct features for performance prediction.

Since HINT graphs often show a three-fold variance in speed over the memory hierarchies, you would expect ordinary benchmarks to incorrectly predict performance by a factor of up to three. This is our experience. However, when we use HINT-AppMap pairs, we typically obtain predictions with a worse case error of 15%. This is less than the experimental variation in much of our experimental data.

We have tested these ideas using over 5,000 performance measurements, and confirmed that the combination is an excellent performance predictor. The ability to predict the ranking correctly and the linearity of the fitting of performance data is strong evidence that we have found a substitute for flops/sec that works. We also have analytical models of HINT that allow a computer designer to examine the effect on a HINT graph of changes to a design (Snell and Gustafson, 1996). This completes the refurbishing of the approach mentioned at the beginning of this paper:

1. measure the AppMaps of the applications in question,

2. use HINT to measure the hierarchical memory performance of a computer,

3. verify that the HINT-AppMap combination predicts actual performance by measuring statistical correlation and maximum deviation from linearity,

4. model the cost of varying the computer features, and

5. vary proposed computer designs with Analytical HINT, AHINT (Snell and Gustafson, 1996), to optimize performance (or performance divided by cost).

1.9 CONCLUSIONS

Advances in computer technology have been dramatic but uneven. Moore's law does not apply to memory speed. Consequently, what used to be a reasonably scientific and quantitative approach to computer design is failing us. We lost our ability to predict performance because data motion became dominant over arithmetic. A possible solution is to analyze computer hardware speed as a function of problem (memory) size, and analyze application requirements as a distribution function of the memory sizes of component operations. Since experiments show that the combination predicts performance, we may have recovered a general model of application-architecture fit. Hence, we may be able to automate the process of optimizing a computer hardware design to run a complex mix of applications.

Acknowledgments

This work is supported by the Applied Mathematical Sciences Program of the Ames Laboratory–U.S. Department of Energy under contract number W-7405-ENG-82.

References

Gustafson, J. and Snell, Q. (1995). HINT: A new way to measure computer performance. *29th Hawaii International Conference on System Sciences*, 2:392–401.

Gustafson, J. and Todi, R. (1998). Benchmarks as a subset of the performance spectrum. *31st Hawaii International Conference on System Sciences*. On CD-ROM.

Hillis, D. (1986). *The Connection Machine*. MIT Press.

Hockney, R. and Jesshope, C. (1988). *Parallel Computers 2: Architecture, Programming, and Algorithms*. Adam Hilger.

Snell, Q. and Gustafson, J. (1996). An analytical model of the HINT performance metric. *Supercomputing'96*. On CD-ROM.

Tichy, W. (1998). Should computer scientists experiment more? *IEEE Computer*, 31(5).

2 SCALABLE SHARED MEMORY AND SHARED MEMORY CLUSTERING

Marc Snir

IBM T. J. Watson Research Center, U.S.A.

snir@watson.ibm.com

Abstract: Users want tightly coupled, shared everything systems that are highly available, scalable, have a single system image, and are cheap. Unfortunately for computer customers (and fortunately for computer engineering research), these goals conflict. The talk will discuss possible design points and trade-offs that can be pursued in an attempt to reconcile these conflicting goals.

We shall survey ongoing research on a scalable shared memory architecture at the IBM T.J. Watson Research Center which explores one such interesting design point, namely shared I/O and shared memory support in an integrated, scalable, distributed OS SP-like system.

3 HIGH-PERFORMANCE COMPUTATIONAL GRIDS

Ian Foster

Mathematics and Computer Science Division
Argonne National Laboratory, U.S.A.

foster@mcs.anl.gov

Abstract:
A computational grid, like its namesake the electric power grid, provides quasi-ubiquitous access to capabilities that cannot easily be replicated at network endpoints. In the case of a high-performance grid, these capabilities include both high-performance devices (networks, computers, storage devices, visualization devices, *etc.*) and unique services that depend on these devices, such as smart instruments, collaborative design spaces, and metacomputations. In my presentation, I review some of the technical challenges that arise when we attempt to build such grids: in particular, the frequent need to meet stringent end-to-end performance requirements despite a lack of global knowledge or control. I then introduce the Globus project, a multi-institutional effort that is developing key grid infrastructure components, for authentication, resource location/allocation, process management, communication, and data management. I conclude with a discussion of our experiences developing testbeds and applications based on Globus components.

For further information on these topics, the reader is referred to the recent book on computational grids (Foster and Kesselman, 1998a), which contains detailed descriptions of grid applications, program development tools, and technologies. Recent articles in a special issue of the Communications of the ACM also contain relevant material (*e.g.* (Stevens *et al.*, 1997)). For more information on Globus, see (Foster and Kesselman, 1998b) or visit www.globus.org.

Keywords: computational grid, high-performance distributed computing, Globus, GUST.

References

Foster, I. and Kesselman, C., editors (1998a). *Computational Grids: The Future of High-Performance Distributed Computing*. Morgan Kaufmann Publishers.

Foster, I. and Kesselman, C. (1998b). The Globus project: A progress report. In *Proceedings of the Heterogeneous Computing Workshop*. To appear.

Stevens, R., Woodward, P., DeFanti, T., and Catlett, C. (1997). From the I-WAY to the National Technology Grid. *Communications of the ACM*, 40(11):50–61.

4 HIGH PERFORMANCE COMPUTING AND COMPUTATIONAL CHEMISTRY IN CANADA—QUO VADIS

Dennis Salahub

Department of Chemistry,
Université de Montréal, Canada

salahub@ERE.UMontreal.CA

Abstract: The talk will illustrate the kinds of biomolecular and materials modeling problems that can be tackled with modern methodology and with available computational and human resources. Examples will be drawn primarily from our own work, because I know them best, but I will also attempt to give a representative sampling from several leading Canadian computational chemists. I will then try to extrapolate into the new millennium (not such a feat anymore...) according to various scenarios for HPC development in Canada.

5 SCALABLE SHARED-MEMORY MULTIPROCESSING: THE FUTURE OF HIGH-PERFORMANCE COMPUTING

Daniel Lenoski

Silicon Graphics Computer Systems, U.S.A.

lenoski@sgi.com

Abstract: Today's options for high-performance computing include message-passing MPPs and clusters, shared-memory multiprocessors, and parallel vector processors. Shared-memory multiprocessors provide the most general-purpose programming model, but generally suffer from limited scalability. New multiprocessor cache coherence protocols together with high-performance switched interconnects enable a new class of machines that scales the shared-memory paradigm.

This talk introduces the scalable shared-memory multiprocessor (SSMP) and SGI's new Origin servers based on this technology. The Origin machines employ an advanced NUMA architecture that allow a single system to grow from a cost-effective desk-side uniprocessor to a supercomputer with hundreds of processors and hundreds of gigabytes of shared-memory. Nodes within the system provide a peak of 50 GB/sec of memory bandwidth and are connected by a fat hypercube network that has a bisection bandwidth of 25 GB/sec. These new capabilities also enable new levels of I/O performance including file servers with sustained bandwidths in excess of 500 MByte/sec and networking interfaces such as Gigabyte System Network (GSN) that can source and sink data at up to 800 MB/sec full-duplex.

We conclude the talk with a look at the future of high-performance computing and predict the merger of HPC computing models onto scalable SMP systems and clusters of these SSMP systems.

6 CHALLENGES OF FUTURE HIGH-END COMPUTING

David H. Bailey

NERSC,
Lawrence Berkeley Laboratory, U.S.A.

dhb@nersc.gov

Abstract: With the recent achievement of one teraflop/s (10^{12} flop/s) on the ASCI Red system at Sandia National Laboratory in the U.S., many have asked what lies ahead for high-end scientific computing. The next major milestone is one petaflop/s (10^{15} flop/s). Systems capable of this level of performance may be available by 2010, assuming key technologies continue to advance at currently projected rates.

This paper gives an overview of some of the challenges that need to be addressed to achieve this goal. One key issue is the question of whether or not the algorithms and applications anticipated for these systems possess the enormous levels of concurrency that will be required, and whether or not they possess the requisite data locality. In any event, new algorithms and implementation techniques may be required to effectively utilize these future systems. New approaches may also be required to program applications, analyze performance and manage systems of this scale.

Keywords: petaflops, teraflops, high-end computers, performance.

6.1 INTRODUCTION

In December 1996, a sustained rate of one teraflop/s (also written as 10^{12} floating-point operations per second) was achieved on the Linpack benchmark, using the ASCI Red system at Sandia National Laboratory in New Mexico, U.S.A. While no one has yet demonstrated a sustained performance rate exceeding one teraflop/s on a real scientific or engineering application, it is expected that this also will be achieved soon.

The next major milestone is a performance rate of one petaflop/s (also written as 10^{15} floating-point operations per second). It should be emphasized that we could just as well use the term "peta-ops", since it appears that large scientific systems will be required to perform intensive integer and logical computation in addition to floating-point operations, and completely non-floating-point applications are likely to be important as well. In addition to prodigiously high computational performance, such systems must of necessity feature very large main memories, between ten terabytes (10^{13} bytes) and one petabyte (10^{15} bytes) depending on application, as well as commensurate I/O bandwidth and huge mass storage facilities. The current consensus of scientists who have performed initial studies in this field is that petaflops systems may be feasible by the year 2010, assuming that certain key technologies continue to progress at current rates (Sterling and Foster, 1996).

If one were to construct a petaflops system today, even with mostly low-cost personal computer components (ignoring for a moment the daunting difficulties of communication and software for such a system), it would cost some 50 billion U.S. dollars and would consume some 1,000 megawatts of electric power.

This paper addresses the hardware and software issues in building and effectively using a petaflops computer.

6.2   APPLICATIONS FOR PETAFLOPS SYSTEMS

The need for such enormous computing capability is often questioned, but such doubts can be dismissed by a moment's reflection on the history of computing. It is well known that Thomas J. Watson Jr., a president of IBM, once ventured that there was a worldwide market of only about six computers. Even the legendary Seymour Cray designed his Cray-1 system on the premise that there were only about 100 potential customers. In 1980, after the Cray-1 had already achieved significant success, an internal IBM study concluded that there was only a limited market for supercomputers, and as a result IBM delayed its entry into the market. In stark contrast to these shortsighted projections, some private homes now have Watson's predicted six systems, and the computational power of each of these personal computers is on a par with, if not greater than, that of the Cray-1. Given the fact that suburban homes have found a need for Cray-class computing, it is not hard to imagine that the requirements of large-scale scientific computing will continue to increase unabated into the future.

Some of the compelling applications anticipated for petaflops computers include the following (Stevens, 1995):

- Nuclear weapons stewardship.

- Cryptography and digital signal processing.

- Satellite data processing.

- Climate and environmental modeling.

- 3-D protein molecule reconstruction.

- Real-time medical imaging.

- Severe storm forecasting.

- Design of advanced aircraft.

- DNA sequence matching.

- Molecular nano-technology.

- Large-scale economic modeling.

- Intelligent planetary spacecraft.

Further, if the history of computing is any guide, a number of exotic new applications will be enabled by petaflops computing technology. These applications may have no clear antecedent in today's scientific computing, and in fact may be only dimly envisioned at the present time.

6.3 SEMICONDUCTOR TECHNOLOGY PROJECTIONS

Some of the challenges that lie ahead to achieving one petaflop/s performance can be seen by examining projections from the latest edition (1997) of the Semiconductor Technology Roadmap (SIA, 1997), as shown in Table 6.1. It can be seen from the table that Moore's Law, namely the 30-year-old phenomenon of relentless increases in transistors per chip, is expected to continue unabated for at least another ten years. Thus the capacity of memory chips is expected to increase as shown. However, the clock rate of both processors and memory devices is not expected to increase at anywhere near the same rate—ten years from now we can only expect clock rates that are double or triple today's state-of-the-art levels. What this means is continued improvements in performance will almost entirely come from a very aggressive utilization of highly parallel designs. It might be mentioned that some of the RISC processors now on the drawing boards will feature advanced features, such as multiple functional units, speculative execution, *etc.* In most cases these features can be seen as merely alternate forms of parallelism.

6.4 LITTLE'S LAW

Before proceedings further, let us review what is known as Little's Law of queuing theory (Bailey, 1997). It asserts, under very general assumptions, that the average number of waiting customers in a queuing system is equal to the average arrival rate times the average wait time per customer. This principle is a simple consequence of Fubini's theorem (the fact that a 2-D integral can be evaluated as an iterated integral in either order). To see this, let $f(t)$ = the cumulative number of arrived customers, and $g(t)$ = the cumulative number of departed customers. Assume $f(0) = g(0) = 0$,

Table 6.1 Semiconductor Technology Roadmap projections.

Characteristic	1999	2001	2003	2006	2009
Feature size (micron)	0.18	0.15	0.13	0.10	0.07
DRAM size (Mbit)	256	1,024	1,024	4,096	16,384
RISC processor (MHz)	1,200	1,400	1,600	2,000	2,500
Transistors (millions)	21	39	77	203	521
Cost per transistor (micro-cents)	1,735	1,000	580	255	100

and $f(T) = g(T) = N$. Consider the region between $f(t)$ and $g(t)$. By Fubini's theorem, $Q T = D N$, where Q is the average length of queue, and D is the average delay per customer. In other words, $Q = (N/T)D$. This is Little's Law.

What does this law have to do with high performance computing? Consider a single processor system with memory. Assume that each operation of a computation requires one word from local main memory. Assume also that the pipeline between main memory and the processor is fully utilized. Then, by Little's Law, the number of words in transit between CPU and memory (*i.e.* the length of memory pipe in a vector architecture, size of cache lines in a cache architecture, *etc.*) is equal to the memory latency times bandwidth. This observation generalizes immediately to multiprocessor systems: concurrency is equal to memory latency times bandwidth, where "concurrency" here is aggregate system concurrency, and "bandwidth" is aggregate system memory bandwidth. This form of Little's Law was first noted by Burton Smith of Tera.

6.5   THE CONCURRENCY CHALLENGE

Several designs have been proposed for a petaflops computer system (Sterling and Foster, 1996). For our purposes here we will mention two such designs. One of these is an all-commodity technology system. It consists of 100,000 nodes, each of which contains a 10 Gflop/s processor. We will assume here that these processors can perform four 64-bit floating-point operations per clock period, and that the clock rate is 2.5 GHz. The interconnection network is constructed from commodity network components that are expected to be available in the 2008 time frame. Another is the "hybrid technology multi-threaded" (HTMT) design, which is now being pursued by a team of researchers headed by Thomas Sterling of CalTech. It features 10,000 nodes, each of which contains a 100 Gflop/s processor, running at 100 GHz. These processors employ a novel technology known as "rapid single flux quantum" (RSFQ) logic, which operates in a cryogenic environment. The system includes multiple levels of memory hierarchy, and an interconnection network based on advanced fiber optic technology (Sterling and Foster, 1996).

Now let us try to calculate, for these two system designs, the overall system concurrency. In other words, calculate the the parallelism assumed in the hardware which a programmer must fully utilize to have a chance of achieving somewhere near the theoretical peak performance of the system. Let us assume here that both systems employ DRAM chips with a latency of 100 nanoseconds for read and write operations. We will also assume that the scientific calculation running on each system is very highly tuned, requires almost no inter-processor communication, but requires one word fetched from memory for each floating-point operation performed. These are obviously rather ideal assumptions.

For the commodity design, we will assume that each processor employs a cache memory system capable of fetching four 64-bit words, one for each floating-point pipe, from main memory each clock period, so that the main memory bandwidth is $4 \times 8 \times 2.5 \times 10^9$ bytes/s $= 40$ gigabyte/s. Observe that at any given time, 250 words are in the pipeline between main memory per pipe (since the latency to main memory is 250 clock periods). Thus the overall system concurrency is 100,000 (nodes) × 4 (pipes/node) × 250 (words in memory pipe) = 10^8.

For the HTMT design, with the ultra-fast clock (100 GHz), the latency to DRAM main memory is now 10,000 clock periods. Thus we will assume that the multi-threaded processor design is capable of maintaining 10,000 outstanding memory fetches simultaneously, so as to enable the processor to sustain somewhere near peak performance. Then a similar calculation as above gives the concurrency as 10,000 (nodes) × 10,000 (words in memory pipe/node) = 10^8.

Note that these two concurrency figures are the same. In fact, either figure could have been calculated without any consideration of the details of the system design, simply by applying Little's Law. If we assume, as above, that one 64-bit word of data is required per flops performed, then the aggregate main memory bandwidth, measured in words/s, is equal to the performance in flops/s. Thus for a one petaflop/s system, Little's Law says that the system concurrency is merely latency (100 ns = 10^{-7} seconds) × bandwidth (10^{15} word/s), which is 10^8.

Needless to say, a system concurrency of 10^8 presents a daunting challenge for future high-performance computing. What this means is that virtually every step of a scientific calculation must possess and exhibit 10^8-way parallelism, or else performance is certain to be poor. Consider, for example, a calculation performed on the commodity system mentioned above, where 90% of the operations performed in this calculation can efficiently utilize 100,000 nodes (*i.e.* exhibits concurrency exceeding 10^8), but that a residual 10% can efficiently only utilize 1,000 nodes (*i.e.* exhibits concurrency not greater than 10^6). Then by Amdahl's Law, the overall performance of the application is limited to 10^{15} / (0.9/105 + 0.1/103) = 9.2×10^{12} flop/s, which is less than one percent of the system's presumed peak performance.

As emphasized above, these analyses were done using highly idealized assumptions. But for computations that require more main memory traffic, or which involve a significant amount of inter-processor communication, the concurrency requirement will be even higher. It is thus clear that a key research question for the field of high performance computing is whether the applications anticipated for future high-end systems can be formulated to exhibit these enormous levels of concurrency. If not, then alternative algorithms and implementation techniques need to be explored.

6.6   THE LATENCY CHALLENGE

It is clear from the discussion so far that latency will be a key issue in future high performance computing. Indeed, even today latency is important in achieving high performance. Consider the latency data in Table 6.2. Note that the local DRAM latency on the SGI Origin is 62 clock periods. On some other RISC systems today the main memory latency exceeds 100 clock periods, and this figure is widely expected to increase further in the years ahead. But these latency figures pale when compared to some of the other latencies in the table.

Recall that in the discussion above on Little's Law, it was assumed that most operations accessed data in main memory. But if algorithms and implementation techniques can be formulated so that most operations are performed on data at a lower level of the memory hierarchy, namely cache or even CPU registers (where latency is lower),

Table 6.2 Latency data.

System	Latency	
	Seconds	Clock Ticks
SGI Origin, local DRAM	320 ns	62
SGI Origin, remote DRAM	1 μs	200
IBM SP2, remote node	40 μs	3,000
HTMT system, local DRAM	50 ns	5,000
HTMT system, remote memory	200 ns	20,000
SGI cluster, remote memory	3 ms	300,000

then there is some hope that Little's Law can be "beaten"—in other words, the level of concurrency that an application must exhibit to achieve high performance can be reduced.

These considerations emphasize the importance of performing more studies to quantify the inherent data locality of various algorithms and applications. Data locality needs to be quantified at several levels, both at low levels, corresponding to registers and cache, and at higher levels, corresponding to inter-processor communication. In general, one can ask if there are "hierarchical" variants of known algorithms, *i.e.* algorithms whose data locality patterns are well matched to those of future high-performance computer systems.

A closely related concept is that of "latency tolerant" algorithms; algorithms whose data access characteristics are relatively insensitive to latency, and thus are appropriate for systems where extra-low latency is not available. One reason this is of interest is the emerging fiber-optic data communication technology, which promises greatly increased bandwidth in future computer systems. This technology raises the intriguing question of whether on future systems it will be preferable to substitute latency tolerant, but bandwidth intensive algorithms, for existing schemes that typically do not require high long-distance bandwidth, but do rely on low latency.

For example, some readers may be familiar with what is often called the "four-step" FFT algorithm for computing the 1-D discrete Fourier transform (DFT). It may be stated as followed. Let n be the size of the data vector, and assume that n can be factored as $n = n_1 \times n_2$. Then consider the input n-long (complex) data vector to be a matrix of size $n_1 \times n_2$, where the first dimension varies most rapidly as in the Fortran language. Then perform these steps:

1. Perform n_1 FFTs of size n_2 on each row of the matrix.

2. Multiply the entry in location (j, k) in the resulting matrix by the constant $e^{(-2\pi i jk)/n}$ (this assumes the indices j and k start with zero).

3. Transpose the resulting matrix.

4. Perform n_2 FFTs of size n_1 on each row of the resulting matrix.

The resulting matrix of size $n_2 \times n_1$, when considered as a single vector of length n, is then the required 1-D DFT (Bailey, 1990). In an implementation on a distributed memory system, for example, it may additionally be necessary to transpose the array before performing these four steps, and, depending on the application, at the end as well. But it is interesting to note that:

- the computational steps are each performed without communication, *i.e.* "embarrassingly parallel";

- the transpose step(s), which are "complete exchange" operations on a distributed memory system, or "block exchange" operations on a cache system, require large global bandwidth, but can be done exclusively with fairly large, contiguous block transfers, which are latency tolerant; and

- the 1-D FFTs required in steps 1 and 4 themselves may be performed, in a recursive, hierarchical fashion, using this algorithm, to as many levels as appropriate.

In short, this algorithm is latency tolerant and hierarchical, so that it is expected to perform well on a variety of future high performance architectures.

A key characteristic of this algorithm is that it relies on array transposes for communication. This same approach, using array transpose operations instead of the more common block decomposition, nearest-neighbor communication schemes, may be effective on future systems for a variety of other applications as well, including many 3-D physics simulations.

6.7 NUMERICAL SCALABILITY

An issue that is related to, but independent from, the issue of parallel scalability is numerical scalability. An algorithm or application is said to be numerically scalable if the number of operations required does not increase too rapidly as the problem is scaled up in size.

It is an unfortunate fact that most of the solvers used in today's large 3-D physical simulations are not numerically scalable. Let n be the linear dimension of the grid in one dimension, so that the other dimensions are linearly proportional to n. Then as n increased, the number of grid points increases as n^3. But for most of the linear solvers used in today's computations, the number of iterations required must also be increased, linearly or even quadratically with n, because the increased grid size increases the condition number of the underlying linear system. Thus the operation count increases as n^4, or even faster in some cases.

This phenomenon is often assumed as an unavoidable fact of life in computational science, but this is not true. Research currently being conducted on domain decomposition and multi-grid methods indicate that many 3-D physical simulations can be performed using techniques that break the link between the number of iterations and the condition number of the resulting linear systems. Thus these methods may yield new computational approaches that dramatically reduce the computational cost of future large-scale physical simulations (Barth, 1997).

6.8   SYSTEM PERFORMANCE MODELING

Developing new algorithms for future computer systems naturally raises the question of how meaningful research can be done years before such systems are available. This suggests a need for effective tools and techniques to model the operation of a future system, thus permitting one to have confidence in the efficiency of an algorithm on such a system.

There are several approaches to perform studies of this sort. One approach is to employ a full-scale computer system emulation program. However, emulators of this scope typically run several orders of magnitude slower than the real systems they emulate. Nonetheless, such simulations may be necessary to obtain high levels of confidence in the performance of an algorithm on a particular design. Along this line, it is interesting to note that accurate simulations of future parallel systems are likely to require today's state-of-the-art highly parallel systems to obtain results in reasonable run times.

A more modest, but still rather effective, approach is to carefully instrument an existing implementation of an algorithm or application (such as by inserting counters), so as to completely account for all run time in a succinct formula. One example of this approach is (Yarrow and der Wijngaart, 1997). These researchers found that for the LU benchmark from the NAS Parallel Benchmark, the total run time T per iteration is given by:

$$T = 485FN^3 2^{-2K} + 320BN^2 2^{-K} + 4L +$$

$$(1 + \frac{2(2^K - 1)}{(N - 2)})(2(N - 2)(279FN^2 2^{-2K} + 80BN^{2-K} + L)) + 953F(N - 2)N^2 2^{-2K})$$

where L = node-to-node latency (assumed not to degrade with large K), B = node-to-node bandwidth (assumed not to degrade with large K), F = floating point rate, N = grid size, and P = 2^{2K} = number of processors.

This formula permitted the authors to determine that the Class C size of the LU benchmark should run with reasonably good efficiently on systems with up to 16,384 processors, assuming the computer system's latency and bandwidth scales appropriately with system size.

6.9   HARDWARE ISSUES

So far we have focused on algorithm and application issues of future high-end systems. Of course there are other issues, which we can only briefly mention here. Among the hardware issues are the following:

1. Should high-end systems employ commodity device technology or advanced technology?

2. How can the huge projected power consumption and heat dissipation requirements of future systems be reduced?

3. Should the processors employ a conventional RISC design, or a more sophisticated multi-threaded design?

4. Should the system be of a distributed memory, or a distributed shared memory architecture?

5. How many levels of memory hierarchy are appropriate?

6. How will cache coherence be handled?

7. What design will best manage latency and hierarchical memories?

8. How much main memory should be included?

With regards to item 2, one advantage of the HTMT design is that the RSFQ processors feature a remarkably low electric power requirement. The HTMT system researchers now estimate that the entire processor complex for a one petaflop/s system would consume only 100 watts. The required super-cooling environment will obviously increase this figure, but it still would be orders of magnitude less than that of conventional RISC processors.

With regards to item 8, some researchers have assumed the "3/4"rule, namely that for typical 3-D physical simulations, memory increases as n^3, while computation increases as n^4. Thus the memory required for future high-end system will scale as the 3/4 power of the performance. Since systems today typically feature roughly one byte of memory per flop/s of computing power, an application of the 3/4 rule implies that petaflops systems will require only about 30 terabytes of main memory.

However, as mentioned above, advances in domain decomposition and multi-grid methods may overturn the rationale underlying the 3/4 rule. If so, then more memory would be required, possibly as much as one petabyte for a one petaflop/s system (on the other hand, larger problems could be performed). Also, an emerging class of "data intensive" problems appear to require, in many instances, larger memories than projected by the 3/4 rule. Finally, real petaflops systems, when they are available, will doubtless be used at least in part for development and multi-user batch processing, especially during daytime hours, even if during they are used off-hours for large, grand-challenge jobs. In an environment where multiple, more modest-sized jobs are being run, a memory-to-flop/s ratio more typical of day's high-end computing will be required. In any event, it is clear that more scaling studies are needed to settle this important issue.

6.10 THE LANGUAGE ISSUE

One important question that needs further study is the issue of programming language. Today, many programmers of high-end system utilize message-passing language systems, such as MPI or PVM, since they tend to obtain the best run-time performance. But message passing languages are difficult to learn, use and debug. They are not a natural model for any notable body of applications, and they appear to be somewhat inappropriate for distributed shared memory (DSM) systems. Further, their subroutine-call format, which unavoidably involves a software layer, looms as an impediment to performance for future computer systems.

Other programmers have utilized array-parallel languages such as HPF and HPC. These language systems are generally easier to use, but their run-time performance lags significantly behind that of similar codes written using MPI. Further, the single-instruction, multiple data (SIMD) model assumed in array-parallel languages is inappropriate for a significant body of emerging applications, which feature numerous asynchronous tasks. The languages Java, SISAL, Linda and others each has its advocates, but none has yet proved its superiority for a large class of scientific applications on advanced architectures.

What will programmers of future systems need? One can dream of the following features:

1. Designed from the beginning with a highly parallel, hierarchical memory system in mind.

2. Includes high-level features for application scientists.

3. Includes low-level features for performance programmers.

4. Handles both data and task parallelism, and both synchronous and asynchronous tasks.

5. Scalable for systems with up to 1,000,000 processors.

6. Appropriate for parallel clusters of distributed shared memory nodes.

7. Permits both automatic and explicit data communication.

8. Permits the memory hierarchy to be explicitly controlled by performance programmers.

It is not known whether such features can be realized in a usable language for high-performance computing. Clearly more research is needed here.

6.11   SOFTWARE ISSUES

There are of course many software issues, including:

1. How can one manage tens or hundreds of thousands of processors, running possibly thousands of separate user jobs?

2. How can hardware and software faults be detected and rectified?

3. How can run-time performance phenomena be monitored?

4. How should the mass storage system be organized?

5. How can real-time visualization be supported?

It is clear that meeting these challenges will require significant advances in software technology. With regards to item 3, for example, the line plots often used today to graphically display the operation of say 16 processors, showing communication between them as time advances, will be utterly inappropriate for a system with 100,000 processors. Completely new approaches for representing performance information, and for automatically processing such information, will likely be required.

6.12 CONCLUSION

For many years the field of high performance computing has sustained itself on a fundamental faith in highly parallel technology, with patience that the flaws in the current crop of systems will be rectified in the next generation, all the while depending on the boundless charity of national government funding agencies. But the result of this informal approach is not hard to observe: numerous high performance computer firms have gone out of business, others have merged, and funding from several federal agencies (at least in the U.S.) has been cut.

What might help, in this author's view, is for researchers in the field to do more rigorous, quantitative studies, replacing the often informal studies that have been made to date. For example, vendors often ask scientists at government laboratories questions such as how much bandwidth is needed at various levels of the memory hierarchy, and how much latency can be tolerated by our applications. On such occasions the author has had little to offer, other than some vague comparisons with some existing systems. Clearly such studies would be both of near-term and long-term value to the field.

Along this line, perhaps the time has come to attempt some quantitative studies in arenas other than raw performance. For example, can the productivity of programmers using various programming systems be quantified? What about the usefulness of various performance tools – can this be quantified and measured? At the least, studies of this sort could help establish the need for continuing work in the field. Let the analyses begin!

References

Bailey, D. (1990). FFTs in external or hierarchical memory. *Journal of Supercomputing*, 4(1):23–35.

Bailey, D. (1997). Little's law and high performance computing. Available from the author.

Barth, T. (1997). A brief accounting of some well known results in domain decomposition. Available from the author.

SIA (1997). The national technology roadmap for semiconductors. Semiconductor Industry Association.

Sterling, T. and Foster, I. (1996). Petaflops systems workshops. Technical Report CACR-133, California Institute of Technology. See also: Sterling, T., Messina, P. and Smith, P. (1995), *Enabling Technology for Petaflops Computers*, MIT Press, Cambridge, MA.

Stevens, R. (1995). The 1995 petaflops summer study workshop. This report and some related material are available from http://www.mcs.anl.gov/petaflops.

Yarrow, M. and der Wijngaart, R. V. (1997). Communication improvement for the LU NAS parallel benchmark: A model for efficient parallel relaxation schemes. Technical report, NAS Systems Division, NASA Ames Research Center.

II Computational Physics

7 PARALLEL PERFORMANCE OF A CODE USING MULTIDIMENSIONAL FFTS

John A. Zollweg

Cornell Theory Center,
Cornell University, U.S.A.

zollweg@tc.cornell.edu

Abstract: Many computationally intensive physics problems spend the majority of their time in multidimensional Fast Fourier Transforms. A case study for the IBM SP is presented based on an algorithm for sound propagation through a finite inhomogeneous medium. Techniques for improving the serial and parallel efficiency of such codes is described, and the use of software tools to aid the process are illustrated.

Keywords: FFT, performance optimization, profiling.

7.1 INTRODUCTION

An initial implementation of an algorithm is typically developed on a workstation using a scaled-down version of the problem. When one does production computation on a real case, the memory requirement or execution time is often so large that the only realistic place to run is on a parallel supercomputer. This article describes how a sound propagation code developed in the laboratory of Professor Robert C. Waag at the University of Rochester was optimized and parallelized for the IBM SP distributed memory machine at the Cornell Theory Center. Many of the techniques that are described here can be applied to the development of efficient parallel codes for other problems in computational physics, especially if they use multidimensional Fast Fourier Transforms (FFTs).

Professor Waag's apparatus contains a circular vessel filled with water that has an object immersed in it. Ultrasonic transducers around the periphery of the vessel serve as both sources and receivers of the ultrasonic waves. By using suitably phased excitation of the transducers, a plane wave can be launched toward the object (see Figure 7.1).

An image of the object can be constructed from ultrasonic signals received at the detectors using the k-space method (Bojarski, 1985; Liu and Waag, 1997). This en-

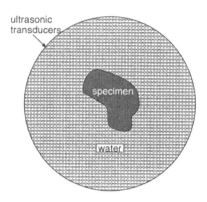

Figure 7.1   Schematic drawing of experimental arrangement.

tails solving a "Forward Problem" for propagation of a plane wave through a model inhomogeneous medium. For purposes of computation, space is divided into cells and the Fourier transforms are calculated over arrays of cells. The object is treated as a region in the medium that has different physical properties (mass density, sound speed, attenuation) from the background. The differential equation that governs sound propagation is solved by a Fourier transformation of the spatial components of the sound field and by using a finite difference scheme on the Fourier components. The inhomogeneous character of the medium is described through Fourier components as well. Because the medium is static, the Fourier components of the model need to be computed only once, but the field components must be recomputed for each step of the propagation.

The object has finite extent but its Fourier representation implies that it has an infinite number of replicas arrayed in the directions along which the transforms have been computed. To avoid interference from the replicas, the computational domain in real space is "padded" with an array of cells that have uniform properties.

This article describes the production of an efficient parallel code for a distributed memory parallel computer. Before any parallel code was written, steps were taken to improve the efficiency of the serial code (Section 7.2). Then a parallel version was produced and further work was done to optimize its performance (Section 7.3).

7.2   OPTIMIZATION OF THE SERIAL CODE

7.2.1   Initial evaluation

The code was developed in Fortran 77 on a Sun workstation. Calls to a two-dimensional FFT routine from the IMSL library were used for the Fourier transforms. The code was first run on the IBM SP without compiler optimization to facilitate debugging. A profile was generated and viewed with *xprofiler*. For a run that took 631 seconds of CPU time, only 129 seconds (20%) was spent in FFT routines. This means that the initial effort to improve the code needed to be focused elsewhere. A sample of the

line	no. ticks per line	source code
499		DO P=1,N
500		DO Q=1,N
501	811	IF((P.LE.N/2).AND.(Q.LE.N/2)) THEN
502	240	TFC(N/2+P,N/2+Q)=VNK(P,Q)/NORM
503	77	ELSE IF((P.GT.N/2).AND.(Q.GT.N/2)) THEN
504	225	TFC(P-N/2,Q-N/2)=VNK(P,Q)/NORM
505	57	ELSE IF((P.GT.N/2).AND.(Q.LE.N/2)) THEN
506	266	TFC(P-N/2,N/2+Q)=VNK(P,Q)/NORM
507	32	ELSE IF((P.LE.N/2).AND.(Q.GT.N/2)) THEN
508	224	TFC(N/2+P,Q-N/2)=VNK(P,Q)/NORM
509		ENDIF
510	41	ENDDO
511	1	ENDDO
512		
513		DO P=1,N
514		DO Q=1,N
515	2077	VNK(P,Q)=TFC(P,Q)
516	36	ENDDO
517	2	ENDDO
518		

Figure 7.2 *xprofiler* window showing annotated source code before optimization.

annotated source for the main program in Figure 7.2 shows that much time was spent copying between arrays, 59 seconds in the two loops shown. Incorrect nesting of the loops is part of the cause of this.

After the code was rearranged to nest the loops correctly, it was compiled with the XL Fortran compiler using flags $-O3$ $-qstrict$ $-qarch=pwr2$. The code shown in Figure 7.2 is *completely gone* in the optimized version. The normalization is done inside the FFT subroutine where it can be combined with other operations for higher efficiency. In the optimized version the transformed arrays are not "unshuffled" (see Section 7.3).

Figure 7.3 shows the profiler output for a similar set of nested loops in the optimized code. Branching has been eliminated by using appropriate index ranges for the loops. The loops are nested correctly. A matrix copy has been eliminated by writing directly to the appropriate matrix. The time for this set of loops is only 0.1 second, but it had been 20.9 seconds for a set of nested loops in the original code in a case where no divisions for normalization were done.

7.2.2 Memory usage

During development, a separate block of memory was used for each different version of an array. After the algorithm was decided upon, it turned out that not all of the arrays were in use at the same time. Thus it would have been possible to eliminate some of the arrays. Instead of eliminating them entirely, arrays that were not in simultaneous use were aliased to each other, improving the readability of the code while saving memory. Another technique that led to substantial memory savings was to make most arrays ALLOCATABLE. This allowed us to overcome the common practice in Fortran 77

line	no. ticks per line	source code
772		
773		DO Q=1,NCA
774		DO P=1,NCA
775	4	WN(N/2+P,N/2+Q)=TFR(P,Q)
776		ENDDO
777		DO P=N-NCA+1,N
778	3	WN(P-N/2,N/2+Q)=TFR(P,Q)
779		ENDDO
780		ENDDO
781		DO Q=N-NCA+1,N
782		DO P=1,NCA
783	2	WN(N/2+P,Q-N/2)=TFR(P,Q)
784		ENDDO
785		DO P=N-NCA+1,N
786	2	WN(P-N/2,Q-N/2)=TFR(P,Q)
787		ENDDO
788		ENDDO

Figure 7.3 $xprofiler$ window showing annotated source code after optimization.

codes of over-dimensioning arrays to accommodate the largest case one might study. It also allows arrays to be DEALLOCATEd when they are no longer needed.

In the time-stepping algorithm, the new Fourier coefficients of the field are calculated from the present and previous coefficients. After this step, the previous coefficients won't be needed, so the new ones can overwrite them. To avoid copying large two-dimensional arrays of coefficients, the array is declared as (: , : , 2) and references to the array are made using a pair of variables that alternate between 1 and 2 at each step. This technique saves both memory and execution time.

7.2.3 Execution time

To find other opportunities for reducing execution time, another run was done with profiling enabled. A portion of the results is shown in Figure 7.4. It showed that there were many calls to exponential and trigonometric routines. The calls to the exponentials are necessary because the incident wave has different values at each time. However, the trigonometric calls are primarily for calculating weighting factors that are invariant. By moving the weight factor calculations outside the time-stepping loop and putting the results into an array, the relatively expensive calls to the trigonometric functions have been reduced to a few at the beginning of a run.

7.2.4 Removing unnecessary FFTs

A major economy in execution time came from reducing the amount of Fourier transformation that was done. Although a two-dimensional (2-D) FFT routine call is very easy to incorporate into an algorithm, in a case where arrays have been padded some transforms need not be done explicitly. Also because the 2-D FFTs represent the bulk of the computation time after the other optimizations have been done (as seen in the profiler output) and are independent of other algorithmic details they provide a good opportunity for introducing parallelism. This opportunity for parallelization has also been described by (Foster and Worley, 1997). Because debugging a serial code is

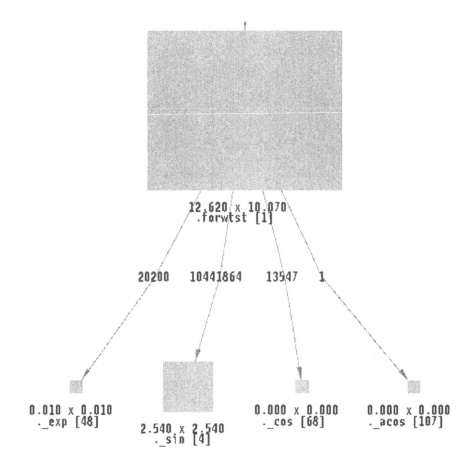

Figure 7.4 *xprofiler* window showing times and call counts for main routine and its descendants. The number on the left below each box is the total time and the one on the right is the exclusive time for that routine. Arcs are labeled with the number of times they are traversed.

much easier than a parallel one, the optimizations to remove the unnecessary FFTs were done on the serial code first.

Before explaining how unnecessary FFTs were avoided, some description of the structure of the data is necessary. Because of the symmetry of the ultrasonic apparatus, it is most natural to have the origin of the coordinate system at the center of the grid on which sound field values are calculated. The FFT routines assume that the origin is in the lower left corner of the computational field, so it is necessary to "shuffle" the data so that the point at the origin is at the beginning of the first column of the data array. Points with negative displacements from the origin are mapped into the upper half of their respective row or column where, by the aliasing property of Fourier

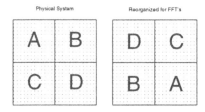

Figure 7.5   "Shuffling" of arrays to shift origin of coordinate system from center to upper left corner.

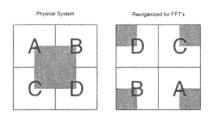

Figure 7.6   Unshaded areas do not need their FFTs calculated.

transforms, they produce the same contributions to Fourier coefficients as if they had a displacement of $2N\Delta x$ less, where N is the number of elements in a row or column of the matrix and Δx is the displacement between elements. A similar alias relationship exists between the Fourier coefficients in the upper calculated half of reciprocal space and coefficients for points with negative reciprocal space coordinates. In Figure 7.5, the left diagram shows the layout of the grid in the physical system and the right diagram shows how the grid points have been "shuffled" before performing the 2-D FFTs. A multidimensional FFT is evaluated by performing sets of one-dimensional FFTs. If we abandon the convenience of calling a 2-D FFT routine and do the 1-D FFTs instead, we can avoid some of the transforms. Evaluation is more efficient if the elements are contiguous in memory, so the first set of one-dimensional FFTs is performed on the columns of the matrix. In Figure 7.6, the unshaded areas are the parts of the matrix where FFTs do not need to be calculated explicitly. However, the first element ($K = 0$) of the transforms that are not calculated does need to be set to $N \times bval$, where $bval$ is the value of each element in the column. In a typical case where the width of the padding is half the width of the physical system, this allows us to avoid calculating half of the FFTs in the first pass.

The second pass of 1-D FFTs needs to be done in the other direction. To bring the elements of each transform into adjacent memory locations, the matrix is transposed. After the transpose, the zero elements from the FFTs that were not done in the first pass are now in the middle section of each column, so no FFTs can be avoided in this pass because of padding. If the first pass was a real to complex transform, though, the columns in the right half of the matrix after transposition are complex conjugates

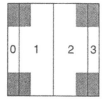

Figure 7.7   Distribution of a data matrix over 4 processors before the first FFT.

of columns in the left half, so only the columns in the left half need to be calculated explicitly. Thus, another $N/2 - 1$ of the 1-D FFTs may be avoided!

7.3   PARALLEL CODE

7.3.1   Parallelization of the Fourier Transforms

Because the intermediate results must be transposed in the middle of a 2-D FFT, this is an excellent opportunity for introducing parallelism into the algorithm with minimal disturbance to the rest of the code. An entire column is necessary for doing a 1-D FFT, so the best way to distribute the data over the processors is by striping. This is shown for the present case in Figure 7.7

The reason for this uneven distribution of data across processors is that one wants to maintain good load balance among all of them. Since most of the computational load comes from doing FFTs, the way to equalize the load is to equalize the number of FFTs each will do. There is no need to use real memory in processors 1 and 2 for the columns that contain only padding. If they are omitted the memory usage is more uniform across processors. It is necessary only to insert the appropriate values into the first column of the transpose when it is done.

After transposition, the columns of the matrix can be distributed equally among processors if there is no complex conjugate symmetry. Otherwise, only the left side of the matrix need be distributed among the processors. Since a column-wise distribution of the matrix after transposition is equivalent to a row-wise distribution beforehand, some elements must be transferred between each pair of processors. An MPI_ALLTOALL call can be used for this purpose. The transfer is more efficient (and simpler to specify) if the elements to be transferred between any pair of processors are contiguous in memory, so it is best for each processor to transpose its piece of the matrix before the call to MPI_ALLTOALL.

If there are several matrices involved in the algorithm, the process of computing 1-D transforms, transposing, transferring, and computing the rest of the 1-D transforms can be repeated for each matrix. The real space and transform space distributions of the matrix are different, so some care has to be exercised in planning other calculations. In particular, if the original matrix had x varying along the rows and y varying along the columns, the transform space matrix will have k_x varying along rows and k_y varying along columns, unless a further transpose is performed.

Table 7.1   Parallel speed of optimized code (small problem, 2000 steps).

Nodes	Time (sec)	Speedup
1	309.3	1.00
2	193.6	1.60
4	125.4	2.47
5	116.6	2.65

Table 7.2   Parallel speed of optimized code (large problem, 100 steps).

Nodes	Time (sec)	Speedup
1	728.9	1.00
2	344.7	2.11
4	155.8	4.68
8	72.3	10.08

In programs that require higher-dimensional FFTs, the decomposition should be performed on only one dimension. The result will be that each processor has slabs with full size in all but one of the dimensions. FFTs can be performed locally in all the dimensions except the one that was decomposed. Then the slabs are transposed and transferred before doing the transforms in the original decomposition direction.

7.3.2   Performance of test case

Test runs were performed on two problems. The small problem had a 512×512 grid, with a computational region that was 200×200. The large problem had a 2048×2048 grid with a computational region of 1024×1024. The results are shown in Tables 7.1 and 7.2, respectively.

All I/O was performed from the master node. For the test runs, there was no significant I/O during the computationally intensive part of the job. The time for distributing the data to the nodes and collecting the results at the end has been included in the timings.

The results for the small problem are typical of a code that is relatively fine-grained. As a problem of fixed size is distributed over more processors, the amount of computation on each node decreases and the amount of communication increases. Both of these changes have the effect of reducing the granularity of the code and the result is often sublinear speedup.

On the other hand, the large problem shows superlinear speedup. A possible explanation is that memory access has not been sufficiently optimized. Since the sections of arrays on each node are larger when fewer nodes are in use, there may be more cache misses when traversing large arrays in this case. This inefficiency can be overcome by

Figure 7.8   Portion of a trace file visualized using NTV.

using "blocking" for some loops. The matrix and vector routines in ESSL make use of this optimization, so substitution of calls to them for DO loops would be the easiest way to improve performance.

7.4 CONCLUSIONS

As a final check on the parallel efficiency of the code, a trace was performed using the tracing facility in the IBM Parallel Environment. A section of the trace file, as visualized by NTV, the NAS Trace Visualization tool, is shown in Figure 7.8.

The portion of the trace file shown covers eleven time-steps for four processors (numbered 0, 1, 2, and 3). The dark regions of each time-line are time spent in computation, and the lighter regions are communication time. (The original figure was color-coded.) Since most of each time-line is dark, the communication overhead is not great.

There are many ways that a code can be optimized to give faster execution. In this case, considerable speedup was gained without even resorting to parallelism. For problems that are large enough, it has been shown that substantial further speedup can be obtained with the parallel code.

References

Bojarski, N. (1985). The k-space formulation of the scattering problem in the time domain: An improved single propagator formulation. *J. Acoust. Soc. Am.*, 77(3):826–831.

Foster, I. and Worley, P. (1997). Parallel algorithms for the spectral transform method. *SIAM J. Sci. Comput.*, 18(3):806–837.

Liu, D. and Waag, R. (1997). Solving the forward scattering problem by the k-space method. report draft.

8 OBJECT-ORIENTED SCIENTIFIC PROGRAMMING WITH FORTRAN 90

Charles D. Norton*

National Research Council, U.S.A

Charles.D.Norton@jpl.nasa.gov

Abstract: Fortran 90 is a modern language that introduces many important new features beneficial for scientific programming. While the array-syntax notation has received the most attention, we have found that many modern software development techniques can be supported by this language, including object-oriented concepts.

While Fortran 90 is not a full object-oriented language it can directly support many of the important features of such languages. Features not directly supported can be emulated by software constructs. It is backward compatible with Fortran 77, and a subset of HPF, so new concepts can be introduced into existing programs in a controlled manner. This allows experienced Fortran 77 programmers to modernize their software, making it easier to understand, modify, share, explain, and extend.

We discuss our experiences in plasma particle simulation and unstructured adaptive mesh refinement on supercomputers, illustrating the features of Fortran 90 that support the object-oriented methodology. All of our Fortran 90 programs achieve high performance with the benefits of modern abstraction modeling capabilities.

Keywords: Fortran 90, supercomputing, object-oriented, adaptive mesh refinement, scientific programming, plasma physics.

8.1 INTRODUCTION

Scientific application programming involves unifying abstract physical concepts and numerical models with sophisticated programming techniques that require patience and experience to master. Furthermore, codes typically written by scientists are constantly changing to model new physical effects. These factors can contribute to long development periods, unexpected errors, and software that is difficult to comprehend, particularly when multiple developers are involved.

---

*Currently in residence at the National Aeronautical and Space Administration's Jet Propulsion Laboratory, California Institute of Technology, U.S.A.

The Fortran 90 programming language (Ellis *et al.*, 1994) addresses the needs of modern scientific programming by providing features that raise the level of abstraction, without sacrificing performance. Consider a 3D parallel plasma particle-in-cell (PIC) program in Fortran 77 which will typically define the particles, charge density field, force field, and routines to push particles and deposit charge. This is a segment of the main program where many details have been omitted:

> dimension part(idimp, npmax), q(nx, ny, nzpmx)
> dimension fx(nx, ny, nzpmx), fy(nx, ny, nzpmx), fz(nx, ny, nzpmx)
> data qme, dt /-1.,.2/
> call push(part, fx, fy, fz, npp, qtme, dt, wke, nx, ny, npmax, nzpmx, ...)
> call dpost(part, q, npp, noff, qme, nx, ny, idimp, npmax, nzpmx)

Note that the arrays must be dimensioned at compile time. Also parameters must either be passed by reference, creating long argument lists, or kept in common blocks and exposed to inadvertent modification. Such an organization is complex to maintain, especially as codes are modified for new experiments.

Using the new features of Fortran 90, abstractions can be introduced that clarify the organization of the code. The Fortran 90 version is more readable while designed for modification and extension:

> use partition_module ; use plasma_module
> type (species) :: electrons
> type (scalarfield) :: charge_density
> type (vectorfield) :: efield
> type (slabpartition) :: edges
> real :: dt = .2
> call plasma_particle_push(electrons, efield, edges, dt)
> call plasma_deposit_charge(electrons, charge_density, edges)

This style of object-oriented programming, where the basic data unit is an "object" that shields its internal data from misuse by providing public routines to manipulate it, allows such a code to be designed and written. Object-oriented programming clarifies software while increasing safety and communication among developers, but its benefits are only useful for sufficiently large and complex programs.

While Fortran 90 is not an object-oriented language, the new features allow most of these concepts to be modeled directly. (Some concepts are more complex to emulate.) In the following, we will describe how object-oriented concepts can be modeled in Fortran 90, the application of these ideas to plasma PIC programming on supercomputers, parallel unstructured adaptive mesh refinement, and the future of Fortran programming (represented by Fortran 2000) that will contain explicit object-oriented features.

8.2 MODELING OBJECT-ORIENTED CONCEPTS IN FORTRAN 90

Object-oriented programming (OOP) has received wide acceptance, and great interest, throughout the computational science community as an attractive approach to address the needs of modern simulation. Proper use of OOP ensures that programs can be written safely, since the internal implementation details of the data objects are hidden. This allows the internal structure of objects and their operations to be modified (to improve efficiency perhaps), but the overall structure of the code using the objects can remain unchanged. In other words, objects are an encapsulation of data and routines.

These objects represent abstractions. Another important concept is the notion of inheritance, which allows new abstractions to be created by preserving features of existing abstractions. This allows objects to gain new features through some form of code reuse. Additionally, polymorphism allows routines to be applied to a variety of objects that share some relationship, but the specific action taken varies dynamically based on the object's type. These ideas are mechanisms for writing applications that more closely represent the problem at hand. As a result, a number of programming languages support OOP concepts in some manner.

Fortran 90 is well-known for introducing array-syntax operations and dynamic memory management. While useful, this represents a small subset of the powerful new features available for scientific programming. Fortran 90 is backward compatible with Fortran 77 and, since it is a subset of High Performance Fortran (HPF), it provides a migration path for data-parallel programming. Fortran 90 type-checks parameters to routines, so passing the wrong arguments to a function will generate a compile-time error. Additionally, the automatic creation of implicit variables can be suppressed reducing unexpected results.

However, more powerful features include derived types, which allow user-defined types to be created from existing intrinsic types and previously defined derived types. Many forms of dynamic memory management operations are now available, including dynamic arrays and pointers. These new Fortran 90 constructs are objects that know information such as their size, whether they have been allocated, and if they refer to valid data. Fortran 90 modules allow routines to be associated with types and data defined within the module. These modules can be used in various ways, to bring new functionality to program units. Components of the module can be private and/or public allowing interfaces to be constructed that control the accessibility of module components. Additionally, operator and routine overloading are supported (name reuse), allowing the proper routine to be called automatically based on the number and types of the arguments. Optional arguments are supported, as well as generic procedures that allow a single routine name to be used while the action taken differs based on the type of the parameter. All of these features can be used to support an object-oriented programming methodology (Decyk *et al.*, 1997a).

8.3   APPLICATION: PLASMA PIC PROGRAMMING ON SUPERCOMPUTERS

Computer simulations are very useful for understanding and predicting the transport of particles and energy in fusion energy devices called tokamaks. Tokamaks, which are toroidal in shape, confine the plasma with a combination of an external toroidal magnetic field and a self-generated poloidal magnetic field. Understanding plasma confinement in tokamaks could lead to the practical development of fusion energy as an alternative fuel source—unfortunately confinement is not well understood and is worse than desired.

PIC codes (Birdsall and Langdon, 1991) integrate the trajectories of many particles subject to electro-magnetic forces, both external and self-generated. The General Concurrent PIC Algorithm (Liewer and Decyk, 1989), which partitions particles and fields among the processors of a distributed-memory supercomputer, can be programmed using a single-program multiple-data (SPMD) design approach. Although the Fortran 77 versions of these programs have been extensively benchmarked and are scalable with nearly 100% efficiency, there is an increasing interest within the scientific community to apply object-oriented principles to enhance new code development.

In the introduction, we illustrated how Fortran 77 features could be modeled using Fortran 90 constructs. In designing the PIC programs, basic constructs like particles (individually and collectively), fields (scalar and vector, real and complex), distribution operations, diagnostics, and partitioning schemes were created as abstractions using Fortran 90 modules. Fortran 90 objects are defined by derived types within modules where the public routines that operate on these objects are visible whenever the object is "used". (The private components of the module are only accessible within module defined routines.)

A portion of the species module (Figure 8.1) illustrates how data and routines can be encapsulated using object-oriented concepts. This module defines the particle collection, where the interface to the particle Maxwellian distribution routine is included.

Some OOP concepts, such as inheritance, had limited usefulness while run-time polymorphism was used infrequently. Our experience has shown that these features, while sometimes appropriate for general purpose programming, do not seem to be as useful in scientific programming. Well-defined interfaces, that support manipulation

```
module species_module
    use distribution_module ; use partition_module
    implicit none
    type particle
        private
        real :: x, y, z, vx, vy, vz
    end type particle
    type species
        real :: qm, qbm, ek
        integer :: nop, npp
        type (particle), dimension(:), pointer :: p
    end type species
contains
    subroutine species_distribution(this, edges, distf)
        type (species), intent (out) :: this
        type (slabpartition), intent (in) :: edges
        type (distfcn), intent (in) :: distf
            ! subroutine body
    end subroutine species_distribution
    ! additional member routines...
end module species_module
```

Figure 8.1   Abstract of Fortran 90 module for particle species.

Table 8.1 3D parallel plasma PIC experiments on the Cornell Theory Center IBM SP? (32 processors, 8M particles, 260K grid points).

Language	Compiler	P2SC Super Chips	P2SC Optimized
Fortran 90	IBM xlf90	622.60s	488.88s
Fortran 77	IBM xlf	668.03s	537.95s
C++	KAI KCC	1316.20s	1173.31s

of abstractions, were more important. More details on the overall structure of the code can be found in (Norton *et al.*, 1995; Norton *et al.*, 1997).

The wall-clock execution times for the 3D parallel PIC code written in Fortran 90, Fortran 77, and C++ are illustrated in Table 8.1. Although our experience has been that Fortran 90 continually outperforms C++ on complete programs, generally by a factor of two, others have performance results that indicate that C++ can sometimes outperform Fortran 90 on some computational kernels (Cary *et al.*, 1997). (In these cases, "expression templates" are introduced as a compile-time optimization to speed up complicated array operations.)

The most aggressive compiler options produced the fastest timings, seen in Table 8.1. The KAI C++ compiler with options "+K3 -O3 -abstract_pointer" spent over 2 hours in the compilation process. The IBM F90 compiler with the options "-O3 -qlanglvl=90std -qstrict -qalias=noaryovrlp" used 5 minutes for compilation. (The KAI compiler is generally considered the most efficient C++ compiler when objects are used. This compiler generated slightly faster executables than the IBM C++ compiler.) Applying the hardware optimization switches "-qarch=pwr2 -qtune=pwr2" introduced additional performance improvements specific to the P2SC processors.

For large problems on supercomputers, we have found Fortran 90 very useful and generally safer, with higher performance than C++ and sometimes Fortran 77. Fortran 90 derived-type objects improved cache utilization, for large problems, over Fortran 77. (The C++ and Fortran 90 objects had the same storage organization.) Fortran 90 is less powerful than C++, since it has fewer features and those available can be restricted to enhance performance, but many of the advanced features of C++ have not been required in scientific computing. Nevertheless, advanced C++ features may be more appropriate for other problem domains (Decyk *et al.*, 1997b; Norton *et al.*, 1997).

8.4 APPLICATION: PARALLEL UNSTRUCTURED ADAPTIVE MESH REFINEMENT

Adaptive mesh refinement is an advanced numerical technique very useful in solving partial differential equations. Essentially, adaptive techniques utilize a discretized computational domain that is subsequently refined/coarsened in areas where additional resolution is required. Parallel approaches are necessary for large problems, but the implementation strategies can be complex unless good design techniques are applied.

```
program pamr
use mpi_module ; use mesh_module ; use misc_module
implicit none
integer :: ierror
character(len=8) :: input_mesh_file
type (mesh) :: in_mesh
call MPI_INIT(ierror)
    input_mesh_file = mesh_name(iam)
    call mesh_create_incore(in_mesh, input_mesh_file)
    call mesh_repartition(in_mesh)
    call mesh_visualize(in_mesh, "visfile.plt")
call MPI_FINALIZE(ierror)
end program pamr
```

Figure 8.2 A main program with selected PAMR library calls.

One of the major benefits of Fortran 90 is that codes can be structured using the principles of object-oriented programming. This allows the development of a parallel adaptive mesh refinement (PAMR) code where interfaces can be defined in terms of mesh components, yet the internal implementation details are hidden. These principles also simplify handling inter-language communication, sometimes necessary when additional packages are interfaced to new codes. Using Fortran 90's abstraction techniques, for example, a mesh can be loaded into the system, distributed across the processors, the PAMR internal data structure can be created, and the mesh can be repartitioned and migrated to new processors (all in parallel) with a few simple statements as shown in Figure 8.2.

A user could link in the routines that support parallel adaptive mesh refinement then, as long as the data format from the mesh generation package conforms to one of the specified formats, the capabilities required for PAMR are available. We now describe the Fortran 90 implementation details that make this possible.

8.4.1 Basic Mesh Definition

Fortran 90 modules allow data types to be defined in combination with related routines. In our system the mesh is described, in part, as shown in Figure 8.3. In two-dimensions, the mesh is a Fortran 90 object containing nodes, edges, elements, and reference information about non-local boundary elements (r_indx). These components are dynamic, their size can be determined using Fortran 90 intrinsic operations. They are also private, meaning that the only way to manipulate the components of the mesh are by routines defined within the module. Incidentally, the remote index type r_indx (not shown) is another example of encapsulation. Objects of this type are defined so that they cannot be created outside of the module at all. A module can contain any number of derived types with various levels of protection, useful in our mesh data structure implementation strategy.

All module components are declared private, meaning that none of its components can be referenced or used outside the scope of the module. This encapsulation adds

```
module mesh_module
use mpi_module ; use heapsort_module
implicit none
private
public :: mesh_create_incore, mesh_repartition, &
        mesh_visualize
integer, parameter :: mesh_dim=2, nodes_=3, edges_=3, neigh_=3
type element
        private
        integer :: id, nodeix(nodes_), edgeix(edges_), &
                neighix(neigh_)
end type element
type mesh
        private
        type(node), dimension(:), pointer :: nodes
        type(edge), dimension(:), pointer :: edges
        type(element), dimension(:), pointer :: elements
        type(r_indx), dimension(:), pointer :: boundary_elements
end type mesh
contains
        subroutine mesh_create_incore(this, mesh_file)
        type (mesh), intent(inout) :: this
        character(len=*), intent(in) :: mesh_file
          ! details omitted...
        end subroutine mesh_create_incore
          ! additional member routines...
end module mesh_module
```

Figure 8.3   Skeleton view of mesh_module components.

safety to the design since the internal implementation details are protected, but it is also very restrictive. Therefore, routines that must be available to module users are explicitly listed as public. This provides an interface to the module features available as the module is used in program units. Thus, the statement in the main program from Figure 8.2:

```
call mesh_create_incore(in_mesh, input_mesh_file)
```

is a legal statement since this routine is public. However the statement:

```
element_id = in_mesh%elements(10)%id
```

is illegal since the "elements" component of in_mesh is private to the derived type in the module.

The mesh_module uses other modules in its definition, like the mpi_module and the heapsort_module. The mpi_module provides a Fortran 90 interface to MPI while the heapsort_module is used for efficient construction of the distributed mesh data structure. The routines defined within the contains statement, such as mesh_create_incore(), belong to the module. This means that routine interfaces, that perform type matching on arguments for correctness, are created automatically. (This is similar to function prototyping in other languages.)

8.4.2   Distributed structure organization

When the PAMR mesh data structure is constructed it is actually distributed across the processors of the parallel machine. This construction process consists of loading the mesh data, either from a single processor for parallel distribution (in_core) or from individual processors in parallel (out_of_core). A single mesh_build() routine is responsible for constructing the mesh based on the data provided. Fortran 90 routine overloading and optional arguments allow multiple interfaces to the mesh_build() routine, supporting code reuse. This is helpful because the same code that builds a distributed PAMR mesh data structure from the initial description can be applied to rebuilding the data structure after refinement and mesh migration. The mesh_build() routine, and its interface, is hidden from public use. Heap sorting techniques are also applied in building the hierarchical structure so that reconstruction of a distributed mesh after refinement and mesh migration can be performed from scratch, but efficiently.

The main requirement imposed on the distributed structure is that every element knows its neighbors. Local neighbors are easy to find on the current processor from the PAMR structure. Remote neighbors are known from the boundary_elements section of the mesh data structure, in combination with a neighbor indexing scheme. When an element must act on its neighbors the neighbor index structure will either have a reference to a complete description of the local neighbor element or a reference to a processor_id/global_id pairing. This pairing can be used to fetch any data required regarding the remote element neighbor. (Note that partition boundary data, such as a boundary face in three-dimensions, is replicated on processor boundaries.) One of the benefits of this scheme is that any processor can refer to a specific part of the data structure to access its complete list of non-local elements.

Figure 8.3 showed the major components of the mesh data structure, in two-dimensions. While Fortran 90 fully supports linked list structures using pointers, a common organization for PAMR codes, our system uses pointers to dynamically allocated arrays instead. There are a number of reasons why this organization is used. By using heap sorting methods during data structure construction, the array references for mesh components can be constructed very quickly. Pointers consume memory, and the memory references can become "unorganized", leading to poor cache utilization. While a pointer-based organization can be useful, we have ensured that our mesh reconstruction methods are fast enough so that the additional complexity of a pointer-based scheme can be avoided.

8.4.3   Interfacing among data structure components

The system is designed to make interfacing among components very easy. Usually, the only argument required to a PAMR public system call is the mesh itself, as indicated in Figure 8.2. There are other interfaces that exist however, such as the internal interfaces of Fortran 90 objects with MPI and the ParMeTiS parallel partitioner (Karypis *et al.*, 1997) which were written in the C programming language.

Since Fortran 90 is backward compatible with Fortran 77 it is possible to link to MPI for inter-language communication, assuming that the interface declarations have been defined in the mpi.h header file properly. While certain array constructs have been useful, such as array syntax and subsections, MPI does not support Fortran 90 directly so array subsections cannot be (safely) used as parameters to the library routines. Our system uses the ParMeTiS graph partitioner to repartition the mesh for load balancing. In order to communicate with ParMeTiS our system internally converts the distributed mesh into a distributed graph. A single routine interface to C is created that passes the graph description from Fortran 90 by reference. Once the partitioning is complete, this same interface returns from C an array that describes the new partitioning to Fortran 90. This is then used in the parallel mesh migration stage to balance mesh components among the processors.

8.4.4 Interfacing among C and Fortran 90 for mesh migration

ParMeTiS only returns information on the mapping of elements to (new) processors; it cannot actually migrate elements across a parallel system. Our parallel mesh migration scheme reuses the efficient mesh_build() routine to construct the new mesh from the ParMeTiS repartitioning. During this mesh_build process the element information is migrated according to this partitioning.

As seen in Figure 8.4, information required by the ParMeTiS partitioner is provided by calling a Fortran 90 routine that converts the mesh adjacency structure into ParMeTiS format (hidden). When this call returns from C, the private mesh_build() routine constructs the new distributed mesh from the old mesh and the new repartitioning by performing mesh migration. Fortran 90 allows optional arguments to be selected by keyword. This allows the mesh_build() routine to serve multiple purposes since a keyword can be checked to determine if migration should be performed as part of the mesh construction process (see Figure 8.5).

This is another way in which the new features of Fortran 90 add robustness to the code design. The way in which the new mesh data is presented, either from a file format or from a repartitioning, does not matter. Once the data in organized in our private internal format the mesh can be reconstructed by code reuse.

```
subroutine mesh_repartition(this)
type (mesh), intent(inout) :: this
    ! statements omitted...
    call PARMETIS(mesh_adj, mesh_repart, nelem, nproc) ! C call
    call mesh_build(this, new_mesh_repart=mesh_repart)
end subroutine mesh_repartition
```

Figure 8.4 A main program with selected PAMR library calls.

```
subroutine mesh_build(this, mesh_file, new_mesh_repart, in_core)
integer, dimension(:), intent(in), optional :: new_mesh_repart
logical, intent(in), optional :: in_core
! statements omitted...
        if (present(new_mesh_repart)) then
            ! perform mesh migration...
        end if
        ! (re)construct the mesh independent of input format...
end subroutine mesh_build
```

Figure 8.5   Skeleton view of the mesh_build() routine.

8.4.5   Design Applications

The AMR library routines have been applied to the finite-element simulation of electromagnetic wave scattering in a waveguide filter, as well as long-wavelength infrared radiation in a quantum-well infrared photodetector. Future applications may include micro-gravity experiments and other appropriate applications. This software runs on the Cray T3E, HP/Convex Exemplar, IBM SP2, and Beowulf-class pile-of-PC's running the LINUX operating system.

8.5   DO SCIENTIFIC PROGRAMS BENEFIT FROM OBJECT-ORIENTED TECHNIQUES?

Many of the benefits of object-oriented programming are probably most suited only for very large programs—typically programs larger than many scientific programmers are involved in—perhaps hundreds of thousands of lines. Nevertheless, most principles can be applied for smaller and medium scale efforts. We have applied these techniques in an experimental way on skeleton programs, but the effort addresses principles that will affect large scale development.

Object-oriented design will not solve every problem, but it does force one to consider issues that normally might be ignored. This includes defining abstractions clearly, their relationships, and organization for extension to new problems. This increases the development time for an initial project, but hopefully reduces the effort in constructing new related projects.

One must question if the highly promoted benefits are real. Since scientific applications contain components that work together to solve complex problems, encapsulation and modularity promote good designs for these programs. This clarifies understanding, allows modifications to be introduced in a controlled manner, increases safety, and supports the work of multiple contributors. Some features, like sub-typing inheritance and dynamic polymorphism are good object-oriented principles, but their general usage in scientific applications was very limited, or non-existent, in our experience. Some research has been performed in measuring the performance effects of constructs used in an object-oriented fashion in C++ and Fortran 90 (Cary *et al.*, 1997; Norton, 1996). However, more work is needed before performance can be a deciding factor in language selection, all other factors being relatively equal. Modern applications are growing more complex, hence object-oriented techniques can clarify their organization, but this does not imply that all aspects of the paradigm are necessary.

We have experienced increased software safety, understandability by abstraction modeling, Fortran 77 level performance, and the modernization of existing programs without redevelopment in a new language. The modern features of Fortran 90 are redefining the nature of Fortran-based programming, although much interest is focussed on comparing Fortran 90 and C++ for scientific programming (Cary *et al.*, 1997; Decyk *et al.*, 1997b; Decyk *et al.*, 1997c; Norton, 1996). New projects may be considering one of these languages, existing projects may reconsider their decision to adopt C++ or Fortran 90, and current projects may consider migration to the "other" language. Most of this activity grew from a realization of the new features Fortran 90 makes available as well as the continual improvement in C++ compiler technology. Scientific programming can benefit from object-oriented techniques.

8.6 CONCLUSION

The use of object-oriented concepts for Fortran 90 programming is very beneficial. The new features add clarity and safety to Fortran programming allowing computational scientists to advance their research, while preserving their investment in existing codes.

Our web site provides many additional examples of how object-oriented concepts can be modeled in Fortran 90 (Norton *et al.*, 1996). Many concepts, like encapsulation of data and routines can be represented directly. Other features, such as inheritance and polymorphism, must be emulated with a combination of Fortran 90's existing features and user-defined constructs. (Procedures for doing this are also included at the web site.) Additionally, an evaluation of compilers is included to provide users with an impartial comparison of products from different vendors.

The Fortran 2000 standard has been defined to include explicit object-oriented features including single inheritance, polymorphic objects, parameterized derived-types, constructors, and destructors. Other features, such as interoperability with C, will simplify support for advanced graphics within Fortran 2000.

Parallel programming with MPI and supercomputers is possible with Fortran 90. However, MPI does not explicitly support Fortran 90 style arrays, so structures such as array subsections cannot be passed to MPI routines. The Fortran 90 plasma PIC programs were longer than the Fortran 77 versions (but more readable), and much shorter than the C++ programs because features useful for scientific programming are not automatically available in C++. Also, the portability of the Fortran 90 parallel adaptive mesh refinement system among various machines and compilers was excellent compared to difficulties experienced with portability of C++ programs among compilers and machines.

Acknowledgments

This work was supported by a National Research Council Resident Research Associateship at the National Aeronautical and Space Administration Jet Propulsion Laboratory, California

Institute of Technology. Additional support was received by the NASA Office of Space Science and the Cornell Theory Center for access to the Cray T3E and IBM SP2 respectively. We also appreciate the support of Tom A. Cwik, Viktor K. Decyk, Robert D. Ferraro, John Z. Lou, and Boleslaw K. Szymanski in this research.

References

Birdsall, C. and Langdon, A. (1991). *Plasma Physics via Computer Simulation*. The Adam Hilger Series on Plasma Physics. Adam Hilger, New York.

Cary, J., Shasharina, S., Cummings, J., Reynders, J., and Hinker, P. (1997). Comparison of C++ and Fortran 90 for object-oriented scientific programming. *Computer Physics Communications*, 105:20–36.

Decyk, V., Norton, C., and Szymanski, B. (1997a). Expressing object-oriented concepts in Fortran 90. *ACM Fortran Forum*, 16(1):13–18. Also as NASA Tech Briefs, Vol. 22, No. 3, pp. 100–101, March, 1998.

Decyk, V., Norton, C., and Szymanski, B. (1997b). How to express C++ concepts in Fortran 90. Technical Report PPG–1569, Institute of Plasma and Fusion Research, UCLA Department of Physics and Astronomy, Los Angeles, CA 90095-1547.

Decyk, V., Norton, C., and Szymanski, B. (1997c). Inheritance in Fortran 90: How to make better stopwatches. Submitted for publication.

Ellis, T. M., I.R.Philips, and Lahey, T. (1994). *Fortran 90 Programming*. Addison–Wesley, Reading, MA.

Karypis, G., Schloegel, K., and Kumar, V. (1997). Parmetis: Parallel graph partitioning and sparse matrix ordering library version 1.0. Technical report, Department of Computer Science, Univ. Minnesota.

Liewer, P. and Decyk, V. (1989). A general concurrent algorithm for plasma particle-in-cell simulation codes. *J. of Computational Physics*, 85:302–322.

Norton, C. (1996). *Object Oriented Programming Paradigms in Scientific Computing*. PhD thesis, Rensselaer Polytechnic Institute, Troy, New York. UMI Company.

Norton, C., Decyk, V., and Szymanski, B. (1996). High performance object-oriented programming in Fortran 90. Internet Web Pages. http://www.cs.rpi.edu/~szymansk/oof90.html.

Norton, C., Decyk, V., and Szymanski, B. (1997). High performance object-oriented scientific programming in Fortran 90. In M. Heath *et al.*, editor, *8th SIAM Conference on Parallel Processing for Scientific Computing*, Minneapolis, MN. On CD-ROM.

Norton, C., Szymanski, B., and Decyk, V. (1995). Object oriented parallel computation for plasma simulation. *Communications of the ACM*, 38(10):88–100.

9 ALTERNATING DIRECTION IMPLICIT METHODS ON DISTRIBUTED AND SHARED MEMORY PARALLEL COMPUTERS

Robert Rankin and Igor Voronkov

Department of Physics,
University of Alberta, Canada

rankin@space.ualberta.ca

Abstract: Global scale simulation codes capable of modeling the interaction of the solar wind with the Earth's magnetosphere are being developed by the international space community. We shall describe different implementations of a 3D alternating direction implicit (ADI) magnetohydrodynamic (MHD) simulation code on distributed and shared memory parallel and parallel/vector computers. The specific architectures considered are the IBM SP and Silicon Graphics Origin, and techniques appropriate to Cray-class computers. On the distributed memory IBM SP, the ADI code was run under PVMe and MPI, and gave good scalability under a single program multiple data (SPMD) implementation. Scalability on the SGI Origin was approximately linear up to the available configuration size of 128 processors, with a change in slope at 32 processors. On Cray-class computers, very high efficiencies are possible when vectorization is combined with parallel processing. It is shown that while this leads to near optimum performance, it is achieved at the expense of increased memory requirements.

Keywords: numerical simulation, parallel processing, vector processing, magnetohydrodynamics.

9.1 INTRODUCTION

The interaction of high speed (400-1,000 km/s) solar wind plasma with the Earth's magnetic field results in a stretched-cigar-shaped magnetic cavity called the magnetosphere. This is a region in which the Earth's magnetic field is compressed on the dayside, and stretched out on the nightside to distances greater than the Moon's orbit. The outer boundary of the magnetosphere prevents the interplanetary magnetic field and solar wind from entering the magnetospheric cavity, and shields biological systems from harmful effects of ionizing radiation and high-energy particles. However, during geomagnetically active periods, solar wind plasma does enter the magnetosphere, particularly during magnetic storms and substorms, which are expected to increase in frequency and intensity as the new cycle 23 solar maximum is approached. A magnetic sub-storm is broadly defined by a growth phase, expansive phase, and a recovery phase. During the growth phase, energy in the form of magnetic fields and particles is stored in the magnetosphere. The expansive phase follows the unloading of field and particle energy into the auroral zones, and is accompanied by enhanced visual auroral displays at high latitudes.

It is the goal of magnetospheric physics to predict magnetic storms and sub storms. Modeling such phenomena places large memory and CPU demands, and the development of new algorithms. While many computational models are at an early stage, progress has been made to the extent that real data can be used to predict certain properties of the magnetosphere. The complexity and scale of the problems to be addressed led to an early decision to exploit large-scale parallel computing systems. From the point of view of a fluid dynamicist, the basic process being modeled is a supersonic flow interacting with a spherical blunt object (the Earth). However, the flow is an ionized plasma rather than a neutral gas, and the interaction is mainly with the Earth's magnetic field rather than with its surface. Further complexity is that electric and magnetic fields are involved, and that flows are often turbulent and subject to instabilities. However, that being said, a standard set of resistive non-ideal MHD equations can be used to model some of the more basic magnetospheric processes.

To model magnetospheric processes within the fluid limit, an alternating direction implicit (ADI) method (Douglas Jr. and Gunn, 1964; Finan III and Killeen, 1981) has been used to solve the general curvilinear resistive 3D magnetohydrodynamic (MHD) equations for a one fluid warm plasma. ADI methods are widely used in engineering and computational fluid dynamics calculations, but until recently were not particularly well suited to 3D problems due to large non-contiguous array access along the third coordinate direction. However, in this article it will be shown that with present day high-speed inter-processor communication, and available large cache memories of 1 MB to 16 MB, high performance with ADI algorithms is possible on shared and distributed memory parallel and parallel/vector computers.

In the material that follows, we shall describe the set of equations that are solved numerically (Section 9.2). We describe the algorithm, implementation and performance of ADI on a Silicon Graphics Origin using the SGI MIPSpro Fortran (Section 9.3). Section 9.4 discusses running ADI on a IBM SP using PVMe/MPI and on Cray-class computers using Cray microtasking. Finally, a summary of the contributions on this paper are presented in Section 9.5.

9.2   PEAR–A 3D PARALLEL MAGNETOHYDRODYNAMIC CODE

The ADI PEAR code (Rankin *et al.*, 1993) is derived from the following set of nonlinear magnetohydrodynamic plasma fluid equations in general curvilinear coordinates:

$$\frac{\partial \rho}{\partial t} + \nabla \bullet (\rho \mathbf{v}) = 0 \tag{9.1}$$

$$\frac{\partial}{\partial t}\rho \mathbf{v} + \nabla \bullet (\rho \mathbf{v}\mathbf{v}) + \nabla P - \frac{1}{c}\mathbf{J} \times \mathbf{B} = 0 \tag{9.2}$$

$$\nabla \times \mathbf{E} = -\frac{1}{c}\frac{\partial \mathbf{B}}{\partial t}, \qquad \nabla \times \mathbf{B} = \frac{4\pi}{c}\mathbf{J} + \frac{1}{c}\frac{\partial \mathbf{E}}{\partial t} \tag{9.3}$$

$$\frac{\partial}{\partial t}\rho \varepsilon + \nabla \bullet \rho \varepsilon \mathbf{v} + P\nabla \bullet \mathbf{v} - \eta J^2 = 0 \tag{9.4}$$

$$\rho \varepsilon = \frac{P}{\gamma - 1}, \qquad \mathbf{E} + \frac{\mathbf{v} \times \mathbf{B}}{c} = \eta \mathbf{J} \tag{9.5}$$

Here, ε is the internal energy per unit mass, $\gamma = 5/3$ is the ratio of specific heats, and η is the plasma resistivity. The above equations can be reduced to a set of eight nonlinear partial differential equations by using equations 9.3 and 9.5 to eliminate the electric field $\mathbf{E}$, current density $\mathbf{J}$, and internal energy per unit mass ε. The resulting equations for the velocity $\mathbf{v}$, magnetic field $\mathbf{B}$, pressure P, and density ρ, can then be expressed in conservation or pseudo-conservation law form. Details are omitted.

To solve the MHD equations, successive calls to three subroutines—*stepx, stepy,* and *stepz,*—are made, where (x, y, z) are the spatial curvilinear coordinates, e.g. (r, θ, ϕ) in spherical coordinates. Each subroutine advances the equations in time along its respective physical coordinate direction. For example, subroutine *stepx* performs a set of independent "sweeps" in the x-direction. During each sweep, a set of block-tridiagonal equations is inverted and iterated to convergence through a Newton-Raphson procedure. If n_x, n_y, and n_z represent the number of nodal points in the three coordinate directions of the numerical finite difference grid, there are $n_y \times n_z$ independent sweeps (involving vectors of length n_x) associated with the execution of subroutine *stepx*. Each sweep or set of sweeps constitutes an amount of parallel work. Likewise, there are $n_x \times n_z$ and $n_x \times n_y$ independent sweep directions associated with execution of subroutine *stepy* and *stepz*, respectively. The advantage of this approach is that a very large 3D problem is reduced to the parallel solution of a large number of grouped together 1D problems, each of which can be solved relatively straightforwardly.

To construct the ADI algorithm, we first define state vectors $\mathbf{U}$ and $\mathbf{u}$ by,

$$\mathbf{u} = \begin{bmatrix} \mathbf{B} \\ \mathbf{v} \\ \rho \\ P \end{bmatrix}, \qquad \mathbf{U} = \begin{bmatrix} \mathbf{B} \\ \rho \mathbf{v} \\ \rho \\ P \end{bmatrix} \tag{9.6}$$

and then express the equations in the form,

$$\frac{\partial \mathbf{U}}{\partial t} + \mathbf{F}(\mathbf{u}) + \mathbf{G}(\mathbf{u}) + \mathbf{H}(\mathbf{u}) = 0 \tag{9.7}$$

where $\mathbf{F}$, $\mathbf{G}$, and $\mathbf{H}$ represent nonlinear MHD flux terms with principal dependencies $\partial/\partial x, \partial/\partial y$, and $\partial/\partial z$, which can be determined by consideration of equations 9.1-9.5. The time-discretized form of Equation 9.7, with separated spatial dependencies, then becomes,

$$\frac{\mathbf{U}^x - \mathbf{U}^n}{\Delta t} + \frac{1}{2}[\mathbf{F}^x + \mathbf{F}^n] + \mathbf{G}^n + \mathbf{H}^n = 0 \tag{9.8}$$

$$\frac{\mathbf{U}^{x+y} - \mathbf{U}^n}{\Delta t} + \frac{1}{2}[\mathbf{F}^x + \mathbf{F}^n + \mathbf{G}^{x+y} + \mathbf{G}^n] + \mathbf{H}^n = 0 \tag{9.9}$$

$$\frac{\mathbf{U}^{x+y+z} - \mathbf{U}^n}{\Delta t} + \frac{1}{2}[\mathbf{F}^x + \mathbf{F}^n + \mathbf{G}^{x+y} + \mathbf{G}^n + \mathbf{H}^{x+y+z} + \mathbf{H}^n] = 0 \tag{9.10}$$

in which superscripts indicate time advancement with respect to the respective spatial coordinate. Thus, in Equation 9.8, $\mathbf{U}^x$ and $\mathbf{F}^x$ are the implicit variables being advanced with respect to the x-direction. After Equation 9.8 is solved, $\mathbf{U}^{x+y}$ and $\mathbf{G}^{x+y}$ become implicit variables in advancing the equations with respect to y, and so on. The remaining task is to linearize equations 9.8-9.10. We illustrate this by considering Equation 9.8, which represents $stepx$ of the PEAR code. Equation 9.8 may be written as,

$$\mathbf{U}^x + \frac{\Delta t}{2}\mathbf{F}^x = \Delta t \left[\mathbf{U}^n - \mathbf{G}^n - \mathbf{H}^n - \frac{1}{2}\mathbf{F}^n \right] = \mathbf{S}^n \tag{9.11}$$

where $\mathbf{S}$ is an effective source term evaluated at the start of a timestep. The $\partial/\partial x$ terms in $\mathbf{F}$ couple adjacent x-coordinate nodal positions $i - 1, i$, and $i + 1$. This dependence is treated implicitly. Mixed or cross-product derivative terms in $\mathbf{F}$ must be treated explicitly since they represent a data dependence between sweep lines. Physically, these terms involve plasma resistivity and are usually of second-order importance. The explicit terms are separated out from $\mathbf{F}$ by writing $\mathbf{F} = \mathbf{F}^{imp}(u^x_{i-1}, u^x_i, u^x_{i+1}) + \mathbf{F}^{exp}$. Since the terms $\mathbf{F}$, $\mathbf{G}$ and $\mathbf{H}$ are nonlinear they are Taylor-expanded about grid point i, leading to the block tridiagonal form,

$$\mathbf{C}_i \delta u^{l+1}_{i-1} + \mathbf{B}_i \delta u^{l+1}_i + \mathbf{A}_i \delta u^{l+1}_{i+1} = \mathbf{D}^l_i \tag{9.12}$$

It is the purpose of subroutine $solvex$ (called successively by subroutine $stepx$) to iteratively solve Equation 9.12 for each sweep line of the finite difference grid. $\mathbf{C}_i$, $\mathbf{B}_i$, and

A_i are formed from Jacobian matrices of the form $\partial F/\partial u$ for $u = u_{i-1}, u_i$, and u_{i+1}, respectively. The Jacobian matrices are a tedious calculation by hand, and are computed symbolically using *Mathematica*, and output in *FortranForm*. The Jacobian 8×8 sub-matrices are the elements of the block-tridiagonal equations. In *stepx* there are $n_y \times n_z$ independent sweep directions each requiring solution of a block-tridiagonal system of the form of Equation 9.12. Note that it is the sweep-directions that are executed in parallel, and not the solution to the block-tridiagonal systems. In general, there is not enough work available to justify parallelizing the block-tridiagonal solvers themselves. Special care is, however, needed, since the solvers are at the deepest level of the code, and determine the overall speed of execution.

9.3   A BASIC PARALLEL VERSION OF THE ADI METHOD

In this section we describe implementation of the PEAR code on the SGI Origin. The Origin is the evolution of the Power Challenge series, and both systems use the MIPS R10000 CPU, with cache sizes varying from 1 MB to 4 MB. The Origin's biggest benefits are scalability and bandwidth. In particular, the Power Challenge was built around a single large 1.2 GByte/s bus. Unfortunately, this bus is a shared resource which must handle all CPU, I/O, and memory transactions. These various transactions must be arbitrated, and this limits the scalability of the system. At some point the benefit of additional processors becomes negligible. On the other hand, the Origin is built around a Module (which can be up to 8 CPUs). Modules are built into a hypercube topology by linking them together via a fast 1.6 GBytes/s CrayLink. The benefit of this approach is that each module comes with its own individual bandwidth, so that as modules are added, bandwidth is correspondingly increased. This means that the Origin is far more scalable. IRIX unifies the physically disparate pools of memory, so that the user sees one logical memory space.

The ADI PEAR code is inherently parallel in that the DO loops which control the solution of the equations in the *x, y,* and *z* directions have no iteration dependencies and are easily made parallel. Figure 9.1 shows the implementation on the SGI Origin running under MIPSpro Fortran.

Similar DO loop structures are involved in subroutines *stepy,* and *stepz*. The SGI $doAcross directive results in the call to *solvex* invoking a child task that solves a set of sweep directions for the unique nodal values j and k. In the above example, the compiler and run time environment parallelize only the outer DO loop, although it is a simple matter to collapse the nested loop structure, and manually set the number

```
C$doAcross local(k,j)
    do k = 1, nz
        do j = 1, ny
            call solvex(j,k)
        enddo
    enddo
```

Figure 9.1   Fragment from the subroutine *stepx*.

```
do k = 1, n-1
  do i = k+1, n
  a(i,k) = a(i,k)/a(k,k)
    do j = k+1, n
    a(i,j) = a(i,j) - a(i,k)*a(k,j)
    enddo
  enddo
enddo
```

```
t10 = a(1,0) / a(0,0)
a(1,1) = a(1,1) - t10*a(0,1)
a(1,2) = a(1,2) - t10*a(0,2)
a(1,3) = a(1,3) - t10*a(0,3)
a(1,4) = a(1,4) - t10*a(0,4)
a(1,5) = a(1,5) - t10*a(0,5)
a(1,6) = a(1,6) - t10*a(0,6)
a(1,7) = a(1,7) - t10*a(0,7)
a(1,0) = t10
t20 = a(2,0) / a(0,0)
a(2,1) = a(2,1) - t20*a(0,1)
. . .
```

(a) Original.   (b) Loop-unrolled.

Figure 9.2   LU-decomposition on $n \times n$ matrix a $(n = 8)$.

of tasks to run within each parallel invocation. The parallel routine *solvex* uses a common block *comMatvP* to store the A_i, B_i, C_i, and D_i matrices for the sweep vector of length n_x. These, and other similar data structures are made task local by including the option "-Xlocal, comMatvP_" on the command line at compilation. This is all that is required for basic parallel execution on a shared memory platform such as the SGI.

One bottleneck in the PEAR code results from the required inversion of large numbers of block-tridiagonal equations. A sweep direction involves a vector of nodal values of length n_x, in the case of *stepx*. The sub-matrices or elements of the block-tridiagonal equations are small matrices that must be inverted using LU-decomposition for each of the vector elements along a sweep line. This results in an inefficient use of the cache memory available on typical RISC processors. In practice, large amounts of loop unrolling are required to alleviate this problem. Figure 9.2 illustrates this with the original LU-decomposition routine in (a) and the loop unrolled routine in (b).

Similar techniques are applied to the many other routines in the solvers, resulting in an increase in the execution speed by a factor varying from 2 to 2.5. Other optimizations have also been made, and source code for the block-tridiagonal solvers is available upon request from the authors. A speedup diagram showing the performance of the PEAR code on up to 128 nodes of an SGI Origin is shown in Figure 9.3. The speedup at 32 processor nodes is around 26.5, with an abrupt change in slope that is followed by a further period of approximately linear speedup. The change in the performance characteristics at 32 nodes probably results from the 'meta-router' which connects the four 32-processor segments of the system that was used. No attempts were made to alleviate this problem, and it remains for future investigation. On 128 nodes, the amount of work per parallel task has decreased by two orders of magnitude, with scalar work now accounting for 10 percent of the total work. This leads us to conclude that Amdahl's law is becoming an issue. However, the ADI code was designed

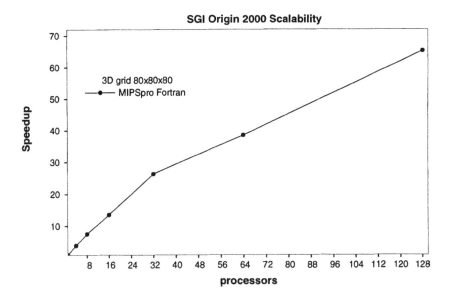

Figure 9.3 Scalability of the PEAR code on up to 128 nodes of a Silicon Graphics Origin 2000. Circles denote data points obtained using SGI MIPSpro Fortran.

for large problems, and the work per task will increase with larger finite difference grid sizes.

9.4 MORE ADVANCED TECHNIQUES

In this section, we describe implementations of the ADI method on distributed and shared memory parallel/vector processing platforms. These techniques require substantial re-writing of the basic algorithm described in the previous section. First we describe the implementation of the PEAR code under the PVMe/MPI environments provided by IBM, and then we briefly describe techniques that offer high efficiency on machines such as the Cray C90.

The IBM AIX PVMe product is an implementation of PVM. PVM (Parallel Virtual Machine) was developed by Oak Ridge National Laboratory, and PVMe maintains the same parallel programming paradigm offered by PVM, so that from the user's point of view no code changes are needed while running a PVM code under the PVMe environment, except that the user has to recompile source code using different libraries and header files. The most important difference between PVM and PVMe is that the latter enables communication through the SP high-speed interconnect or switch. IBM has in fact supplemented PVMe by MPI, which is now the preferred communications package of choice. However, the user implementation for MPI is very similar to PVMe, and the latter is described here.

In MIMD-based parallel environments, two programming models are commonly used. One is master/slaves, and the other is single program multiple data (SPMD).

Master/slaves uses one master program to control the main stream of procedures, and to dispatch tasks or data to slaves running in parallel. In the SPMD model, the same program is replicated on a number of processors, and each processor operates on statements and data assigned to it using processor identifying labels 0, 1, 2...N-1, where N is the total number of processors. In adapting the ADI method for the SP2, we used both a SPMD implementation and a master/slaves model. In the master/slaves model, the master spawns N-1 slaves, handles serial code execution, and executes 1/N of the parallel work left behind.

Load balancing is a key factor affecting parallel efficiency. Under PVMe, the most efficient division of work among processors involved cutting the domain in the z−direction for the x and y solvers, and then cutting the domain along the y−direction when moving to the z−solver. Reduction of communication overhead is also a key issue. In our original PVM implementation of the ADI method, 3D explicit flux terms $\mathbf{F}^n$, $\mathbf{G}^n$ and $\mathbf{H}^n$, and the state vector, $\mathbf{U}^n$, were computed serially at the start of a timestep and then broadcast to the *stepx* and *stepy* solvers even though they both use the same domain cut. This introduced an unnecessary communication related to the x and y solvers. This was modified so that data arrays containing the field variables $\mathbf{b}^n$, $\mathbf{v}^n$, ρ^n and P^n at the start of a timestep were broadcast before *stepx* was called, and the calculation of the explicit flux terms was moved inside each sub-domain. This meant that no communication was needed when changing from the x to the y direction. All iteration and timestep controls were also done within each sub-domain.

Figure 9.4 shows the master/slaves scalability of the PEAR ADI code on a model problem with $n_x \times n_y \times n_z = 80 \times 48 \times 48$, respectively, and for up to eight nodes of an IBM SP2. For comparison, we show PVMe and PVM implementations that use the switch and normal Ethernet communication protocols. Clearly, there is almost no speedup while running PVM on Ethernet, and better scalability when using PVM and the SP2 switch. The best result is obtained with PVMe on 8 nodes, giving a 5.2-fold speedup, or 65% parallel efficiency. However, it is clear that all of the PVM results are strongly influenced by communication time overhead, which is due to the fact that parallelism in the ADI code is reasonably fine grained. This prompted us to re-program the algorithm using a SPMD implementation, with results summarized in Figure 9.5. The difference here is that the entire data is partitioned among nodes at the out-set, and the only communication required is after *stepx* and *stepy* when data must be broadcast to the partition that achieves the z−direction solution. In the SPMD model, near linear speedup is achieved on 16 SP2 nodes, with elapsed times comparable to those observed on the SGI Origin on the same number of processors.

On machines such as the Cray T3D or C90, parallel and vector processing is possible. However, the basic ADI method is not efficiently vectorizable at its deepest level, namely at the level of the block-tridiagonal solvers. As described in Section 9.3, loop unrolling on scalar processors can alleviate problems with short (length 8 or nested 8) DO loops, but the performance gains are limited. On vector processors, short loops suffer similar or worse performance hits. To circumvent these problems on parallel/vector machines, and considering only *stepx* of the PEAR code, we can divide the number of independent sweep lines $n_y \times n_z$ into n_{cpu} chunks (normally the number of processors available). Then we stripe the number of nodal points within a chunk into vectors of

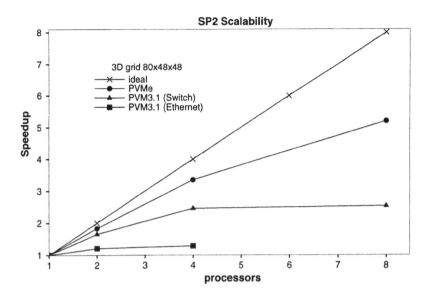

Figure 9.4   Master/slaves implementation of the PEAR ADI code on up to 8 nodes of a "thin-node" IBM SP2.

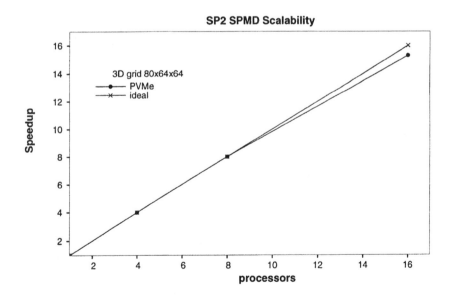

Figure 9.5   SPMD implementation of the PEAR ADI code on up to 16 nodes of a "thin-node" IBM SP2.

```
isize = (ny*nz+ncpu-1)/ncpu
do parallel np = 1, ncpu
     call solvex((np-1)*isize,min(isize,ny*nz-(np-1)*isize))
enddo
```

Figure 9.6   Fragment from subroutine *stepx*.

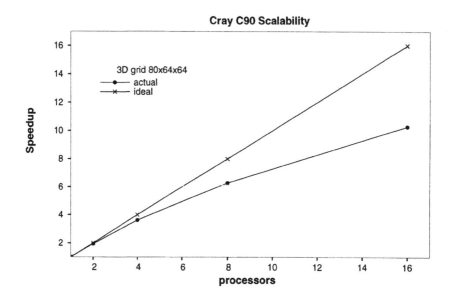

Figure 9.7   Implementation of the PEAR ADI code on up to 16 nodes of a Cray C90.

the required length (64 on Cray-class machines). We can view the vectorization part as solving all lines (or sweeps) within a parallel segment simultaneously. The code segment in Figure 9.6 summarizes the procedure.

The call to *solvex* in Figure 9.6 cuts the domain into chunks along the z-direction. Then, within subroutine *solvex*, striping over the number of nodal points *isize* is done for the optimum vector length. The advantage is that almost all of the pipelined operations will take place on vectors of optimum length. Additionally, short DO loops in the solvers will now also have inner loops of optimum vector length. This is, however, achieved at the expense of increased memory, since the $\mathbf{A}_i$, $\mathbf{B}_i$, $\mathbf{C}_i$, and $\mathbf{D}_i$ arrays now have the vector length as an additional parameter: arrays which were originally dimensioned (n, n, n_x) are now dimensioned as $(vec_{length}, n, n, n_x)$. Results are shown in Figure 9.7, with the 16 processor result corresponding to approximately 4 gigaflops. Although the scalability in Figure 9.7 is flattening out, diagnostics show that this is likely due to not enough work being available for the given problem size. Again, the reason for using parallel processing is to address very large problems. The implementation on the Cray C90 very likely approaches optimum performance with higher computation speeds than either of the implementations described earlier.

9.5 CONCLUSIONS

In this article, we have shown that 3D ADI methods can be implemented efficiently on distributed and shared memory parallel and parallel/vector processors. We have described implementations that use message passing libraries, such as PVM and MPI, and vendor supplied parallel Fortran compilers. Under PVM, the most efficient implementation was obtained for a SPMD version of the PEAR ADI code. Each processor then executes virtually identical statements for its unique partition of the 3D finite difference grid. Communication is kept to a minimum, and is required during a single timestep only when switching to the third dimension of a 3D problem. Shared memory ADI applications are easily implemented, although with block-tridiagonal systems, extensive loop unrolling is necessary to achieve optimum execution speed. Shared memory parallel/vector processors give superb performance on the PEAR ADI code, and also offer good portability across platforms. It is anticipated that extremely large problems can be addressed with current technology, thus opening up the possibility of new knowledge that would otherwise be unattainable.

Acknowledgments

Research for this project has been supported by the Canadian Space Agency and the Natural Science and Engineering Research Council of Canada, NSERC. The authors would also like to thank SGI, IBM Canada, and Cray Research for invaluable discussions related to this work.

References

Douglas Jr., J. and Gunn, J. (1964). A general formulation of alternating direction methods. *Numerische Mathematik*, 6:428.

Finan III, C. and Killeen, J. (1981). Solution of the time dependent, three dimensional resistive magnetohydrodynamic equations. *Comput. Phys. Commun.*, 24:441.

Rankin, R., Harrold, B., Samson, J., and Frycz, P. (1993). The nonlinear evolution of field line resonances in the Earth's magnetosphere. *J. Geophys. Res.*, 98:5839.

10 SIMULATING GALAXY FORMATION ON SMPS

Robert Thacker[1], Hugh Couchman[2] and Frazer Pearce[3]

[1]Theoretical Physics Institute,
University of Alberta, Canada
[2]Department of Physics and Astronomy,
University of Western Ontario, Canada
[3]Department of Physics,
University of Durham, UK

thacker@phys.ualberta.ca
couchman@coho.astro.uwo.ca
f.r.pearce@dur.ac.uk

Abstract: We discuss the implementation of HYDRA, a combined hydrodynamic and gravity N-body particle integrator, on shared memory architectures with a symmetric multi-processor configuration (SMP). Parallelization is achieved using the parallelization directives in the MIPSpro Fortran 77 instruction set. We examine the performance of the code and find excellent scaling up to 8 processors. The maximum number of process elements is limited by race conditions which are difficult to overcome. The new code takes advantage of the considerable processing power of mid-range machines and permits simulations of a scale lying between those attempted on workstations and massively parallel supercomputers.

Keywords: numerical methods, hydrodynamics, cosmology, galaxy formation.

10.1 INTRODUCTION

Computational cosmology has come far since the first analog simulations by (Holmberg, 1941), which used light bulbs and galvanometers to model the gravitational forces between an ensemble of particles. The interaction of N-bodies under gravity is a computationally demanding problem as simple pairwise algorithms require N^2 operations to calculate the forces for every particle in the ensemble. However, provided a small but quantifiable integration error is acceptable, alternative methods of solution may be used. By accepting this small error it is possible to progress to larger and larger N.

The large-scale dynamics of our Universe are dominated by a component which only interacts gravitationally and hence is undetectable by traditional (electromagnetic) means. This 'dark matter' is widely believed to be in the form of relic subatomic particles, such as massive neutrinos, which interact via particle mechanisms that were common a few seconds after the 'Big Bang' but have become progressively more dormant as the Universe evolves into the low energy state we see today. The actual amount of dark matter in the Universe is not known. Dynamical estimates suggest there may be thirty times as much dark matter as baryonic (protons and neutrons, for example), although a more realistic estimate is in the range 8 to 15. As a consequence of this, cosmologists can study the very large-scale structures in the Universe by simulating the interaction of purely gravitational particles. This field has developed from initial studies on the gravitational clustering of 32^3 particles (Efstathiou and Eastwood, 1981) to the state of the art 1024^3 simulations now possible on the Cray T3E (Evrard, 1998). In addition to the hardware development that has allowed this increase in simulation size, there has also been a corresponding improvement in the performance of simulation algorithms. The advent of combined mesh/particle methods (Hockney and Eastwood, 1981) and tree methods (Barnes and Hut, 1986) provides a toolkit for numerical cosmological studies that can cover almost all physical scenarios from the galactic scales all the way to superclusters and beyond (MacFarland *et al.*, 1997).

Purely gravitational N-body simulations do not address the problem of galaxy formation for which an account of the gas dynamics is necessary. Convergent motions in the gas are dissipated during the formation of protogalaxies and this can only be modeled by including hydrodynamics. This increases the computational requirements by an order of magnitude. Not only must the collisional gas component be added, the number of time-steps is typically higher since gas dynamics involve shock processes. Memory requirements are also more than tripled due to additional arrays that must be stored. However, the goal of understanding galaxy formation provides sufficient motivation to overcome these obstacles. Because galaxies are the yardstick by which we measure the Universe, it is vital to have a good understanding of their formation process. Analytic models provide insight, but cannot accurately address details of dynamic evolution and morphology. This is because it is necessary to make simplifying assumptions, such as spherical symmetry, to make the problem analytically tractable. Simulation quantifies the three-dimensional dynamics and morphology. This is of vital importance since galaxy formation is believed to occur via a hierarchical merging process in which small sub-galactic objects merge over time to form the final galaxy.

To study galaxy formation at very high resolution, massively parallel supercomputing is necessary. The reason for this is that doubling the linear resolution—achieved by increasing the particle number by eight—increases the computational task by a factor of approximately 20. Not only does the solution time per time-step increase because of the additional particles by a factor of approximately 10, but the Courant condition asserts that the number of time-steps must be doubled due to the higher resolution. Thus the progression from a simulation with 64^3 particles to one with 256^3 results in a problem that is roughly 400 ($2 \times 2 \times (256/64)^3 \times 3 \times (\log 256 - \log 64)$) times more computationally expensive. This is close to the performance scaling difference between workstations and supercomputers.

The complexity of designing and implementing a high-performance application on a supercomputer makes code development inherently slow. Since most of these machines have distributed memory, message-passing algorithms are necessary. However, the message-passing paradigm is cumbersome to use. Any program that uses global solution methods (MacFarland *et al.*, 1997), as opposed to local finite difference methods, will require extensive effort to rewrite. Global addressing makes the task inherently easier, as can be seen in the implementation of the HYDRA simulation code on the Cray T3D using CRAFT by (Pearce and Couchman, 1997). In view of this, the global addressing of the shared memory programming model renders parallelization a simple task. Additional motivation for development of a shared memory code can be had from examining the computational scaling of the program. As indicated previously, doubling the linear resolution leads to a task 20 times as computationally intensive. Thus shared-memory SMP machines, which typically have 8-32 CPUs, provide an extremely important resource for problems that fall midway between workstations and supercomputers. These machines are becoming more commodity based, reducing the cost significantly while maintaining the performance. Many academic researchers now have access to the gigaflop computing that these machines can supply.

The organization of this paper is as follows: Section 10.2 details the coupled hydrodynamic and gravitational equations that must be solved. An overview of the algorithm is presented in Section 10.3. Section 10.4 gives some performance results and algorithm optimizations. We conclude in Section 10.5 with a brief review.

10.2   THEORY

The dominant physical process on galactic scales is gravitation. Rather than solving the 2-body force equation for all the bodies in the simulation, we convolve to a field representation and solve Poisson's Equation,

$$\nabla^2 \phi = 4\pi G \rho, \tag{10.1}$$

where ϕ is the gravitational potential, G is Newton's constant and ρ is the mass density. Using Green's functions and a Fourier convolution, we can solve this equation and recover the gravitational potential. Given the potential, we may then extract the force and consequently the acceleration for any particle in the simulation.

As mentioned in the introduction, modern cosmology theory dictates that most of the mass of the Universe is in the form of gravitationally interacting dark matter. However, the remaining matter is a self-gravitating gas (an admixture of Hydrogen and Helium), part of which condenses to form stars. The gas component is a low density plasma and the hydrodynamic equations for the fluid elements i are,

- mass continuity,

$$\frac{d\rho_i}{dt} + \rho_i \nabla.\mathbf{v}_i = 0 \tag{10.2}$$

- equation of motion,

$$\frac{d\mathbf{v}_i}{dt} = -\frac{1}{\rho_i}\nabla P - \nabla\phi + \mathbf{a}_i^{visc} \tag{10.3}$$

- internal energy equation,

$$\frac{du_i}{dt} = -\frac{P}{\rho_i}\nabla.\mathbf{v}_i, \tag{10.4}$$

where d/dt is the convective directive of Lagrangian mechanics, P pressure, ρ density and $\mathbf{a}_i^{visc}$ is an artificial viscosity which is required to resolve shocks in the flow. The relationship between pressure, density and temperature is provided by the perfect gas equation of state.

We have presented the simplest representation of these equations. They can be generalized to include the effects of electric and magnetic fields as well as radiative cooling, which is a particularly important phenomenon in galaxy formation. The solutions to these equations are found on an individual particle basis using a technique known as Smoothed Particle Hydrodynamics (SPH) (Gingold and Monaghan, 1977; Lucy, 1977). Briefly, SPH solves the equations by assuming that the particles in the simulation are a Monte-Carlo representation of the underlying fluid distribution. The internal energy equation, for example, is solved by constructing a Monte-Carlo estimate of the velocity divergence, multiplying by the local temperature (and factors relating to the adiabatic index) and finally integrating over the time-step.

Initial conditions are provided by a set of adiabatic fluctuations in the matter density. The precise spectrum of perturbations has to be calculated according to the cosmology of interest. At very large scales, data from the COBE experiment (Smoot *et al.*, 1992) constrains the acceptable amplitude of these fluctuations.

10.3   ALGORITHM

10.3.1   Efficiency and Overview

When attempting to solve equations 10.1-10.4 for a large number of particles, algorithm efficiency is of vital importance. Solution of the equations of motion for gravity can be found using a pairwise approach, but this is extremely slow for a large number of particles (N greater than 10^4), since the calculation is $O(N^2)$. As mentioned earlier, an alternative approach is to use a field representation for the density of the particle data. In converting to a field representation on a mesh we have to perform an operation of $O(N)$. Once we have done this we can achieve a rapid solution to Poisson's equation using a fast Fourier transform (FFT), which is $O(L^3 \log L)$, where L is the size of the Fourier mesh. The total execution time using this method scales as $\alpha N + \beta L^3 \log L$, where α and β are constants. This technique is widely known as the "Particle-Mesh" (PM) method (Hockney and Eastwood, 1981).

The shortcoming of PM is that it is unable to resolve below a length scale set by the resolution of the mesh which is Fourier transformed (corresponding to the Nyquist frequency of the mesh). It is, however, possible to supplement this PM force with

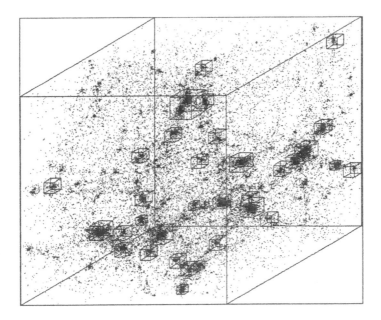

Figure 10.1   Refinements placed in a box 1/5 the size of the total simulation volume.

an additional short-range force calculated from local particles on a pairwise basis, yielding a force which is accurate below the resolution of the mesh. This method is known as "Particle-Particle, Particle-Mesh" (P^3M) (Hockney and Eastwood, 1981) and the execution time scales in proportion to $\alpha N + \beta L^3 \log L + \gamma \sum N_{pp}^2$, where γ is a constant and N_{pp} corresponds to the number of particles in the short-range force calculation within a specified region. The summation is performed over all the PP regions, which are identified using a chaining mesh. P^3M suffers the drawback that under heavy gravitational clustering the short-range sum used to supplement the PM force slows the calculation down dramatically—the N_{pp}^2 term dominates as too many particles contribute to the short-range sum. Although acutely dependent upon the particle number and relative clustering in a simulation, the algorithm may slow down by a factor between 10-100 or possibly more. One may go to finer and finer meshes, but memory limitations rapidly make this prohibitive and, additionally, computation becomes wasted on areas that do not require the higher resolution.

Adaptive P^3M (AP^3M) (Couchman, 1991) solves this problem by isolating regions where the N_{pp}^2 term dominates and solving for the short-range force in these regions using FFT methods supplemented by a smaller number of short-range calculations. This process is a repeat of the P^3M algorithm except that the FFT is now calculated with isolated boundary conditions. At the expense of a little additional bookkeeping, this method circumvents the slow down dramatically. Dependent upon clustering, AP^3M is often 10 times or more faster than P^3M. See Figure 10.1, for an example of where refinements are placed.

When implemented in an adaptive form, SPH is an $O(N)$ scheme. It fits well within the P^{3}M method since the short-range force supplement for the mesh force must coincidentally search to find the particles which are used in the SPH calculation. Hence, once a list of particle neighbors has been found, it is simple to sort through this and establish which particles are to be considered for the gravitational calculation and the SPH calculation. Details of the exact solution method for SPH may be found in (Gingold and Monaghan, 1977). The modifications to make the scheme adaptive are found in (Hernquist and Katz, 1989). To incorporate the adaptive SPH into AP^{3}M we simply have to make coordinate scalings and do minor bookkeeping. The combined adaptive P^{3}M-SPH code, HYDRA, in serial Fortran 77 is available on the World Wide Web from http://coho.astro.uwo.ca/pub/hydra_consort/hydra.html.

The solution cycle of one time-step may be summarized as follows,

1. Assign mass to the Fourier mesh.

2. Convolve with Green's function using the FFT method to get potential. Difference this to recover mesh forces in each dimension.

3. Apply mesh force and accelerate particles.

4. Identify regions where the standard short-range $O(N^2)$ algorithm is inefficient. Where so, place a new sub-mesh and evaluate the short-range forces via the further use of $O(N \log N)$ Fourier methods.

5. Accumulate the gas forces (and state changes) as well as the short-range gravity for all positions not in a sub-mesh.

6. Repeat steps 1-5 on the sub-meshes until the forces on all particles in the simulation have been accumulated.

7. Update the time-step and repeat.

Note that the procedure for placing meshes is hierarchical in that a further sub-mesh may be placed inside a sub-mesh. This procedure can continue to an arbitrary depth but, typically, speedup only occurs to a depth of six levels of refinement.

10.3.2   SMP Fortran Implementation

Fortran remains a popular language among scientists. Although primarily due to the large amount of legacy code, this popularity is also maintained by the continued high performance of Fortran compilers. Indeed, high performance is especially important in large N-body simulations where the total number of calculations borders on a petaflop. A further reason for the popularity is the ongoing evolution of the language to include modern programming paradigms, such as dynamic allocation of memory in the Fortran 90 standard.

From the programming viewpoint, if algorithmic development is an ongoing project (as opposed to having a predefined production code) then it is extremely useful to keep the program code as simple as possible. To this end the addition of parallel extensions, similar in nature to the Parallel Computing Forum (PCF) standard or the later X3H5 proposal (Leasure, 1994), simplifies the task of maintaining parallel codes. Further, the shared memory environment allows global addressing which saves the trouble of implementing explicit message passing.

The code which we discuss was first developed on an SGI Power Challenge and then later on an Origin 2000 system using MIPSpro Fortran (SGI, 1997). We have made extensive use of their dynamic load scheduling option, which has proven particularly useful for load balancing some difficult situations.

10.3.3   Data Geometry and Parallel Scheme

Particle-grid codes of this kind are difficult to parallelize. The fundamental limitation is the degree to which the problem may be subdivided. Given that the computation-to-communication cost is relatively low compared with many other simulation and modeling codes (*e.g.* finite element codes), we must ensure that the computational granularity is sufficiently large that communication does not dominate the calculation and destroy the algorithm scalability. For our code, this corresponds to maintaining a low surface-area-to-volume ratio for the spatial blocks into which the particles and grid cells are decomposed. This sets an upper bound on the degree to which the problem can be subdivided which, in turn, limits the number of processors that may be used effectively for a given problem size. The code is a good example of Gustafson's conjecture: a greater degree of parallelism may not allow arbitrarily increased execution speed for problems of fixed size, but should permit larger problems to be addressed in a similar time.

On SGI S^2MP machines, where data assignment across nodes may be specified, we chose a block distribution for the global particle data arrays (*e.g.* positions, velocities). The motivation for distributing the data is derived from the fact that a large amount of the computation proceeds using a linked list which is ordered in the z-direction. For efficient memory access we wish to keep the data organization closely tied to the ordering of the linked list. While there is the possibility of a load imbalance, the homogeneous nature of the matter distribution which is being modeled, relative to the computational slab decomposition, usually prevents serious problems. However, for a large number of processors—greater than 32—this does become significant.

The code uses two meshes for force evaluation: the main Fourier mesh used in the representation of the particle density data and a coarser chaining mesh used to create a linked list of particles for the short-range force. Typical sizes for the meshes are 256^3 and 116^3, respectively, for a simulation involving $N = 2 \times 128^3$ particles. We do not face memory limitations with this code as the storage requirements for these arrays and the particle data amounts to 18N words for the serial code and 30N words for the parallel version due to local replication of data on processor elements (PEs).

We denote the process of interpolating the particles onto the Fourier mesh as *mass assignment*. This process involves a race condition since if two processors have particles which occupy the same position in space then they may both attempt to write to the same grid position simultaneously. As mentioned, the particles are ordered along the z-axis. Hence if each PE is given a simple ordered block of particles to assign (of size $N/\#PEs$), then this problem is unlikely to occur because of the homogeneous nature of the particle distribution. Should we need to study problems with a less homogeneous distribution then a different algorithm is necessary. We admit that this solution does not remove the race condition. However tests in both clustered and unclustered states show that, provided the number of PEs does not go above $L/16$, we do not encounter problems. This is the only simple algorithm that guarantees load balancing since each PE must perform the same number of calculations. To completely avoid race conditions a complicated system of ghost cells must be used, and a cyclic data distribution is necessary to provide load balancing. Clearly, our method is a simple way of avoiding this complexity. We are also investigating the possibility of using software locks to allow us to use alternative algorithms. The solution to this problem in the CRAFT implementation is to use the *atomic update* facility which is a lock–fetch–update–store–unlock hardware primitive that allows fast updating of arrays where race conditions are present.

The short-range forces are also calculated over the chaining mesh in a similar fashion. In this case there is a loop over all the chaining mesh cells with the outermost loop distributed across the processor space (leading to a slab decomposition). The race condition is less prevalent here since the work quantum is large and evenly distributed. Furthermore, instead of using particle indexes to schedule the computation, it proceeds over the chaining cells which have a fixed geometry. We have tested this algorithm and have found no failure provided that the number of PEs is less than $Ls/8$, where Ls is the size of the chaining mesh along one dimension. Note that if we place a barrier at the halfway point of each slab then the race condition would be exactly removed, but we have not implemented this as it is not necessary. Again this condition is avoided in the CRAFT version by the use of the atomic update primitive. As in the mass assignment problem, we are also investigating the use of software locks to allow the implementation of other algorithms which provide better load balancing and more PEs to be used.

The adaptive part of the algorithm is dealt with in two ways. Firstly the algorithm may place sub meshes that have half the number of cells along one dimension of the base Fourier mesh. If so, the calculations for these refinements are performed in parallel across the maximum number of PEs allowed by the race condition avoidance criteria. This avoids a possibly serious load imbalance that could occur if one very large refinement was placed and calculated on a single PE. Next, smaller sub-meshes are distributed as a task farm amongst the processors. As soon as one processor becomes free it is immediately given work from a pool. Guidance for this process is provided by the load scheduler in the compiler (SGI's dynamic scheduling option). By dynamic scheduling we decrease the possibility of a load imbalance.

10.4 PERFORMANCE AND OPTIMIZATION

The initial goal set for optimization was to achieve a 60 second cycle time for the top-level base mesh in a 2×128^3 simulation (on an 8-processor SGI Origin 2000). Although the final goal is to achieve a code that scales as best as possible, this was the acceptable limit (given the development time available) for the first production code. This compares to 40 seconds on a 200 MFlop serial workstation for a simulation with 2×64^3 particles.

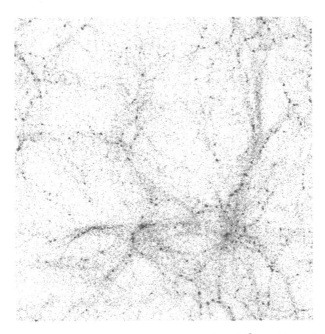

Figure 10.2   The z-projection of a simulation with 2×64^3 particles. Box size is 16 comoving Mpc, greyscale is representative of gas density and redshift is z=3.0.

As the particles in the simulation become clustered, the serial code slows down by a factor of 3-5 depending on the problem. This is acceptable since without refinements the code is 40 times slower in the clustered state. For a 2×128^3 particle simulation with 2,000 time-steps, the runtime is roughly seven days. While this may seem excessive, such large runs are only performed a few times a year. Most questions can be answered using smaller simulations and then, if more resolution is needed, a progression to larger particle number is made. The gains to be had from the increase in resolution may be seen by examining Figures 10.2 and 10.3.

Parallel speedup for the code on four nodes is 3.3 (83% efficiency) increasing to a 5.4-fold speedup on eight nodes (68% efficiency). This is due to approximately 6.5% of the code being inherently difficult to parallelize in a simple way, thus limiting the scalability. Given that 6.5% of the code is serial, the maximum theoretical speed up on eight processors is 5.5, indicating that we are achieving high efficiency in the parallel sections of our code.

A significant part of the serial execution time time (80%) is spent in the routine responsible for controlling the placement of the adaptive refinements. The workaround used on the T3D code is to update the placement of the largest adaptive meshes every 20 (for example) time-steps. This is a workable solution since the configuration of the particles does not change appreciably from one time-step to the next. We will be implementing this in our SMP code in the near future, which should markedly increase the parallel efficiency. Further, a parallel version of this algorithm is under development which we will incorporate when a production version is completed. Calculation

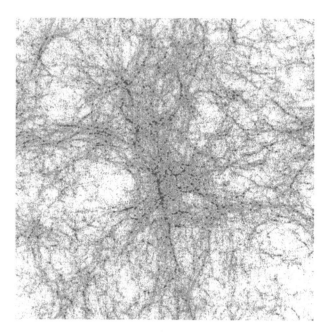

Figure 10.3   The z-projection of a simulation with 2×144^3 particles. The box size is 16 comoving Mpc at epoch z=3.8. The figure shows the gas distribution only, with the greyscale indicative of logarithmic density. Compared to Figure 10.2 (which has different realization of the initial conditions), the increase in mass resolution shows up filaments and void detail.

shows that removal of the serial overhead due to placing refinements can lead to a theoretical maximum speedup of 7 (88% efficiency) on eight nodes.

The remainder of the inherently scalar code (which amounts to 1.5% of the total run time) calculates Green's functions, as well as performing loading and unloading of data. We note that most routines are amenable to parallelization to the same degree as the parts already completed. With our collaborators, we have thus been able to show that it is possible to perform large particle calculations on massively parallel super-computers (MacFarland *et al.*, 1997). However, at present the benefit of incorporating these new routines is minimal given that there is room for improvement elsewhere.

The particle ordering that we enforce to alleviate race conditions helps to improve data locality. Locality reduces the probability of a cache miss, and as much as a 20% speed up improvement occurs as a result of this. Since we only reorder the array every 40 steps the small overhead paid for doing this serial computation is more than recouped.

Repackaging the mass array into the position array (as a fourth element) also improves performance, primarily due a large number of memory calls which ask for the positions and then the mass. A similar approach can be applied to the packaging of the velocity and acceleration arrays, although the performance gains are less significant.

The Fourier transform is used extensively and we should perhaps go to a math library implementation. However, the convolution routine is quite complex and it is a significant rewrite to change this for different data arrangements in the FFT. Further, the execution time spent calculating FFT's is small (less than 10%).

We have also yet to investigate the enormous number of possible compilation options that effect code performance. It is conceivable that aggressive pre-fetching may well speed up our code significantly. Also cache optimizations may well have a significant impact.

10.5 SUMMARY

We have presented the methodology behind our parallel implementation of the combined adaptive P^3M-SPH code, HYDRA, on shared memory SMP hardware using X3H5 compatible parallel Fortran extensions.

The global shared memory model allows simple array addressing, thereby removing the need for message passing, which in turn speeds up development time. The use of PCF compatible directives makes the code portable to a number of different shared-memory SMP architectures.

Given the difficulty inherent in parallelizing particle–grid codes in which the Lagrangian map suffers extreme distortion as gravitational clustering occurs, the scaling achieved on coarse-grained parallel machines is excellent. This judgement results from the only relevant criterion for success: the utility of the code for scientific experiments.

Development of this code fills a niche that exists between simulations that can be performed on workstations and those which require massively parallel supercomputers. Given the range of problems for which modest (2×128^3 particle) hydrodynamic simulations are necessary, and given the arms-length nature of supercomputing, this version of the code has been developed to take advantage of the wide range of modestly parallel midrange computers that are now available. The increased resolution that the parallelized code provides should allow in-depth investigations at a mass resolution ten times that possible on workstations. Alternatively, volumes eight times larger may be simulated to assess the role of the cosmological environment more fully.

Postscript: The new OpenMP standard was announced at Supercomputing'97. The acceptance of this standard by a number of vendors inspires confidence that a cross-platform shared-memory programming standard will be adopted. We shall endeavor to convert our code to this standard when compilers become available. Further, the inclusion of an atomic primitive in the instruction set will remove the race condition problem, thus opening the door to a higher number of PEs.

Acknowledgments

The authors thank NATO for providing Collaborative Research Grant CRG 970081 which facilitated their interaction.

References

Barnes, J. and Hut, P. (1986). A hierarchical O(N log N) force-calculation algorithm. *Nature*, 234:446–449.

Couchman, H. (1991). Mesh refined P^{3}M: A fast adaptive n-body algorithm. *Astrophysical Journal Letters*, 368:L23.

Efstathiou, G. and Eastwood, J. (1981). On the clustering of particles in an expanding universe. *Monthly Notices of the Royal Astronomical Society*, 194:503–526.

Evrard, A. (1998). In preparation.

Gingold, R. and Monaghan, J. (1977). Smoothed particle hydrodynamics: Theory and application to non-spherical stars. *Monthly Notices of the Royal Astronomical Society*, 181:375–389.

Hernquist, L. and Katz, N. (1989). TREESPH: A unification of SPH with the hierarchical tree method. *Astrophysical Journal Supplements*, 70:419–446.

Hockney, R. and Eastwood, J. (1981). *Computer Simulation Using Particles*. McGraw Hill, New York.

Holmberg, E. (1941). *Astrophysical Journal*, 94:385.

Leasure, B., editor (1994). *Parallel Processing Model for High Level Programming Languages*. Draft Proposed American National Standard for Information Processing Systems.

Lucy, L. (1977). A numerical approach to the testing of the fission hypothesis. *Astronomical Journal*, 82:1013–1024.

MacFarland, T., Couchman, H., Pearce, F., and Pichlmeier, J. (1997). A new parallel P^{3}M code for very large-scale cosmological simulations. http://xxx.lanl.gov/abs/astro-ph/9805096.

Pearce, F. and Couchman, H. (1997). HYDRA: A parallel adaptive grid code. *New Astronomy*, 2:411–427.

SGI (1997). *MIPSPro Fortran 77 Programmer's Guide*. Silicon Graphics document 007-0711-060.

Smoot, G. *et al.*(1992). Structure in the COBE differential microwave radiometer first-year maps. *Astrophysical Journal*, 396:L1.

III   Database Systems

11 DATA PARALLELISM IN JAVA

Michael Philippsen

Computer Science Department,
Universität Karlsruhe, Germany
phlipp@ira.uka.de

Abstract: Java supports threads and remote method invocation, but it does not support either data-parallel or distributed programming. This paper discusses Java's shortcomings with respect to data-parallel programming and presents workarounds. The technical contributions of this paper are twofold: a source-to-source transformation that maps forall statements into efficient multi-threaded Java code, and an optimization strategy that minimizes the number of synchronization barriers needed for a given program.

The transformation, the optimization, and a distributed runtime system have been implemented in the JavaParty environment. In JavaParty, code compiled with current just-in-time compilers runs merely a factor of 4 slower than equivalent Fortran 90 or HPF, on both a shared-memory parallel machine (IBM SP/2) and a distributed-memory parallel machine (SGI Origin 2000). Furthermore, better compiler technology is on the horizon, which will further narrow the performance gap.

Keywords: data parallelism, Java, elimination of synchronization barriers.

11.1 INTRODUCTION

Java has been adopted by application programmers so quickly, in part, because of its portability. Similarly, Java's portability is one of the key issues that might make Java one of the main languages for scientific programming as well. Unfortunately, Java has neither been designed for parallel programming nor for distributed programming; hence, it does not offer adequate support for either of them. Sun has not indicated whether it will alter Java to address these concerns.

High Performance Fortran (HPF) is a procedural language that maintains much of the original Fortran syntax and semantics. From the software engineering point of view, Java has some advantages over HPF (Koelbel *et al.*, 1994), since it allows for clearly designed, reusable, and maintainable code. However, HPF combines the collective knowledge on data-parallel programming. HPF offers forall statements, directives to distribute array data in a distributed memory environment, and statements for collective communication. Java does not offer language support for any of these. There are at least three different approaches for adding HPF expressiveness to Java:

A) Dual-Language Approach. To use both Java and HPF at the same time, libraries and wrapper functions are added to Java. Libraries communicate array data between Java and HPF, while wrappers allow the invocation of HPF code. This approach aims at a gradual move from HPF programs to Java programs using HPF functionality. It preserves investments made in HPF programs but does not prevent well-designed object-oriented programs from taking over in the future.

Mixing Java and HPF causes several problems. Since different HPF compilers behave differently, the portability gained by using Java is lost in the HPF subroutines. At the Java-to-HPF interface, problems are caused both by the different sets of primary types in each language (in particular by Java's lack of complex numbers), and by Java's inability to access array substructures efficiently (such as dimensions or their subsections). Finally, since Java's floating point operations are overly restrictive with respect to IEEE floating point standards, different sections of a dual-language program may compute incompatible results.

B) Language and ByteCode extensions. The syntactic elements of HPF are added to the Java language; compilers that implement Java plus HPF semantics are constructed.

All HPF features can be covered with this approach. In a compiler for the new language, costly array bound checks can be avoided, the format used for floating point I/O can be chosen in a way that best suits state-of-the-art parallel computers, other Java features can be omitted to avoid difficult interactions with HPF semantics (*e.g.* there may be no need to implement arrays and their sub-dimensions as objects), and the memory layout of array data can be hidden from the user. The disadvantage is that the special purpose compiler cannot be as portable as the JDK. For instance, since array accesses are explicit in ByteCode, the decision to avoid them requires a specialized ByteCode. For efficient access to sub-sections of arrays, the memory layout must be visible and the garbage collector must understand these sub-arrays. Therefore, one of the key advantages of Java, namely portability, is lost in this approach.

C) Pre-Compiler and Library. A third approach is to extend the expressive power by means of libraries for data distribution, collective communication, and `forall` statements. This approach has been taken for various parallel extensions of C++, *e.g.* (Chandy and Kesselman, 1993; Grimshaw, 1993). However, since Java does not support templates, macros, and generic classes, some restrictions apply.

Both data distribution and collective communication are best handled by libraries that manipulate and partition arrays on a distributed parallel machine. Since Java does not support parameterized classes, replicated versions of these libraries must exist for all primary types and for all objects. Because of the lack of generic classes, these libraries need to be replicated for every primitive type and for all objects. For objects, type casts must be used to down-cast elements to the proper type by the programmer. One of Java's strengths, the strong static type checking, is lost.

Section 11.2 shows that the `forall` statement is *not* a candidate for macro or template processing. Therefore, the `forall` semantics must either be offered by means of another library or be implemented by a pre-compiler's source-to-source transformation that accepts `forall` statements in an extended Java and generates multi-threaded Java equivalents for them.

JavaParty consists of a pre-compiler and a distributed runtime system, and offers data-parallel programming by taking the third approach.

In this paper, we discuss the design ideas of JavaParty with respect to data-parallel programming. In Section 11.2 it is argued that `forall` statements cannot be handled by macros or templates. The key issues of JavaParty's source-to-source transformation are presented in Sections 11.3 to 11.5. The transformation techniques that enable transparent distribution of JavaParty objects in a parallel hardware environment can be found elsewhere (Philippsen and Zenger, 1997). Section 11.6 discusses the related work. Benchmark results are presented in Section 11.7 and compared to equivalent HPF and Fortran 90 implementations.

11.2 FORALL MACROS DO NOT WORK IN JAVA

Java's threads express control parallelism. Threads are objects inheriting from a standard Thread class. This class offers a `run` method that returns immediately to the caller but is in fact executed concurrently. There is no other concept of concurrency available in Java. In particular, data parallelism cannot be expressed directly.

This section argues that `forall` statements are too complex for macro processing:

Too-Simplistic Macro.   The Java documentation supports the belief that threads can easily be used for data parallelism as well. Consider the `forall` statement (left column) in Figure 11.1. With inner classes, the JDK documentation suggests a multi-threaded equivalent (right column) (Rose, 1996).

Although the code fragment is hard to read at first, one can get used to the pattern. In the body of the first `for` loop, an anonymous inner class is declared that inherits from Thread (lines 4–8). The `run` method of class Thread is overwritten by a specific `run` method that calls `bar(ii)` (lines 5–7). Since Java does not allow regular local variables to be used in inner classes, one has to declare, initialize, and use a final helper variable instead (`ii`, lines 3 & 6). To execute `bar(ii)` concurrently, an object of that anonymous class is created and stored in an array of workers before its `start` method is called (lines 4 & 9). `Start` calls `run` which concurrently executes `bar(ii)`. The main program can only continue after all concurrent invocations of `bar(i)` have been completed. This is implemented in lines 11–16. A `join` is called

```
1   Thread[] worker = new Thread[100];
2   for (int i=0; i<100; i++) {
3       final int ii = i;
4       worker[i] = new Thread() {
5           public void run() {
6               bar(ii);
7           }
8       }
9       worker[i].start();
10  }
11  try {
12      for (int i=0; i<100; i++)
13          worker[i].join();
14  } catch (InterruptedException e) {
15      //some complicated cleanup code
16  }
```

```
forall (int i=0; i<100; i++)
    bar(i);
```

Figure 11.1   A `forall` and its naïve implementation using a simplistic macro.

for each worker in the array (line 13). Since one of the threads might have been interrupted, the synchronization needs to be in a `try-catch`.

This informal macro, transforming `forall` statements into multi-threaded Java, is incomplete, superficial, and inefficient with respect to the demands of data-parallel programs. `Forall` transformations require semantic knowledge and are much more complex.

Lack of Semantic Knowledge. Independent of the approach—there are alternatives to using inner class—the body of the `forall` is moved during the transformation. The surrounding state (as defined by the contents of local variables) must be captured at runtime and conveyed to the body's new position. Since Java's semantics distinguish between primary and reference types, and between local and instance variables, capturing this information is type dependent. Hence, `forall` transformation schemes depend on the types of surrounding variables.

The above simplistic macro is incomplete if local variables are involved. Different transformations are needed depending on whether local variables or instance variables are used in the original `forall`. Assume a `forall` statement that uses local variables for the "iterations" to store results.[1] The transformation of such a `forall` requires that additional helper instance variables be declared at the class level and are used instead of the local variables in the transformed body. After the end of the `forall`, the contents of the helpers must be assigned back to the local variables.

The transformation is thus non-local. It needs semantic knowledge about the types of variables involved. Since macros and templates operate locally and without semantic knowledge, the transformation cannot be achieved by them.

Although the macro approach cannot work, it is useful for understanding the simplistic macro's other disadvantages.

Lack of Generality. The simplistic macro suffers from limited applicability and generality. Instead of a simple function call, there can be statements that alter the flow of control: the enclosing method can be left by a `return` statement, an exception can be thrown, a surrounding loop can be the target of a `break` or `continue` statement, *etc.* The semantics of these statements inside of a `forall` are similar to their semantics when used inside of a regular `for` statement. All these constructs cannot easily be used inside of the inner class. A generally applicable transformation scheme needs to be much more complicated.

Lack of Efficiency 1: Object creation cost. Object creation in Java is expensive since each object needs some memory that must be initialized and registered for garbage collection. An additional lock object must either be created by the Java virtual machine or is provided by the operating system through a costly kernel entry.

---

[1] For example, imagine a parallel search for a target value in a two dimensional array. Each "iteration" of the `forall` searches for a different row of the array. If it is known that the target value either appears once or not at all, it is sufficient if the finder, if any, writes the target's coordinates into two index variables. No additional synchronization is needed.

Hence, for performance reasons, numerous or frequent creation of thread objects must be avoided.

Unfortunately, the simplistic macro creates threads *en mass* and frequently. A thread object is started for every single "iteration" of the original `forall`. Moreover, for each `forall`, a new set of worker threads is created, started, and joined. After joining, the threads are useless and disposed of by the garbage collector. They are useless because subsequent `foralls` have different bodies and require different `run` methods and because Java threads cannot be restarted.

For recursion and nested parallelism the above transformation scheme is not just costly, but also is useless, for it easily results in a number of thread objects that overwhelms the capabilities of either the Java virtual machine or the virtual memory system. To avoid these inefficiencies and to make a transformation work for nested parallelism as well, threads must be reused and standard virtualization loop techniques (Quinn and Hatcher, 1990) must be used. Both will increase the complexity of the transformation, since reuse of threads is difficult to achieve in Java, and since code for index set splitting and additional boundary checks for the first/last thread need to be added.

Lack of Efficiency 2: Fan-out and fan-in restrictions. The bottleneck caused by sequential start and join of threads is unacceptably costly, especially since data-parallel programs have too many and too small `foralls`.

When threads are recycled, `start` and `join` can no longer be used. Without them, however, tree-structured restart and barriers are more difficult to implement. The reason for this is that Java's weak memory consistency model (there is no sequential consistency (Lea, 1996)) is coupled to unsuitable or too costly synchronization operations. (It is not guaranteed that a thread sees the effects of write operations performed by different threads, unless synchronization operations have been executed earlier.)

Monitors and critical sections are too heavy weight to implement restarts and barriers since they often cause slow kernel entries in current JVMs. Other mechanisms are awkward to use. For example, `wait` and `notify` suffer from race conditions since `notify` operations are not buffered but are lost if no thread is waiting at an object. They are also difficult to use for (distributed) load-balancing, since the JVM or the operating system may decide at will which out of several waiting threads is continued upon notification.

An efficient transformation template thus needs a tree structure for start-up and barriers. It must be capable of dealing with the specific characteristics of Java's synchronization primitives. Both requirements add complexity.

Lack of Barrier Support. Pure SIMD semantics demand that all "iterations" proceed in perfect unison, *i.e.* no "iteration" starts to process an expression before the previous expression is completed by all "iterations". Because of polymorphism, dynamic dispatch, and aliasing, there is no way a macro can transform all methods that are called in the body of a `forall` into a form that is amenable for completely synchronous execution. Moreover, it is not even promising to use compile-time data de-

```
forall (int i=0; i<100; i++) {          forall (int i=0; i<100; i++)
  bar(i);                                 bar(i);
  sync;                                  forall (int i=0; i<100; i++)
  gee(i);                                 gee(i);
}
```

Figure 11.2   Explicit synchronization implemented by splitting `forall`s.

pendence analysis techniques for that task (Philippsen and Heinz, 1995; Tseng, 1995), since the same reasons render those techniques either too costly or too weak.

The only way out is to force that the programmer to detect potential data dependencies and split the `forall` into parts so that the sources and targets of all dependencies are located in different `forall` statements. Consider the code fragments in Figure 11.2. On the left, the necessity of a synchronization barrier is indicated by `sync`. On the right are the split `forall`s.

Since, in general, a macro cannot determine the proper placements of barriers, and since a (pre-)compiler can only determine them in very special cases, this task is left to the programmer. Section 11.5 discusses compiler support for splitting `forall`s. Compilers can even apply optimization techniques to minimize the number of barriers to improve performance.

11.3   PRE-COMPILING FORALLS DOES WORK

A pre-compiler can avoid the disadvantages of the macro approach. It has the necessary semantic knowledge and can modify given classes as a whole without distorting the clarity of the original code. A pre-compiler can use complex transformations, *i.e.* it can do its job completely and still achieve efficiency.

This is the design idea of the `forall` treatment implemented in JavaParty: an additional data structure is created that holds a pool of worker threads. Instead of creating, starting, joining, and discarding new threads for every single data-parallel section of the code, worker threads from that pool alternate between being blocked and executing: they are blocked and wait for work to arrive, execute the work, and return to a blocked state again. To allow for non-sequential start-up and join, the work packets are in a form which enables the worker thread to decide whether to execute the packet entirely (= virtualization loop) or to split it up and work only on a part. The worker's strategy takes into account the current load, the number of available threads, and the size of the packet.

Based on this design idea, Section 11.4 presents the details of JavaParty's source-to-source transformation and of the required libraries.

11.4   FORALL IMPLEMENTATION IN JAVA PARTY

The JavaParty pre-compiler accepts Java enhanced with `forall` statements and generates efficient multi-threaded pure Java code. Instead of a back-end that emits Java source code, a regular ByteCode back-end can be used to speed up compilation by avoiding additional file I/O and repeated semantic analysis.

```
class ForallThread extends Thread {
  private WorkPile wp;
  ...
  public void run() {
    while(true) {
      (->void) fkt = wp.getWork();
      fkt();
      wp.doneWork();
    }
  }
}
```

(a) ForallThread class.

```
public class WorkPile {
  private static WorkPile wp = new WorkPile();
  private static SplitStrategy splstrat = new DefaultSplitStrategy();
  ...
  public static void doWork(ForallController fac, (->void) fkt) {
    if (!wp.AddWork(fac, fkt))
      fkt();
  }

  synchronized boolean AddWork(ForallController fac, (->void) fkt) {
    if (full()) return false;
    else {
      totalWorkInQ++;
      fac.addWork();
      enqueue(new Work(fkt));
      notify();
      return true;
    }
  }

  synchronized (->void) getWork() {
    while (isEmpty()) {
      try {
        wait();
      } catch (Exception e) { /* ignore */ }
    }
    totalWorkInQ--;
    totalBusyThreads++;
    return dequeue().fkt;
  }
}
```

(b) WorkPile class.

Figure 11.3   ForallThread and WorkPile.

For now, we deliberately restrict the presentation to a shared memory architecture using Java's threads and a common address space. Section 11.5 presents the optimized implementation of synchronization barriers.

ForallThread and WorkPile.   A central component in JavaParty's forall library is the ForallThread. Upon program startup, a fixed number of these worker threads are started. Their run method is an endless loop that gets work from a work pile, executes that work, indicates completion, and starts over (see Figure 11.3(a)).

JavaParty uses Pizza's closures (Odersky and Wadler, 1997; Pizza, 1997) to capture state and to construct functions. We discuss the intermediate Pizza form here (with closures) because it is easier to understand than the Java code that results from Pizza's transformation. In Figure 11.3(a), wp.getWork returns a function from the work pile that neither takes arguments nor returns a result. In Pizza terminology, this function is of type (->void).

The method doWork puts new work onto the pile (see Figure 11.3(b)). If the pile is full and AddWork returns false, the current thread executes the work rather than

```
try {
    ForallController fac = new ForallController();
    WorkPile.doWork(fac, makeCl(0, 100, 1, fac));
    fac.finalBarrier();
} catch (RuntimeException _exc) {
    throw _exc;
} catch (Error _exc) {
    throw _exc;
}
```

(a) Replacement for the `forall` statement.

```
1  void() (int, int, int, ForallController)
2  makeCl = fun(int lb, int ub, int step, ForallController fac) {
3      void() forallCl = fun() {
4          if (! fac.breakFlag) {
5              if (! WorkPile.split(fac, makeCl, lb, ub, step)) {
6                  for (int i = lb; i <= ub; i += step) {
7                      if (fac.breakFlag) break;
8                      try {
9                          bar(i);
10                     } catch (Throwable e) {
11                         fac.caughtException = e;
12                         fac.breakFlag = true;
13                     }
14                     Thread.currentThread().yield();
15                 }
16             }
17         }
18         fac.workDone();
19     }
20     return forallCl;
21 }
```

(b) The additional `makeCl()` method.

Figure 11.4 `Forall` transformation pattern.

postponing it for another `ForallThread` to do the work. The `SplitStrategy` implements the semantics of `full`.

In addition to `doWork`, there is the synchronized method `AddWork` that either returns `false` if the work pile is full, or the work is added to the pile. An arbitrary `ForallThread` that is blocked in `getWork` is notified and resumes. It is irrelevant which `ForallThread` resumes, hence, Java's awkward notification mechanisms does not interfere.

A `ForallThread` waits inside of the `getWork` method when there is no work on the pile. Otherwise the work is dequeued and returned to the thread. Since a `while` loop is used to implement the blocking, a lost notification does not hurt. Since the method is synchronized, no other `ForallThreads` can interfere.

Forall Transformation Pattern. We now describe the code that results from the simple `forall` statement from Figure 11.1. The part of the resulting code that textually replaces the given `forall` is discussed first. The body of the original `forall` is moved to a closure declaration (discussed below) that precedes the code fragment in Figure 11.4(a). A try-catch block replaces the original `forall`. Inside of it, a `ForallController` is created that is unique to an execution of the whole `forall`. The work pile is asked to `doWork`. It is handed the `Forall-Controller` and the result of `makeCl` (Figure 11.4(b)). This function returns a closure describing the `forall` range and containing the original body.

Before we discuss the details of the closure, let us look into the rest of the `try`. The `finalBarrier` makes sure that all work packets that belong to the `forall` have been completed before execution continues. Since an unknown number of ForallThreads will each work on an unknown number of work packets, and since ForallThreads do not terminate but continue to serve, a sequential join phase no longer works. Instead, the ForallController keeps track of the number of packets to be completed. If that number reaches zero, execution can continue past the barrier. The catch clauses are necessary to pass exception objects from the body of the forall to the caller. If one of the "iterations" of the original forall throws an exception, this exception is now thrown inside of a ForallThread, *i.e.* at a different position in the code. To simulate the intended behavior, the exception object is caught in the ForallThread, buffered in the ForallController, and is then re-thrown inside of finalBarrier. The JavaParty compiler generates catch clauses for RuntimeExceptions and Errors since those need not be declared by the programmer. In addition (not shown in the example), a catch clause is generated for every named exception that can be thrown in the original body.

The function makeCl (Figure 11.4(b)) is a closure that has four parameters and returns a void function without parameters. That function is called forallCl; it is defined in lines 3–19. A ForallThread will later execute forallCl. The forall closure first checks that no other thread has encountered a break (line 4). The call of the split strategy, which checks whether the work packet must be split into parts before being executed (line 5), implements a high fan-out start. If no split is necessary, the work is performed in the virtualization loop (lines 6–15). (The upper bound check is simplified for readability; the correct test depends on the sign of step.) In each iteration, the loop checks the break flag again and executes bar(i). Any potential exception is caught and registered with the ForallController. To work on Java virtual machines without preemptive scheduling as well ("green threads"), the current thread is asked to yield. The thread indicates completion at the ForallController so that the finalBarrier can work as expected. When splitting work packets, forallCl calls makeCl. This is why two nested closures are required.

ForallController. The ForallController now works as expected (see Figure 11.5). It has a counter to keep track of the number of work packets that are

```
public class ForallController {
    public boolean breakFlag;
    public Throwable caughtException;
    int counter;
    ...
    public synchronized void workDone() {
        if ((--counter)<=0)
            notifyAll();
    }
    public synchronized void finalBarrier()
    throws Throwable {
        try {
            while (counter > 0)
                wait();
        } catch (Exception e) { /* ignore */ }
        if (caughtException != null)
            throw caughtException;
    }
}
```

Figure 11.5 ForallController class.

being processed or are on the pile. If `counter` reaches zero, all waiting threads are notified to continue. An exception that might have been registered with the `Forall-Controller` during the execution of the body is thrown to the caller at the original position in the code.

Split Strategy.   The strategy can access the `ForallController`, can call the closure `makeCl` that makes a `forall` closure, and knows the range of the virtualization loop in a given work packet. Moreover, the strategy knows about the number of active and blocked threads and the number of packets on pile. The routine is defined as follows:

```
static boolean split(ForallController fac,
                     (int,int,int,ForallController->(->void)) mCl,
                     int lwb, int upb, int stp) {
    ...
}
```

Several strategies can be implemented and selected dynamically. The JavaParty environment provides some default strategies that work well with nested `forall` statements and do not cause any deadlocks. These strategies can be refined by the user.

11.5   OPTIMAL BARRIERS IN JAVA PARTY

In the implementation of `foralls` presented so far, the programmer is in charge of splitting `forall` statements if data dependencies demand synchronization. However, as has been mentioned above, the compiler can assist the programmer with the splitting, especially if control flow statements are to be split. This section gives a motivating example first, before presenting the central idea of the restructuring that results in a minimal number of barriers.

Example of Barrier Elimination.   Consider the left hand side of Figure 11.6. The two necessary synchronizations are indicated by `sync`. The first `sync` refers to those conceptual threads that entered the `if` branch; it does not affect the other threads (and vice versa for the second `sync`.)

One synchronization barrier is sufficient for both the synchronization of the `if` branch and the synchronization of the `else` branch. The right hand side of Figure 11.6 shows the result of the transformation. An indication of which branch has been selected is stored temporarily in helper arrays `tmp_if` and `tmp_else`. The actual implementation in JavaParty uses bit vectors for performance reasons.

```
forall (int i=1; i<100; i++) {          boolean tmp_if[]   = new boolean[100];
  if (cond(i)) {                         boolean tmp_else[] = new boolean[100];
    bar(i);                              forall (int i = 1; i < 100; i++) {
    sync;                                  if (cond(i)) {
    gee(i);                                  bar(i);
  } else {                                   tmp_if[i] = true;
    bar(i/2);                              } else {
    sync;                                    bar(i/2)
    gee(i/2);                                tmp_else[i] = true;
  }                                        }
}                                        }
                                         forall (int i = 1; i < 100; i++) {
                                           if (tmp_if[i])   gee(i)
                                           if (tmp_else[i]) gee(i/2)
                                         }
```

Figure 11.6   Barrier elimination and transformation pattern for if-statements.

The straightforward approach is to use a single helper array to store the evaluation of cond(i), and to use that helper array in the second forall to decide for each thread which branch to enter. But that does not work in the general case of arbitrary control flow statements, especially when an individual thread k leaves the flow of control early in the first forall (*e.g.* because of a break statement). In such cases, the use of a single helper would result in a unintended execution of "iteration" k in the second forall. Thus, the general solution is to use a branch specific helper variable to register the fact that a thread k should continue in a later forall.

Forall Restructuring Pattern. JavaParty provides automatic and efficient forall splitting and thus simplifies the programmer's task. The central idea of the restructuring algorithm is to label individual statements in the body of a forall with *synchronization ranks* and sort them topologically according to these ranks. Two statements in the body of a forall that are separated by sync have synchronization ranks that are at least one apart. In Figure 11.6, the condition and both occurrences of bar have synchronization rank 1. The two occurrences of gee must be separated from the preceding invocations of bar by a barrier, hence they have synchronization rank 2. Statements that alter the flow of control, like the if statement, are attributed with an interval of synchronization ranks defined by the smallest and the largest synchronization rank of any statement in their bodies. The if statement of the example has the synchronization interval [1:2] since in both branches there are statements with ranks 1 and 2.

The synchronization ranks define an order that is used during the restructuring: for each rank, a separate forall statement is constructed. All those statements are placed in the body of the forall of rank x that either have synchronization rank x, or have an interval of synchronization ranks including x. Control flow statements can thus reappear a few times, albeit with changing conditions. In Figure 11.6, there are if statements in both foralls.

Before the first forall is generated that belongs to the synchronization interval of a control flow statement, as many helpers are created as there are branches in the control flow statement. Since the if of Figure 11.6 has two branches, two helpers are created. For loops, a single helper array is sufficient. The default initialization (false) is sufficient, except for while loops. For all loop constructs, there is the additional problem that the transformation has to pull the loop out of the forall statement. This is shown in Figure 11.7 where the forall on the left hand side runs until the while loop has been terminated for each value of i. The result of the transformation is shown in the right column.

In a first forall statement, the helper array is initialized according to cond(i). The while loop runs until no element of the helper array is true. In the body of the while loop there are two foralls, one for each synchronization rank. At the end of the last forall, the helper variable might be set to false if for a certain index the original while loop would terminate.

The JavaParty implementation includes transformation rules for all control flow constructs. In particular, JavaParty can transform nested forall statements with arbitrary many sync points into a sequence of nested forall statements without

```
forall (int i=1; i<100; i++) {        forall (int i=1; i<100; i++)
  while (cond(i)) {                      tmp_while[i] = cond(i);
    bar(i);                            while (atLeastOneTrue(tmp_while)) {
    sync;                                forall (int i=1; i<100; i++) {
    gee(i);                                if (tmp_while[i])
  }                                          bar(i);
}                                        }
                                         forall (int i=1; i<100; i++) {
                                           if (tmp_while[i]) {
                                             gee(i);
                                             if (!cond(i)) tmp_while[i]=false;
                                           }
                                         }
                                       }
```

Figure 11.7   Barrier elimination and transformation pattern for while-statements.

any internal sync. The details of the other transformation rules are omitted here for reasons of brevity.

11.6 RELATED WORK

An advantage of JavaParty over other projects aimed at adding data parallelism to Java is portability. JavaParty programs run on workstations, on shared memory parallel computers, and in truly distributed environments. Moreover, JavaParty's forall statement allows for a single program approach that is independent of the underlying topology. And finally, JavaParty programs come close to the native HPF and Fortran 90 performance.

(Hummel *et al.*, 1997) worked on SPMD programming in Java and faced some of the same problems we have discussed in Section 11.2. Their main contributions were classes for efficient synchronization and for regular and irregular collective communications. Such classes are useful for JavaParty. The system runs on an IBM SP/2, but in contrast to JavaParty's 100% pure Java approach, it uses a runtime system written in C that interacts with a native MPI communication library.

In the HPJava project at Syracuse, (Carpenter *et al.*, 1997) use wrappers to interface to native HPF code and library-based extensions of Java. They offer useful classes for distribution of array data and for collective communication as well. Whereas Java-Party has the capability of transforming a single program with forall statements into a distributed program that communicates by means of message passing, HPJava requires the implementation and invocation of several node programs. Such node programs know about the processor topology, the arrays to be used and their distribution across the machines. Virtualization loops get their boundaries from library calls. Since the programmer is in charge of distributing the threads (one per processor) several of the disadvantages of a macro transformation are avoided. However, HPJava's approach cannot handle nested parallelism nor does it offer any support for the efficient implementation of barriers. HPJava uses native routines and MPI calls to implement the communication as well.

11.7 RESULTS

We have studied large-scale geophysical algorithms to evaluate the efficiency of Java-Party's forall. Veltran velocity analysis and Kirchhoff migration are basic algo-

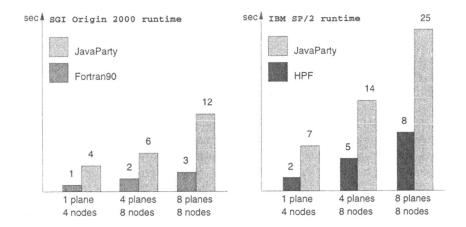

Figure 11.8   Performance numbers.

rithms used in geophysics for the analysis of the earth's sub-layers by means of sound wave reflection caused by explosions. Since it can take terabytes of input data to cover a reasonable area, the performance of these algorithms is crucial. The geophysics and the details of the benchmarks are discussed in (Jacob, 1997).

In cooperation with the Stanford Exploration Project (Clearbout and Biondi, 1996), we have implemented these algorithms in JavaParty, HPF, and Fortran 90. We then benchmarked the programs on up to 8 nodes of an IBM SP/2 distributed memory parallel computer and on 8 nodes of an SGI Origin 2000 shared memory machine. The JavaParty implementation uses JDK 1.1.4 with just-in-time compilation on the IBM SP/2. On each node, a separate JVM is started. The JVMs communicate by means of RMI. For the HPF measurements, version 2.2 of the Portland Group High Performance Fortran compiler was used. On the SGI Origin 2000, we used a beta version of JDK 1.1.5 with just-in-time compiler. Since this beta version offers native thread support, JavaParty did not use any RMI communication. SGI's standard parallelizing Fortran 90 compiler is used for the Fortran measurements.

On both platforms, optimizing Fortran compilers have been used. Java's performance was not attained by compiled and optimized native code but instead relied on interpreters with just-in-time compilation features.

As shown in Figure 11.8, on the SGI our JavaParty implementation is slower than the equivalent Fortran 90 program by a factor of about 4. On the SP/2, JavaParty faces a slowdown by 3. In part, the slowdowns are due to Java's mandatory and implicit array boundary checking.

The JavaParty program automatically adapts both to the number of planes[2] to be processed and to the number of nodes available. The Fortran programs did not have

---

[2]A plane is a two-dimensional matrix that holds the sensors' measurements for a single explosion over time. Informally, the evaluation of a plane can show the structure of the Earth's sub-layers on a straight line from the point of the explosion towards the center of the Earth.

the same adaptability. Instead, we had to change some constants and recompile the Fortran code for each of the measurements. Without the manual changes and recompilation, the performance of the most general and slowest program would have shown up repeatedly. For example, to process one plane on four nodes with the general program, the Origin 2000 needs 3 seconds and the SP/2 needs 8 seconds.

We expect significant performance increases for Java in the near future because of two main reasons. First, we had to use JDK 1.1.x, because later releases are not yet available for our hardware platforms. Later versions (JDK 1.2, HotSpot) have increased performance (especially RMI performance, improved native thread support and just-in-time compilation) on Solaris and Wintel platforms and are likely to show the same effect in our environments. Second, compilers producing optimized native code like IBM's High Performance Java Compiler (IBM, 1997) are on the horizon. These compilers will approach Fortran performance because they can apply much more sophisticated optimization techniques than current just-in-time compilers.

11.8 CONCLUSION

This paper has discussed alternatives to add data parallelism to Java, has reasoned that `forall` statements in general cannot be implemented by means of standard macro processing, has presented JavaParty's source-to-source transformation that turns `forall` statements into efficient multi-threaded equivalents, has sketched an optimization technique to avoid unnecessary synchronization barriers, and has demonstrated that JavaParty's data parallelism can achieve performance that comes close to HPF and Fortran 90 code and is expected to improve.

The JavaParty system is freely available for non-commercial projects. For details and downloading see http://wwwipd.ira.uka.de/JavaParty.

Acknowledgments

I would like to thank the JavaParty group, especially Matthias Zenger, for their support of the JavaParty environment. Matthias Jacob implemented the geophysical algorithms. Christian Nester pointed out several synchronization bugs in the first version of the forall transformation (Winkel, 1997). Furthermore, we want to express gratitude to Maui High Performance Computing Center as well as Karlsruhe Computing Center for the access to the IBM SP/2 and SGI Origin 2000.

References

Carpenter, B., Chang, Y.-J., Fox, G., Leskiw, D., and Li, X. (1997). Experiments with "HPJava". *Concurrency: Practice and Experience*, 9(6):579–619.

Chandy, K. and Kesselman, C. (1993). CC++: A declarative concurrent object-oriented programming notation. In Agha, G., Wegner, P., and Yonezawa, A., editors, *Research Directions in Concurrent Object-Oriented Programming*, pages 281–313. MIT Press.

Clearbout, J. and Biondi, B. (1996). Geophysics in object-oriented numerics (GOON): Informal conference. In *Stanford Exploration Project Report 93*.

Grimshaw, A. (1993). Easy to use object-oriented parallel programming. *IEEE Computer*, 26(5):39–51.

Hummel, S., Ngo, T., and Srinivasan, H. (1997). SPMD programming in Java. *Concurrency: Practice and Experience*, 9(6):621–631.

IBM (1997). High performance Java. http://www.alphaWorks.ibm.com.

Jacob, M. (1997). Implementing Large-Scale Geophysical Algorithms with Java: A Feasibility Study. Master's thesis, Computer Science Department, Universität Karlsruhe.

Koelbel, C., Loveman, D., Schreiber, R., Steele, G., and Zosel, M. (1994). *The High Performance Fortran Handbook*. MIT Press.

Lea, D. (1996). *Concurrent Programming in Java – Design Principles and Patterns*. Addison-Wesley.

Odersky, M. and Wadler, P. (1997). Pizza into Java: Translating theory into practice. In *24th ACM Symposium on Principles of Programming Languages (POPL)*, pages 146–159, Paris, France.

Philippsen, M. and Heinz, E. (1995). Automatic synchronization elimination in synchronous foralls. In *Frontiers'95: The 5th Symposium on the Frontiers of Massively Parallel Computation*, pages 350–357, McLean, VA.

Philippsen, M. and Zenger, M. (1997). JavaParty: Transparent remote objects in Java. *Concurrency: Practice and Experience*, 9(11):1225–1242.

Pizza (1997). http://www.cis.unisa.edu.au/~pizza.

Quinn, M. and Hatcher, P. (1990). Data-parallel programming on multicomputers. *IEEE Software*, 7(5):69–76.

Rose, J. (1996). *Inner class specification: Further example: Multi-threaded task partitioning*. JavaSoft.

Tseng, C.-W. (1995). Compiler optimizations for eliminating barrier synchronization. In *5th ACM SIGPLAN Symposium on Principles and Practice of Parallel Programming (PPoPP)*, pages 144–155, Santa Barbara, CA.

Winkel, M. (1997). Erweiterung von Java um ein FORALL. Technical report.

12 A COMPARATIVE STUDY OF SOFT REAL-TIME TRANSACTION SCHEDULING POLICIES IN PARALLEL DATABASE SYSTEMS

Sonia Takkar* and Sivarama P. Dandamudi

Centre for Parallel and Distributed Computing, School of Computer Science,
Carleton University, Canada

stakkar@ca.newbridge.ca
sivarama@scs.carleton.ca

Abstract: Real-time database systems support transactions with timing constraints such as deadlines. Real-time transactions, in addition to preserving the consistency of the database as in traditional transactions, have to meet the deadlines. Scheduling real-time transactions has received considerable attention in centralized and distributed databases. However, real-time transaction scheduling in parallel database systems has not received much attention. This paper focuses on real-time transaction scheduling in shared-nothing parallel database systems. We evaluate the performance of a new priority-based scheduling policy in which all scheduling decisions are made locally by each node. In contrast, several other algorithms, proposed for the distributed systems, require communication among the nodes to globally synchronize their local block and abort decisions. Such synchronization can deteriorate performance as real-time transactions will have to meet deadlines. We use miss ratio and average lateness as the performance metrics and show that, in general, the new policy provides a superior performance for the workload and system parameters considered in this study.

Keywords: earliest deadline scheduling, databases, real-time databases, parallel databases, shared-nothing systems, soft real-time transactions, performance evaluation, transaction scheduling.

---

*Currently with Newbridge Networks, Kanata, Canada.

12.1   INTRODUCTION

Real-time database systems support transactions with timing constraints such as deadlines. Example applications include computer-integrated manufacturing, banking, financial markets, the stock exchange, command systems, and control systems (*e.g.* threat analysis systems (Abbott and Garcia-Molina, 1992)) (Ulusoy and Belford, 1993). The transactions in such applications will have to be completed within a specified deadline. In addition to the deadline, real-time transactions, like traditional transactions, have to maintain the ACID (Atomicity, Consistency, Isolation, and Durability) properties (Gray and Reuter, 1993). Real-time transaction processing has received a lot of attention during the past several years (Abbott and Garcia-Molina, 1992; Harista, 1991; Hong *et al.*, 1993; Kuo and Mok, 1996; Lam and Hung, 1995; Ramamritham, 1993; Son and Park, 1994; Ulusoy and Belford, 1993; Yu *et al.*, 1994).

In conventional databases, the execution time and the data to be accessed are not known at the time of transaction submission. However, for most well-defined real-time transactions, an estimate of the execution time and the locks required by the transaction are known *a priori* (Lam and Hung, 1995). In fact, several previous studies have assumed this information in order to use a static locking approach. In the static locking method, either all locks are granted to the transaction or none is granted. Such an approach is appropriate as it avoids deadlocks in the system. Even if the precise lock information is not possible to get, it is possible for well-defined real-time transactions to get a list of data items touched by the transaction that is an overestimate. Such an overestimate is preferred to a dynamic locking scheme, in which locks are requested as needed, because it avoids the need for deadlock checking. Detecting deadlocks in the dynamic locking scheme can potentially affect the completion times of all transactions and hence their ability to meet the deadlines. In this paper, as in the literature, the static locking approach is assumed. However, our new policy proposed in (Sonia and Dandamudi, 1998a) can be easily extended to handle dynamic locking.

Real-time transactions can be further categorized into hard and soft transactions, depending on the flexibility the application can grant in the completion of the transactions (Ramamritham, 1993). Hard transactions are unyielding about satisfying their timing constraints and a failure to satisfy the deadlines can lead to catastrophic results. On the other hand, soft transactions are flexible about their timing constraints and a delay in completion of these transactions can be tolerated. Our policy is applicable to both hard and soft transaction scheduling. However, the focus of this paper is on the performance evaluation of the new policy for soft real-time transactions. The results for hard real-time transactions are available in (Sonia, 1997; Sonia and Dandamudi, 1998b).

Most of the existing work in the area of real-time transaction scheduling has been focused on centralized databases. We study the behavior of real-time transactions in parallel databases. We report the performance of a new priority-based policy for scheduling soft real-time transactions in parallel database systems.

Parallel database systems provide an opportunity to significantly improve performance (DeWitt and Gray, 1992). This policy depends on the current priority of the transaction and involves minimal communication among various processors. Continuous need for communication among the processors for synchronization (*e.g.* to block all sub-transactions) is an important factor in the deterioration of performance in parallel systems and, more importantly, in real-time systems where deadlines of transactions have to be met. We consider the shared-nothing parallel database architecture with soft real-time transactions.

The shared-nothing architecture has high scalability, low interference among processors and a simple interconnection network, as most of the processing is done locally at individual processors (Stonebraker, 1986). These factors have led to the shared-nothing approach to be the most popular of the existing architectures. Some commercial systems using this approach are available from Teradata, Oracle/nCube and Tandom. Examples of experimental systems include Bubba (Boral *et al.*, 1990) and Gamma (DeWitt *et al.*, 1990).

Section 12.2 describes two synchronized scheduling policies for parallel databases that are adapted from the distributed database area. Our adaptive highest priority first (AHPF) policy is described in Section 12.3. To test this policy, we construct a model in Section 12.4 and present simulation results based on it in Section 12.5. Section 12.6 presents a summary of the results.

12.2 SYNCHRONIZED SCHEDULING POLICIES FOR PARALLEL DATABASES

Most of the existing real-time transaction scheduling policies have been designed for centralized databases (Son and Park, 1994). These policies schedule transactions according to the priorities assigned to them. Transaction priority can be calculated in a variety of ways (Buchmann *et al.*, 1989). These priorities could be a function of the arrival time transactions, their deadlines or their slack time (slack time is defined as the difference between the deadline and the estimated execution time).

Most of the scheduling algorithms for centralized databases have been modified and adapted for distributed databases (Lam and Hung, 1995). Distributed databases are prone to distributed deadlocks (Bernstein and Goodman, 1987).

In this section, we describe two synchronized scheduling policies for parallel databases that are adapted from the distributed database area.

12.2.1 Synchronized Scheduling Policy

The transactions entering the system have associated with them a deadline, an approximate execution time and a list of data items they will access. The priority of the transaction is determined by one or a combination of these parameters. Depending on the location of the data, the query manager splits the transaction into several sub-transactions. At most one sub-transaction of a transaction is present at a processor. The sub-transactions, at each processor, are queued in decreasing order of priority. If the new sub-transaction has a higher priority than a currently executing sub-transaction, and these sub-transactions access data in a conflicting mode, then the lower-priority transaction is restarted. This restart is carried out globally at the transaction level. That is, all the sub-transactions of that transaction, present at other processors, are restarted. When the sub-transaction gets scheduled, it requests locks from the local lock manager.

If all the locks could not be granted, all sub-transactions of that transaction at all the processors block and are queued in the respective block queues for a predetermined amount of delay. Thus, if a transaction cannot get all the locks at a processor, the transaction blocks at all the processors. After the expiration of the blocking delay time, the sub-transactions at all the processors attempt to get the locks again. The amount of time for which the transaction blocks is doubled (up to a ceiling) each time the transaction get blocked. This is similar to the exponential back-off policy used in Ethernets to resolve collisions.

After a sub-transaction obtains locks on all the required data items, it reads the data from the disk and processes them. If the data is modified, the modified data will have to be written back to the disk after the transaction commits. The two-phase commit protocol is used to commit a transaction.

In this policy, all the decisions involved are made globally. If a transaction blocks at one processor, it will be blocked at all the other processors, regardless of whether it could proceed at these processors or not. Similarly, in the case of sub-transaction restart transaction at one processor, the decision is communicated to all the other processors and all sub-transactions of the transaction are restarted at all the other processors. Continuous communication among the processors is required in this policy to synchronize blocking and restarts, which deteriorates its performance (see Section 12.5 for details).

12.2.2 Modified Synchronized Scheduling Policy

This policy is an advanced version of the synchronized scheduling policy. It is similar to the previous policy except for the manner in which it handles lock conflicts when the priority of the sub-transaction requesting a lock is higher than the priority of the transaction holding the lock. In this case, the remaining execution time of the sub-transaction holding the lock is compared with the slack time of the requesting sub-transaction. If more than one sub-transaction have locked the data item (*i.e.* read lock), the maximum remaining execution time among all the sub-transactions is considered as the remaining execution time of the data item. If the slack time of the requesting sub-transaction is greater than the remaining execution time of the data item, a local decision is made to block the requester. Otherwise, the holding transaction is aborted.

This policy inherits some of the performance problems associated with the basic synchronization policy. That is, as in the basic synchronization policy, local abort or block decisions are required to be implemented at the transaction-level rather than at the local sub-transaction-level.

12.3 ADAPTIVE HIGHEST PRIORITY FIRST POLICY

The adaptive highest priority first (AHPF) policy takes into consideration the problems associated with the policies described in the previous section. The major advantages of this policy over the previous ones are:

- The processing decisions at each processor are purely local. They do not involve communication among the processors, thereby, eliminating the communication delay.

■   Global synchronization of abort or block decisions made locally is avoided by the new policy.

At each node, the sub-transactions are queued in decreasing order of their priority. Each node maintains a status queue that keeps track of the state of all sub-transactions at that node. Sub-transactions can be in one of three states: *Scheduled, Blocked* or *Ready*. A sub-transaction is in *Ready* state when it enters the node. A sub-transaction is scheduled on the basis of the scheduled and blocked sub-transactions at that node. The status queue keeps track of all the sub-transactions at a node in decreasing order of priority.

In this policy, the sub-transactions can go ahead only if they get locks on all the data items to be accessed, thereby eliminating the problem of local deadlock. Before a sub-transaction can request locks, some checks have to be made. These checks are done locally at each node, with no communication among the nodes. These checks are performed on the status queue, thereby considering all the sub-transactions at that node. The two basic rules to be followed in this policy are as follows:

Backward Rule: This rule checks all sub-transactions that are enqueued behind the newly arrived sub-transaction (*i.e.* all lower-priority sub-transactions). If any of these lower-priority sub-transactions have been scheduled and they access data in a conflicting mode with the new, higher-priority sub-transaction, then they release all their locks on the data and are restarted. Note that only the local sub-transaction is restarted.

Forward Rule: This rule checks all the sub-transactions enqueued in front of the newly arrived sub-transaction (*i.e.* all higher-priority sub-transactions). If any of these sub-transactions are accessing data in a conflicting mode with the new, lower-priority sub-transaction, the new sub-transaction is blocked. Again only the local sub-transaction is blocked.

On getting all the required locks, the sub-transaction reads the data from the disk and the processes. If the values of the data items read have been modified, the sub-transaction maintains a local copy of the modified value, thus eliminating the need to rollback the sub-transaction in case of an abort. The only situation where communication among the different nodes is required in this policy is at the time a transaction commits. The two-phase commit protocol is considered in this case, which requires communication among the nodes. Unlike in the synchronized policies, restarting and blocking are done locally at the sub-transaction-level with no effect on the other nodes. A more precise description of these rules is given in (Sonia and Dandamudi, 1998a).

To maintain the current priority of the sub-transactions at each processor, each time a new sub-transaction arrives at a processor, the priorities of the existing sub-transactions at that processor are updated. During this period, if a transaction has missed its deadline, it is immediately aborted as hard transactions are being considered. In addition to maintaining a priority-based queue to determine if a sub-transaction can be scheduled, processing at the disk and the CPU is also priority-based, thus emphasizing the real-time property of the sub-transactions.

Correctness of the database is maintained by this policy as the Forward Rule and the Backward Rule force all the processors to maintain a global order of execution of the transactions. Note that this policy avoids deadlocks at the local level as well as at the global level. Local deadlocks are avoided as a transaction cannot proceed until it obtains locks on all the required data items. To avoid global deadlock, the global serializability of transactions is maintained at all processors. Thus, the order in which transactions will be scheduled is the same at all the processors. A detailed discussion of these properties is given in (Sonia and Dandamudi, 1998a). It can be observed that some delay is involved in the continuous calculation of priority and in the elaborate checks conducted, but all these are localized with no communication among the processors. All these delays are incorporated into our simulation model.

12.4 SIMULATION MODEL

To assess the performance of AHPF, we constructed a simulation model. This section describes the workload, system and database components of the model.

12.4.1 Workload Model

Transactions are characterized by the amount of data they access, the nature of the accesses (read or write), the priority assigned to them, their deadline and so on. Transaction arrival process is assumed to be Poisson with parameter λ (*i.e.* inter-arrival time is exponentially distributed with a mean $1/\lambda$). The deadline of the transactions, $deadline_T$, is calculated using the following function (Ulusoy and Belford, 1993):

$$deadline_T = arrival_T + slack_T + service_T$$

where $arrival_T$ is the arrival time, $slack_T$ is the slack time, and $service_T$ is the estimated execution time of transaction T. In the simulation experiments, the slack time is uniformly distributed between min_slack_time and max_slack_time. The estimated processing time of transaction T, $service_T$ is calculated as follows:

$$service_T = lock\_grant\_delay_T + processing\_delay_T + io\_delay_T$$

where $lock\_grant\_delay_T$ is the delay for granting the required locks to transaction T, $processing\_delay_T$ is the delay before processing the data, and io_delay_T is the delay to access the disk to read the data required by the transaction.

The locking delay depends on the number of data items to be locked by the transaction. Similarly, the processing and io delays also depend on the number of data items to be processed and read from the disk, respectively. The priority of a transaction is determined by the deadline of the transaction.

12.4.2 System Model

The system model characterizes the behavior of the actual parallel database system. The model used in this study is shown in Figure 12.1. Upon arrival, transactions are enqueued in the ready queue ordered by their priority. Depending on the location of the data, the query manager splits the transaction into as many sub-transactions as the

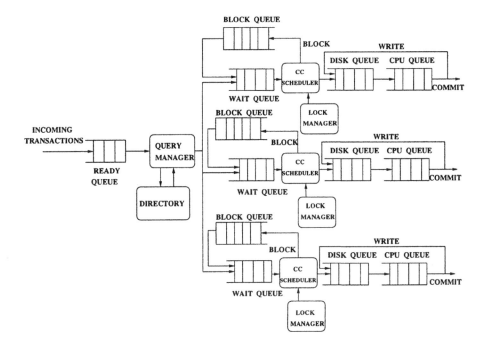

Figure 12.1 High-level block diagram of the system model.

number of nodes where the data to be accessed are located. Note that at most one sub-transaction of a particular transaction is present at a node.

At each node, the incoming sub-transaction is enqueued in the wait queue, also ordered by decreasing priority. The status of the sub-transaction is maintained in the status queue, which keeps track of the status of all sub-transactions at that processor, regardless of their actual position. Depending on the two rules of the AHPF policy, the decision whether the sub-transaction can be scheduled or blocked is made. If the sub-transaction is scheduled, it requests locks on the data items it wants to access.

The local lock manager is responsible for granting the locks on the data items stored at that node. Unless a sub-transaction can obtain all the locks, it is not allowed to proceed. If the sub-transaction is unable to get all the required locks, it blocks. A blocked sub-transaction can only be unblocked if it can receive all the required locks.

12.4.3 Database Model

The model of the database considered consists of a number of relations, each containing a large number of tuples. Each relation is horizontally fragmented across different disks. The degree of fragmentation or declustering of relations can be varied to facilitate experimentation. The number of partitions into which a relation is divided is:

$$number\_of\_partitions = frac\_partition * N$$

where N is the number of processors and $frac\_partition$ is the degree of declustering. A value of 1 implies full declustering with a partition on every disk.

Complete information on the partitioning of each relation and the position of each fragment in the disk is maintained by the query manager. The locking granularity of the database is at the page level. In addition to declustering of data, static and dynamic skew effects are also taken into consideration.

Static skew is observed when data is *unequally distributed* across a subset of disks which can lead to the respective nodes becoming potential bottlenecks. Dynamic or execution skew is observed due to *unequal data accesses* across the different nodes by transactions. The node that is accessed most often can also become a potential bottleneck at high loads. To investigate the effects of both dynamic and static data skew on the performance of the system, the database is designed in the following manner (Hua and Lee, 1991).

Let
T = size of the database (in pages),
P_S = number of pages in the skewed partition of a relation,
P_U = number of pages in each of the unskewed partitions of a relation,
N = number of processors in the system, and
σ = the degree of data skew, defined as follows:

$$\sigma = \frac{|P_S - P_U|}{P_S}.$$

In the model, only one skewed node is considered, which can be a potential bottleneck in the system. Then, the following can be deduced.

$$P_S + (N - 1) * P_U \;=\; T$$
$$\Rightarrow P_S \;=\; \frac{T}{1 + (N - 1) * (1 - \sigma)}$$
$$\Rightarrow P_U \;=\; \frac{(1 - \sigma) * T}{1 + (N - 1) * (1 - \sigma)}$$

Depending on the values of P_S and P_U, the size of the skewed partition of each relation is determined and stored at the predetermined skewed node.

12.5 SIMULATION RESULTS

In this section, the performance of the AHPF policy is compared with the two synchronized policies. The default values of the transaction parameters used in the simulation are summarized in Table 12.1. The number of nodes is fixed at 10. *page_mean* is the mean number of pages accessed by all the transactions. *conflict_delay* is the delay

Table 12.1 Some simulation parameters and their default values.

Parameter	Meaning	Default value
B	Number of batches	30
N	Number of nodes	10
db_size	Size of the database	400 pages
λ	Mean transaction arrival rate	36 ms
$page\_mean$	Mean pages accessed by a transaction	13 pages
cpu_time	CPU time per page	8 ms
io_time	I/O time per page	20 ms
$lock\_time$	Time to check and lock a page	2 ms
$block\_delay$	Time a transaction remains blocked before retrying	4 ms
$conflict\_delay$	Time to check Forward and Backward Rules	0.1 ms
$cost\_delay$	Time to globally block a transaction	2 ms
$communicate\_delay$	Communication delay	1 ms
$slack\_time$	Slack time	unif[200···400]
$frac\_partition$	Degree of declustering of a relation	1
$write\_percent$	% of pages written by a transaction	25%
num_trans	Number of transactions per batch	2000

experienced to check for both the Forward and Backward Rule of the AHPF policy. $lock\_delay$ is the time required to check and set the lock on a page accessed by a transaction. $write\_percent$ is the percentage of write operations each transaction will carry out. The slack time is distributed uniformly between 200 and 400 (except in the section that discusses the impact of the slack time).

Due to the small size of the database and the relatively large number of pages accessed by each transaction, the possibility of a conflict is high in this system. Therefore, the experiments can be said to have been conducted on a heavily loaded system. This is very important as real-time database systems have applications in critical environments, where the performance of the system is important, especially at high system loads. All these experiments were carried out for 30 batches of 2,000 transactions each and with a $write\_percent$ of 25%.

The primary performance metrics used in analyzing the results of the experiments are the *miss percent* of the system, which is defined as the percentage of transactions that do not complete before their deadlines (Son and Park, 1994) and *average lateness* of the transactions, which is defined as the average amount of time by which the transactions have missed their deadlines.

12.5.1 Sensitivity to Inter-arrival Time

Inter-arrival time determines the load of the system. The system load increases as the inter-arrival time decreases. The performance of the three policies as a function of the

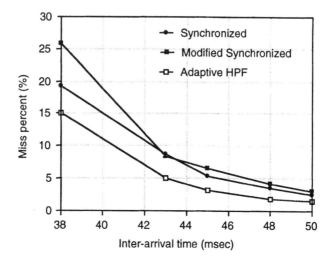

Figure 12.2   Sensitivity of miss percent to inter-arrival time.

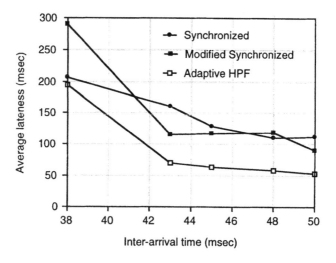

Figure 12.3   Sensitivity of average lateness to inter-arrival time.

inter-arrival time is shown in Figure 12.2. It can be observed that even at higher loads, the percentage of transactions missing their deadlines is much lower in the AHPF policy than in the other two policies. The AHPF policy also provides better average lateness as shown in Figure 12.3.

The performance superiority of the AHPF policy is due to the following reasons. The main drawback with the synchronized policies is that, whenever a sub-transaction gets blocked or aborted, all sibling sub-transactions of the transaction are blocked or aborted. This strategy wastes resources, which are particularly valuable at high system loads. This point is illustrated by means of an example. Assume that there is a local conflict and a sub-transaction of T_x is aborted at node 1. Suppose further that there is no local conflict at any of the remaining nine nodes. The synchronized policies abort all ten sub-transactions of T_x, even though it is not necessary to abort the other nine sub-transactions. This is precisely what the AHPF policy avoids. Therefore, its performance is superior to that of the synchronized policies. Additionally, since the AHPF policy localizes all decisions, a blocked sub-transaction is unblocked only when the required locks are available. There is no need to repeatedly try to get the locks as is necessary in the two globally synchronized policies.

12.5.2   Sensitivity to Data Skew

Data skew modifies the distribution of data and sub-transactions across the nodes. The skewed node has more data than the other nodes and, thus, more sub-transactions are scheduled at the skewed node. This can lead to the skewed node becoming a bottleneck and the degree of data skew can drastically affect the performance of the system.

The performance sensitivity of the three policies to the degree of data skew is shown in Figures 12.4 and 12.5. The value of inter-arrival time used is 45 ms. As the degree of data skew increases, the load on the predetermined skewed node increases and can lead to an increase in the number of transactions missing their deadlines. It can be observed that the AHPF policy is relatively less sensitive to the degree of data skew. It maintains a low miss percentage even when the disk utilization on the skewed node is about 95%.

On the other hand, both the synchronized scheduling and the modified synchronized policies exhibit similar behavior and are sensitive to data skew. A further increase in skew can easily saturate the systems for both the synchronized policies. The reasons for their performance deterioration are discussed in the last section. As the data skew increases, the requirement of synchronized block and abort imposes an additional penalty as the skewed node typically operates at a higher load than the other nodes in the system. This causes more transactions to miss their deadlines.

12.5.3   Sensitivity to Slack Time

Slack time is the maximum amount of time a transaction can spend waiting without missing its deadline. Slack time provides a transaction with some extra time to allow it the opportunity to complete before its deadline. If the slack time of a transaction is reduced, the waiting time of the transaction reduces and, thus, probability of a transaction missing its deadline increases.

The performance of all three policies is shown in Figures 12.6 and 12.7. It can be seen that the performance of the three policies tend to converge when the slack time is too small or too large. At these boundary cases, the scheduling policy is not as influential. At very high slack times, all the three policies converge to 0 as the slack time is more than sufficient for almost all the transactions to complete within their deadlines independent of the scheduling policy used. On the other hand, too small a slack time causes most transactions to miss their deadline because of the tightness of

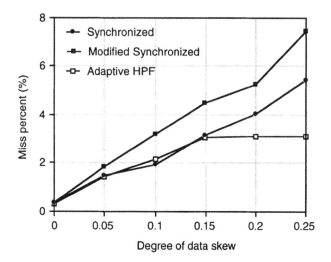

Figure 12.4   Sensitivity of miss percent to data skew.

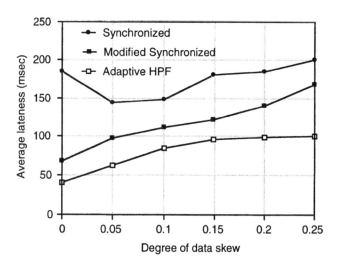

Figure 12.5   Sensitivity of average lateness to data skew.

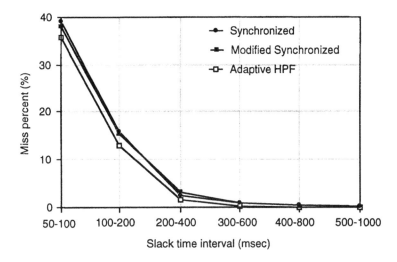

Figure 12.6   Sensitivity of miss percent to slack time. Note that the slack time is uniformly distributed between the minimum and maximum vales shown for each data point.

the deadline. The actual scheduling policy used makes a difference at slack time values that are intermediate between these two extremes. The data presented in Figure 12.6 indicates that the AHPF policy retains its performance advantage over the other two policies. The value of miss percent in AHPF policy converges fairly early to 0 with the average lateness rapidly decreasing as can be observed from Figure 12.7. The value of average lateness in the two synchronized policies increases initially as transactions spend more time in the system due to their global blocking and aborting strategies.

12.5.4   Sensitivity to Transaction Size

The size of a transaction is the number of data items the transaction accesses. If the average number of data items accessed by a transaction increases, the transactions require more locks on data items, and also spend more time in the system. An increase in transaction size tends to increase the number of conflicts and consequently increases the percentage of transactions missing their deadlines in the system.

The performance of the three policies as a function of the transaction size is shown in Figure 12.8. When the size of the transactions is small, all the three policies perform identically, but as the size of transactions increases, the percentage of transactions missing their deadlines in both the synchronized and modified synchronized policies increases sharply. The performance difference between the two synchronized policies increases with the transaction size. As the transaction size increases, the system operates at higher system loads. As we have discussed before, relative performance difference tends to increase with the system load. As can be observed from Figure 12.9, the average amount of time by which the transactions are missing their deadlines increases with transaction size. However, the average lateness of transactions in the AHPF pol-

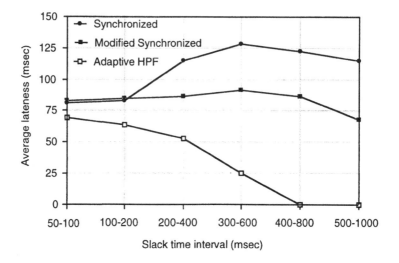

Figure 12.7 Sensitivity of average lateness to slack time. Note that the slack time is uniformly distributed between the minimum and maximum vales shown for each data point.

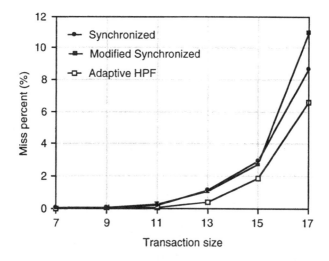

Figure 12.8 Sensitivity of miss percent to size of the transaction.

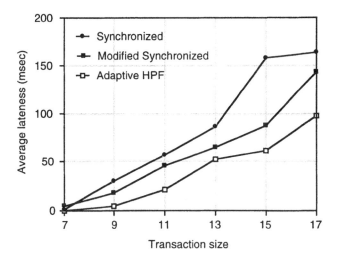

Figure 12.9   Sensitivity of average lateness to size of the transaction.

icy remain considerably lower than the other two policies even when the transaction size is large.

12.5.5   Other Results

We have also conducted detailed experiments to study the performance sensitivity to other parameters like the degree of declustering and write percent. Due to space constraints, these results are not presented here. These results generally support the observations made here. The only exception is that when the write percent is high, performance of the AHPF policy is worse than the synchronized policies. Similar behavior has been observed in scheduling hard real-time transactions. See (Sonia and Dandamudi, 1998b) for a discussion of the reasons for this behavior.

12.6   CONCLUSIONS

Scheduling of real-time transactions involves maintaining the consistency of the database like traditional transactions, while taking into consideration the timing constraints of the transactions. The scheduling policies proposed in the literature for a distributed environment involve communication to synchronize abort and block actions at the transaction level. This leads to a degradation in the performance of the policies.

In this paper, we presented the performance of a new priority-based policy that is purely local in nature and requires minimal communication among the processors during the decision-making process. The new policy avoids both local and global deadlocks as well as priority inversion. Performance results presented here suggest that the proposed policy performs better than the synchronized policies (except when there is a large fraction of pages read are modified by transactions).

We have also investigated the performance of this new policy to schedule hard real-time transactions (Sonia and Dandamudi, 1998b). In this environment, a transaction missing its deadline is aborted unlike in the soft real-time transaction environment. The transactions, therefore, do not consume resources after missing their deadlines. Our results presented elsewhere (Sonia, 1997; Sonia and Dandamudi, 1998b) indicate that the AHPF policy performs substantially better than the two synchronized policies even in this environment. In fact, its performance is better than that in the soft real-time transaction environment.

Acknowledgments

The authors gratefully acknowledge the financial support provided by the Natural Sciences and Engineering Research Council of Canada (NSERC) and Carleton University.

References

Abbott, R. K. and Garcia-Molina, H. (1992). Scheduling real-time transactions: A performance evaluation. *ACM Trans. on Database Systems*, 17:513–560.

Bernstein, P. A. and Goodman, N. (1987). Concurrency control in distributed database systems. *Computing Surveys*, 13:185–221.

Boral, H., Alexander, W., Clay, C., Copeland, G., Danforth, S., Franklin, M., Hart, B., Smith, M., and Valduriez, P. (1990). Prototyping Bubba, a highly parallel database system. *IEEE Trans. on Knowledge and Data Engineering*, 2:4–23.

Buchmann, A. P., McCarthy, D. R., Hsu, M., and Dayal, U. (1989). Time-critical database scheduling: A framework for integrating real-time scheduling and concurrency control. In *5th IEEE International Conference on Data Engineering*, pages 470 480.

DeWitt, D., Ghandeharizadeh, S., Schneider, D., Bricker, A., Hsiao, H.-I., and Rasmusson, R. (1990). The Gamma database machine project. *IEEE Trans. on Knowledge and Data Engineering*, 2:44–61.

DeWitt, D. and Gray, J. (1992). Parallel database systems: The future of high performance database systems. *Communications of the ACM*, 35:85–98.

Gray, J. and Reuter, A. (1993). *Transaction Processing: Concepts and Techniques*. Morgan Kaufmann Publishers.

Harista, J. R. (1991). Earliest deadline scheduling for real-time database systems. In *12th Real-Time Systems Symposium*, pages 232–242.

Hong, D., Johnson, T., and Chakravarthy, S. (1993). Real-time transaction scheduling: A cost conscious approach. *ACM SIGMOD Record*, 22:197–206.

Hua, K. A. and Lee, C. (1991). Handling data skew in multiprocessor database computers using partition tuning. In *17th International Conference on Very Large Data Bases*, pages 525–535.

Kuo, T. and Mok, A. K. (1996). Real-time database – Similarity semantics and resource scheduling. *ACM SIGMOD Record*, 25:18–22.

Lam, K. and Hung, S. (1995). Concurrency control for time-constrained transactions in distributed database systems. *The Computer Journal*, 38:704–716.

Ramamritham, K. (1993). Real-time databases. *International Journal of Distributed and Parallel Databases*, 1:199–226.

Son, S. H. and Park, S. (1994). Scheduling transactions for distributed time-critical applications. In *Distributed Computing Systems*, pages 592–619, IEEE Computer Society Press.

Sonia, T. (1997). Scheduling real-time transactions in parallel database systems. Master's thesis, School of Computer Science, Carleton University, Ottawa, Canada.

Sonia, T. and Dandamudi, S. (1998a). An adaptive scheduling policy for real-time parallel database systems. In *International Conference on Massively Parallel Computing Systems*, pages 8.19–8.27.

Sonia, T. and Dandamudi, S. (1998b). performance of hard real-time transaction scheduling policies in parallel database systems. In *International Symposium on Modeling, Analysis and Simulation of Computer and Telecommunication Systems (MASCOTS-98)*.

Stonebraker, M. (1986). The case for shared nothing. *Database Engineering*, 9:4–9.

Ulusoy, O. and Belford, G. G. (1993). Real-time transaction scheduling in database systems. *Information Systems*, 18:559–580.

Yu, P., Wu, K., Lin, K., and Son, S. (1994). On real-time databases: Concurrency control and scheduling. *Proceedings of the IEEE*, 82:140–157.

13 AN AGE-THRESHOLD ALGORITHM FOR GARBAGE COLLECTION IN LOG-STRUCTURED ARRAYS AND FILE SYSTEMS

Jai Menon and Larry Stockmeyer

IBM Almaden Research Center, U.S.A.

menonjm@almaden.ibm.com

stock@almaden.ibm.com

Abstract: A new algorithm for choosing segments for garbage collection in Log-Structured Arrays and Log-Structured File Systems is proposed and studied. The basic idea of our algorithm is that segments that have been recently filled by writes from the system should be forced to wait for a certain amount of time (the age-threshold) before they are allowed to become candidates for garbage collection. Among segments that pass the age-threshold, we select ones that will yield the most amount of free space. We show, through simulation, that our age-threshold algorithm is more efficient at garbage collection (produces more free space per garbage-collected segment) than previously known greedy and cost-benefit algorithms. It is also simpler to implement a scalable version of the age-threshold algorithm than to implement a scalable version of the cost-benefit algorithm. The performance of the age-threshold algorithm depends on a good choice of an age-threshold; therefore, we suggest methods for choosing a good age-threshold.

Keywords: log-structured array, log-structured file system, garbage collection, segment cleaning, RAID.

13.1 INTRODUCTION

In this paper, we propose and study a new algorithm for garbage collection in Log-Structured File Systems (LFS), Log-Structured Disks (LSD) and Log-Structured Arrays (LSA). LSD and LSA are relatively new types of disk architectures (Menon, 1995) which are important because they can improve write performance over standard disk architectures, and because they can support disk compression.

Log-Structured Disks (LSD) and Log-Structured Arrays (LSA) borrow heavily from the log-structured file system (LFS) approach pioneered at UC Berkeley (Rosenblum and Ousterhout, 1992). Our results apply equally well to LSD, LSA and to LFS systems; however our focus in this paper is on LSA. The LSA technique combines LSD and RAID and is typically executed in an outboard disk controller rather than in the file system. Because LSA is implemented in an outboard controller which has no understanding of files, it is more appropriate to think of LSA as a log-structured track manager rather than a log-structured file system. In this respect, LSA has some similarities to Loge (English and Stepanov, 1992) and to Logical Disk (de Jonge *et al.*, 1993), both of which are implemented below the file system.

In LSA, updated data is written into new disk locations instead of being written in place. Large amounts of updated data are collected in controller memory and written together to a contiguous segment on the disks. Parity on this data is also computed in controller memory and written to the segment at the same time. This technique avoids the standard write penalty of RAID arrays (Patterson *et al.*, 1988), but requires a process called garbage collection to continually create new empty segments to substitute for the ones that got filled during the writing process. When data that went into a segment is rewritten by the system, the data will be placed into a new segment, and a hole or empty space forms in the segment which originally had the data. Garbage collection is the process of compacting some number of partially empty segments into a fewer number of full segments, thus creating some completely empty segments.

The focus of this paper is on garbage collection in LSA and LFS systems. Specifically, we look at the impact on performance of the algorithm used to decide which segments should be chosen for garbage collection. Two algorithms have been previously proposed in the paper that introduced the log-structured approach (Rosenblum and Ousterhout, 1992): a *greedy* algorithm which selects segments that will yield the most amount of free space; and a *cost-benefit* algorithm which selects segments based on both how much free space it has and how long it has been since the segment was last filled with a bunch of data by the controller (its age). The latter algorithm includes the age of the segment in the selection criterion, because the expectation is that younger segments will have data which is likely to be rewritten shortly, so if we wait a little longer, these segments will yield even more free space in the future. This argues for not selecting young segments. Although the amount of free space and age are clearly important criteria to use in deciding which segments should be chosen for garbage collection, it is less clear how to combine these two criteria into a single criterion which can be used to order segments according to their desirability for garbage collection. A particular *cost-benefit* method for combining amount of free space and age into a single number is suggested in (Rosenblum and Ousterhout, 1992); the details of this method are discussed later in the paper.

We introduce a new class of algorithms for selecting segments for garbage collection based on an age-threshold. The basic idea is that segments which have been recently filled by writes from the system should be forced to wait for a certain amount of time (the age-threshold) before they are allowed to become candidates for garbage collection. This way, we give the system a reasonable amount of time to rewrite any of the data in the segment, before we make it a candidate for garbage collection. The expectation is that if the age-threshold is properly chosen, segments that have reached the age-threshold are unlikely to get significantly emptier due to future rewrites. Among segments that pass the age-threshold and become candidates for garbage collection, we select ones that will yield the most amount of free space. We show, through simulations both on synthetic traces and on an actual I/O trace, that our age-threshold algorithms are better in performance than greedy or cost-benefit, where performance is measured by the average amount of free space produced per garbage-collected segment. This means that designs using age-threshold will give better system performance than designs using greedy or cost-benefit. The age-threshold algorithms are also simpler to implement than the cost-benefit algorithm, if we want a scalable design that continually maintains an ordering of the segments according to their desirability for garbage collection. The performance of the age-threshold algorithm depends on a good choice for the age-threshold. A mathematical analysis can be used to choose an optimal age-threshold under certain workload assumptions. We also suggest how to choose good age-thresholds when nothing is known about the workload.

A more detailed version of this paper, including details of the analysis and additional simulations, is available (Menon and Stockmeyer, 1998).

13.2   OVERVIEW OF LSA

We begin by describing LSA which uses a parity technique similar to that used in RAIDs for improving reliability and availability. A Log-Structured Array (LSA) consists of a disk controller and $N + 1$ physical disks, where each disk is divided into large consecutive areas called *segment-columns*. A segment-column consists of some number of consecutive sectors on the disk. Corresponding segment-columns from the $N + 1$ disks constitute a *segment*. The array has as many segments as there are segment-columns on a disk in the array. One of the segment-columns of a segment contains the parity (XOR) of the remaining N segment-columns of the segment. *Logical devices* are mapped and stored in this Log-Structured Array, and host programs access logical tracks stored on logical devices. A *logical track*, or simply track, which we will define for this paper as the smallest unit writable by the host system, is stored entirely within some segment-column of some physical disk of the array; many logical tracks can be stored in the same segment-column. The location of a logical track in an LSA changes each time the track is rewritten by a host program. A directory maintained by the LSA in the controller, called the LSA directory, indicates the current physical location of each logical track in the disk array. Given a request to read a logical track, the controller examines the LSA directory to determine its physical location, and then reads the relevant data from the relevant disk.

Writes to the array operate as follows, using a section of controller memory, logically organized as $N + 1$ segment-columns called the *memory segment*. It consists of $N + 1$ memory segment-columns: N data memory segment-columns and 1 parity memory segment-column. When a logical track is updated by the system, the entire logical track is written into one of the N data memory segment-columns (and the host is told that the write is done). When the memory segment is full (all data memory segment-columns are full), we XOR all the data memory segment-columns to create the parity memory segment-column, and then all $N + 1$ memory segment-columns are written to an empty segment on the disk array. All logical tracks that are written to disk from the memory segment must have their entries in the LSA directory updated to reflect their new disk locations. Note that writing to the disk is more efficient in LSA than in RAID-5, where 4 disk accesses are needed for an update (Patterson *et al.*, 1988). However, LSA needs to do a process called *garbage collection*, since holes (garbage) form in segments that previously contained one or more of the logical tracks that were just written. To ensure that we always have an empty segment to write to, the controller garbage collects segments in the background. All logical tracks from a segment selected for garbage collection that are still in that segment (are still pointed to by the LSA directory) are read from disk and placed in a memory segment. These logical tracks will be written back to disk when this memory segment fills. In this paper, we assume that this memory segment is separate from the memory segment described above that is filled with track writes. Thus, written tracks and tracks found by the garbage collector are placed into separate segments.[1] Empty segments produced through garbage collection are returned to the empty segment pool and are available when needed.

13.3 OVERVIEW OF GARBAGE COLLECTION

As we mentioned before, we call the smallest unit of data that is written a *track*. In LSA, these tracks are organized into *segments*. At any time, each track is *live* in exactly one segment. A basic operation of the system is to *write* a particular track, that is, change the contents of the track. We can imagine that the tracks being written are placed into (*i.e.* become live in) an initially empty segment s_0. While s_0 is being filled, s_0 resides in controller memory. If track k is written and if track k was previously live in some other segment s before this write, then k becomes *dead* in s, and k becomes live in the segment s_0 being filled. This continues until s_0 is filled to capacity, at which point s_0 is *destaged*, *i.e.* written to the disk (or disk array). Then another empty segment is chosen to be filled. As writing proceeds, storage becomes fragmented; there can be many segments that are partially filled with live tracks. At any point, the *utilization* of a segment is the fraction of the segment containing live tracks, *i.e.* if the segment contains L live tracks and if segment capacity is C tracks, then its utilization is L/C. For simplicity, in our simulations we assume that all the data (tracks) are loaded into the LSA at initialization time, and no new tracks are added or existing tracks deleted following this initial loading or initialization phase. That is, we assume that tracks are not added or deleted, just rewritten; therefore, the *average segment utilization* (ASU) of the system is constant. If there are S segments, if each segment has a capacity of C tracks, and if there are T tracks, then

$$ASU = T/CS.$$

---

[1]We also investigated placing these two types of tracks into the same segment. Briefly, we found that it is better to keep them separate. The intuition behind this is that a segment filled with hot recently-written tracks has more potential for rapidly becoming emptier than does a segment filled with both hot written tracks and cold garbage-collected tracks.

Since the writing process will eventually deplete the empty segments, garbage collection is done to create empty segments. This is done by choosing a certain number of segments and compacting the live tracks in these segments into a fewer number of full segments, so some empty segments are created. For example, if we collect 3 segments, each having utilization 2/3, the live tracks can be reorganized into 2 full segments, thus creating a net of one new empty segment. The main subject of this paper is algorithms for deciding which segments to choose for garbage collection.

We measure the performance of a garbage collection algorithm by the average utilization of the garbage collected segments. We call this measure *garbage collection utilization* (GCU). This seems a reasonable measure of performance since small GCU means that the garbage collector is doing a good job of collecting free space, at an average of $C(1 - \text{GCU})$ per collected segment. GCU is related to other performance measures that have been considered elsewhere. The *write cost* of (Rosenblum and Ousterhout, 1992) is equal to $2/(1 - \text{GCU})$, and the *moves per write* of (McNutt, 1994) is equal to $\text{GCU}/(1 - \text{GCU})$. Both of these are increasing functions of GCU, so that the relative ordering of different algorithms by GCU is the same ordering that would be obtained using either of the other measures. Previous studies of garbage collection algorithms show (not surprisingly) that GCU increases as the average utilization ASU increases, and that ASU should be somewhat less than 1 to allow some build-up of dead tracks, since garbage collection converts dead tracks to free space.

13.4 MODELS OF TRACK WRITING

The GCU of a particular garbage collection algorithm depends on the sequence of track writes. We begin by considering a synthetic model of track writing, used in (Rosenblum and Ousterhout, 1992), that assumes a random choice of the track to be written, allowing a certain fraction of the tracks (the "hot tracks") to be more likely to be chosen than the others (the "cold tracks"). More precisely, in the *hot-and-cold* model the degree of "hotness" is specified by two numbers h and p with $0 < h \leq p < 1$, where h is the fraction of tracks that are hot and p is the probability of choosing from among the hot tracks. A particular subset $\mathcal{H}$ of the tracks is designated as the "hot tracks", such that $\mathcal{H}$ contains a fraction h of the tracks, *i.e.* $|\mathcal{H}| = hT$. Whenever a track choice k is needed, then with probability p choose k at random from $\mathcal{H}$, or with probability $1 - p$ choose k at random from the tracks not in $\mathcal{H}$.

In Section 13.10 we compare various garbage collection algorithms on an actual I/O trace, the "snake" trace (Ruemmler and Wilkes, 1993).

13.5 THE SIMULATION PROCEDURE

When simulating a particular garbage collection algorithm, the details of the simulation depend, of course, on the particular algorithm. Some details of the simulations, however, are common to all of our simulations. For simulations done with the hot-and-cold model described above, we used $S = 3000$ segments, each having capacity $C = 300$ tracks. As it turns out, however, our analysis suggests that the results are relatively independent of the specific numbers we selected here. Other values for S and C are used in Section 13.10. Having chosen S and C, the number T of tracks is chosen to give a desired ASU ($T = \text{ASU} \times CS$). For hot-and-cold track choice, the set $\mathcal{H}$ of hot tracks is initially chosen to be a random set of the correct size to give a desired h ($|\mathcal{H}| = hT$). Another parameter in the simulation is *max-empty*, the number of empty segments produced during each phase of garbage collection. It is sometimes convenient to express max-empty as a fraction of the number of segments. Define the *normalized max-empty*, m, as

$$m = \text{normalized max-empty} = \text{max-empty}/S.$$

We typically took max-empty to be 5% of the segments, *i.e.* $m = .05$.

At the start of the simulation, the tracks are packed into $\lceil T/C \rceil$ segments in sequential order. Thus, at the start, $\lfloor T/C \rfloor$ segments are full (utilization 1) and $S - \lceil T/C \rceil$ are empty (utilization 0). To begin the simulation, there is an initial filling process where track writing is invoked to fill each empty segment, in turn, until there are no empty segments. Then simulation of the particular garbage collection algorithm is started. This is done by repeatedly executing the following steps:

S1. Run the garbage collector to create max-empty empty segments.

S2. Fill the empty segments, one at a time, by track writing.

It is useful to draw a distinction between two types of segments: (1) segments that were last filled by track writing in step S2 (call these *TW-filled segments*); and (2) segments that were last filled with live tracks by the garbage collector in step S1 (call these *GC-filled segments*).

Our basic simulation uses a simple procedure for the garbage collector to pack live tracks into segments: it packs the tracks taken from the selected segments in the order in which these segments were selected. For some cases, described later, we investigated the effect of grouping tracks by age.

For synthetic (randomly generated) traces, the simulations were run until GCU was observed to stabilize, indicating that the simulation had reached steady state. Reported GCU values are the average utilization of the segments selected by the garbage collector, averaged over several tens of thousands of segment selections in steady state.

13.6   THE AGE-THRESHOLD ALGORITHM

We now describe the basic age-threshold algorithm. The general rationale for the age-threshold algorithm is discussed in Section 13.1. An important parameter in this algorithm is the *age-threshold*, which places a lower bound on the time that a segment must wait, after it is filled with track writing, before it can be selected by the garbage collector.

The age-threshold algorithm selects segments based on smallest utilization, although it considers a segment for selection only if its age exceeds the age-threshold. The age of a segment is defined as follows.

Definition of segment age. There is a *destage clock*, initially 0. Whenever a TW-filled segment s is filled by track writing, the timestamp of s is set to the current value of the destage clock, and the destage clock is incremented by 1. Whenever a GC-filled segment s is produced, the timestamp of s is set to the largest timestamp of any of the segments that contributed tracks to s. At any point, the age of s is the difference between the current value of the destage clock and the timestamp of s. (Thus, the age of a GC-filled segment is initially set to the age of the youngest segment that contributed tracks to s.)

Among the segments whose age exceeds the age-threshold, the algorithm first selects segments with the smallest utilization. As a tie-breaker among segments with the same (smallest) utilization, it first selects segments with the largest age. In the age-threshold algorithm, the age of a GC-filled segment has a secondary effect. In fact, in the more efficient implementation described in Section 13.9, the timestamp of a GC-filled segment can simply be set to zero. The age of a GC-filled segment is more relevant in one of the cost-benefit algorithms described later.

A GC-filled segment immediately passes the age-threshold when it is created, because it was filled by tracks from segments that passed the age-threshold. In effect, only the TW-filled segments must wait to pass the age-threshold. We refer to this version of the algorithm as the *TW-age* version. We considered another version, the *all-age* version, where all newly filled segments must wait to pass the age-threshold before they can be selected for garbage collection. A convenient way to make all the segments wait about the same amount of time is to place each newly filled segment (both TW-filled and GC-filled) into a FIFO queue, called the *waiting list*. Whenever the destage clock is incremented (which still happens only when a TW-filled segment is destaged) we look at the head of the queue to see if the segment there passes the age-threshold. If so, we remove segments from the head of the queue as long as they pass the age-threshold; these removed segments are now available to the garbage collector.

It is often convenient to express the age-threshold as a fraction of the number of segments. Define the *normalized age-threshold*, t, as

$$t = \text{normalized age-threshold} = \text{age-threshold}/S.$$

13.7 RESULTS OF ANALYSIS AND SIMULATION

In this section we describe results of both analysis and simulation that show how the GCU of the age-threshold algorithm depends on the normalized age-threshold t in the hot-and-cold model. The actual analysis is omitted due to space limitations; details are in the full paper (Menon and Stockmeyer, 1998). The analysis can be used to compute the optimal age-threshold, under the hot-and-cold model. Other relevant parameters are $a = \text{ASU}$, $m = $ normalized max-empty, and the parameters h and p of the hot-and-cold model. Figure 13.1 shows how GCU of the TW-age-threshold algorithm depends on normalized age-threshold from the analysis (dotted line) and simulation (small circles), with ASU $= .8$, $m = .05$, and degree of hotness $h = .1$, $p = .9$. There is an intuitive explanation for the shape of the curve. For a range of age-thresholds sufficiently near to 0, the age-threshold algorithm is essentially the same as the greedy algorithm, since the greedy algorithm will not select a segment based on smallest utilization until the age of the segment has passed the age-threshold anyway. At some value of age-threshold (the value at which GCU starts to decrease), the age-threshold will "protect" a segment that the greedy algorithm would have selected, allowing its

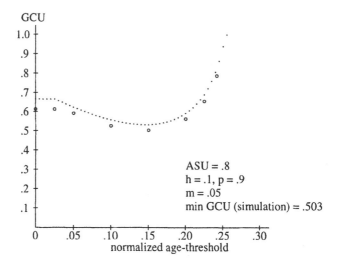

Figure 13.1   GCU vs normalized age-threshold for the TW-age-threshold algorithm. The dotted line was obtained from the analysis. Small circles are points obtained from simulation.

utilization to decrease further. But as the age-threshold continues to increase, we eventually reach a point of diminishing returns because too many low-utilization segments are being protected, and this causes GCU to increase.

Varying m, we found that the optimal normalized age-threshold decreases as m increases. We also plotted GCU against normalized age-threshold for a lower degree of hotness ($h = .1, p = .7$), and observed that the curve becomes flatter.

13.8   COMPARISON WITH COST-BENEFIT SELECTION CRITERION

In the paper that introduced log-structured file systems (Rosenblum and Ousterhout, 1992), the *cost-benefit criterion* is used to decide which segments to select for garbage collection. In this section, we compare the age-threshold criterion with the cost-benefit criterion. The two criteria have in common that they use the utilization and the age of a segment to order the segments, but they do the ordering in different ways. The age-threshold method keeps a segment out of the ordering until it passes the age-threshold; the segments that do pass are ordered by utilization (with smaller being better). The cost-benefit method orders the segments by their benefit/cost ratio (with larger being better): if a segment has age A and utilization u, its benefit/cost ratio is

$$\frac{\text{benefit}}{\text{cost}} = \frac{(1 - u)A}{1 + u}.$$

A segment is good to select if a large fraction of the segment is dead and if the segment is old. The benefit (numerator) is taken to be the product of the dead fraction $(1 - u)$ and the age A. The denominator represents the cost of selecting the segment, since a whole segment is read and a fraction u (the live tracks) is written back to disk.

(Rosenblum and Ousterhout, 1992) uses a definition of segment age different than the one we use: the age of a segment is the age of the youngest track in the segment,

Table 13.1 GCU values obtained by simulation for various garbage collection algorithms, where ASU $= .8$, $h = .1$, $p = .9$, and $m = .05$.

Segment-Age	Selection-Criterion	Age-Grouping? Yes	No
segment	greedy	.610	.612
track	cost-benefit	.688	.535
track2	cost-benefit	.580	.610
segment	cost-benefit	.528	.526
segment	TW-age-threshold	.501	.503
segment	all-age-threshold	.457	.457

where the age of a track is the elapsed time since the track was last modified. For simulation, we take the age of a track k to be the number of track writes that have occurred since k was last written. To distinguish the two definitions, call this definition the *track-based* definition, and call our definition given in Section 13.6 the *segment-based* definition. (Rosenblum and Ousterhout, 1992) also suggests that the garbage collector group the collected live tracks by age before packing them into segments, the idea being that tracks last written at around the same time have some temporal affinity, so it could be advantageous to have them together in the same segment.

13.8.1 Comparison of GCU

By simulation, we compared the cost-benefit selection criterion using both the segment-based and the track-based definition of segment age, the age-threshold criterion with the segment-based definition, and the greedy algorithm. For the age-threshold criterion, both the TW-age and the all-age versions were considered. In each case, the simulation was done both with and without age-grouping. Simulations were done with ASU $= .8$, degree of hotness $h = .1$ and $p = .9$, and $m = .05$. The age-thresholds used for the age-threshold algorithms were the optimal values obtained from the analysis. When age-grouping was in effect, the tracks were sorted by age in batches of about 3000. We also considered a modified version of the track-based definition of segment age, where the age of a segment is reset to zero when any live track in the segment is written, even though this write causes the track to become dead in that segment; call this the *track2-based* definition.

Table 13.1 shows the GCU values obtained in each case. Age-grouping had little effect in the cases using the segment-based definition of segment age. The age-threshold algorithms gave the best GCU. Since the performance of these algorithms depends on a good choice for the age-threshold, it is also worth mentioning that the TW-age version is still better than the next better competitor (.526 for the cost-benefit criterion and segment-based age) for a range of normalized age-thresholds between roughly .1 and .175. The all-age version has GCU better than .526 for a larger range of age-thresholds between roughly .07 and .19.

13.8.2   Example scenarios to illustrate differences between algorithms

A scenario where the age-threshold algorithm outperforms the cost-benefit algorithm is one like the hot-and-cold model, where very old segments containing only (or almost only) cold tracks become emptier very slowly as cold tracks are rewritten. Before the utilization of these segments has decreased very far, their age becomes large, causing their benefit/cost ratio to become large enough for them to be the best choice for selection, even though they have high utilization. Age-grouping just exacerbates this effect.

On the other hand, a reason to garbage collect very-old high-utilization segments is to reclaim the free space in these segments. This consideration suggests a scenario where cost-benefit outperforms age-threshold (this was verified by simulation): 90% of the tracks (the "frozen" tracks) are *never* rewritten; 10% of the tracks are written uniformly; initially each full segment contains 90% frozen tracks; ASU = .9; and $m = .01$. Cost-benefit does better than age-threshold at reorganizing the frozen tracks into full segments, so it has better GCU after the reorganization is done. To handle this situation, the age-threshold algorithm could be augmented by another process that collects free space from very-old high-utilization segments. Since this process has relatively high overhead and relatively low priority, it could be done during idle periods. In the context of log-structured file systems, there is existing work on scheduling garbage collection during idle periods (Blackwell *et al.*, 1995).

13.8.3   Implementation issues for scalable designs

In our opinion, the cost-benefit algorithm is more difficult to implement and consumes more resources than the age-threshold algorithm, provided that a list of the segments (actually, segment names) is to be maintained with the segments sorted according to their desirability for garbage collection. To implement a garbage collection algorithm, one of the following needs to be done:

1. keep all segments sorted according to their desirability for garbage collection (*e.g.* by the cost-benefit criterion in the cost-benefit algorithm), so when segments are needed for garbage collection, they can be easily found; or

2. sort all segments according to desirability whenever we need to do garbage collection.

However, method (2) may require significant CPU resources for large systems with an extremely large number of segments. This means that method (2) is not scalable, since we may get reasonable performance in small systems, but poorer performance in larger systems when the sort may take considerable resources. Remember that, in many controllers, the CPU resource in the controller is often the bottleneck to performance; method (2) makes this situation worse. Therefore, we may want to consider method (1), where the segments are always maintained in sorted order. However, it is difficult to maintain segments in sorted order by the cost-benefit criterion. Because of the way this criterion is defined, it is possible that the positions of two segments in the list may need to be reversed even if there was no activity to either of those two segments (or to any other segment), just because the two segments each got a little older. For the age-threshold algorithm, on the other hand, a segment changes position in the list only if some live track in the segment is written, causing its utilization to decrease. In the next section, we show some very attractive implementations of the age-threshold algorithm.

13.9 MORE EFFICIENT IMPLEMENTATIONS OF AGE-THRESHOLD ALGORITHMS

Using the definition of the age-threshold algorithm given in Section 13.6, the garbage collector needs an ordered list of the segments available to it (*i.e.* those outside the waiting list) where the segments are sorted by utilization. To distinguish this algorithm from the variation described below, call it the *full-sort* algorithm. For reasons described in Section 13.8.3, suppose that we want to maintain this ordering as track writing occurs. Whenever a live track in segment s is written, segment (name) s might change its position in the list because its utilization has decreased.

One way to decrease the number of repositionings in the list is to use coarse-grained accuracy for the utilizations. The range of possible utilizations is divided into intervals of the same length. All segments having utilizations in the same interval are placed together in the same "bucket". For example, we could have 10 buckets, one bucket containing segments whose utilizations lie between 0 and 0.1, a second bucket for segments whose utilizations are between 0.1 and 0.2, and so on. In the *bucket-sort* algorithm, each bucket is organized as a FIFO queue. The age-threshold is implemented as in Section 13.6 by a *waiting list*, which is a FIFO queue. Each FIFO queue has a "tail" where segments (actually, segment names) enter, and a "head" where segments are removed. The bucket-sort algorithm has low overhead and is easy to implement. After a TW-filled segment is filled with written tracks, it enters the tail of the waiting list. Segments at the head of the waiting list that pass the age-threshold are removed and enter the appropriate bucket list depending on their utilization. In the TW-age version, a GC-filled segment enters the highest-utilization bucket. In the all-age version, a GC-filled segment enters the waiting list. In either case, the timestamp of a GC-filled segment is set to zero. If segment s is outside the waiting list and a write to a live track in segment s causes the utilization of s to cross an interval boundary, s is removed from its current bucket and enters the tail of the next lower one. Whenever the garbage collector needs a segment, it finds the lowest-numbered non-empty bucket, and removes the segment from the head of that list. If all the buckets are empty, it removes the segment from the head of the waiting list.

We simulated these algorithms with parameters ASU = .8, degree of hotness $h = .1$ and $p = .9$, and two values of m, .05 and .01. 10 buckets were used for the bucket-sort algorithm. In each case, we attempted to locate a good age-threshold, guided by the analysis. Two conclusions from these results are: (1) the all-age version does better than the TW-age version in all cases examined; and (2) for both the TW-age and all-age versions, we do not pay a price in increased GCU by using the more efficient bucket-sort algorithm with 10 buckets, instead of the full-sort algorithm.

13.10   SIMULATIONS USING AN ACTUAL I/O TRACE

In this section, we compare garbage collection algorithms using an I/O trace collected
from a running system. We used the "snake" trace (Ruemmler and Wilkes, 1993). The
trace was collected on a 3 gigabyte file server over a period of two months. It contains
about 6.8 million writes (only the writes are relevant to the GCU of a garbage collection algorithm). Further information about the trace can be found in (Ruemmler and
Wilkes, 1993). Each write I/O was converted to a sequence of one or more track writes
depending on the disk, starting address, and length of the I/O. The track size was taken
to be 32 kilobytes. This yielded a total of $T = 98224$ tracks in the system. Of these,
36607 tracks were written (at least once) in the trace. To achieve ASU = .8, the simulations were done with $S = 2455$ segments of capacity $C = 50$. Max-empty was
122, $i.e.$ 5% of the segments. To better simulate the situation that track writing and
garbage collection can occur concurrently, during each phase of garbage collection
(started whenever the number of empty segments reached zero) the simulation alter-
nately created 10 new empty segments and filled 5 empty segments until the number
of empty segments reached max-empty. We allowed the system to warm-up from its
initial configuration for 1 million writes, and then calculated GCU over the remaining
5.8 million writes. (GCU was also computed over the entire trace; these GCU numbers were slightly lower but generally followed the same pattern as those computed
only for the last 5.8 million writes.)

The following five algorithms were simulated:

A1. the bucket-sort TW-age-threshold algorithm with 10 buckets;

A2. the bucket-sort all-age-threshold algorithm with 10 buckets;

A3. the greedy algorithm;

A4. the cost-benefit algorithm with segment-based segment age and no age-
grouping;

A5. the cost-benefit algorithm with track-based segment age and age-grouping.

For cost-benefit algorithm A5, the I/O start times were used to determine the age of
a track. Results are shown in Figure 13.2. Since the same track might be written many
times over a short time interval, we used an LRU write cache in the simulation, with the
cache size equal to one segment's worth of tracks (50). A track is placed in the segment
being filled only when it is pushed out of the write cache. Two differences between
these results and results obtained for the hot-and-cold model with $h = .1, p = .9$ are
that the optimal normalized age-threshold is smaller for the trace results, and the TW-
age and all-age versions have virtually identical GCU for all reasonably small values
of the age-threshold. The likely reason for both differences is that young segments
empty more quickly in the trace simulations. Thus, a smaller age-threshold suffices,
and GC-filled segments have less "hot" data in them.

Another experiment was done where the system contains only the tracks that are actually written, and each written track is loaded the first time it is written. Since 36607
tracks are written, we took $C = 50$ and $S = 915$ to achieve ASU = .8. The results
were similar to those obtained above. The age-threshold algorithm at normalized age-

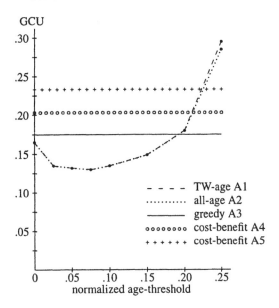

Figure 13.2   Results of simulations on the snake trace with ASU = .8 and m = .05.

threshold t = .075 had the best GCU (.124), followed by greedy (.169), cost-benefit A4 (.174), and cost-benefit A5 (.199).

13.11   AGE-THRESHOLDS FOR UNKNOWN WORKLOADS

Our analysis can be used to select an optimal age-threshold for a workload that follows the hot-and-cold model of track writing. In the absence of any knowledge of the workload, we propose a dynamic learning algorithm for choosing an age-threshold. We start the system with an age-threshold of 0, and set the value of a tuning knob called "direction" to *up*. The algorithm then repeatedly performs the following steps:

1. Measure GCU over a long period of time.

2. If the measured GCU is larger than that computed during the previous period (and this is not the first iteration), then change the value of direction (for example, from *up* to *down*); otherwise keep it the same.

3. Increase or decrease the value of age-threshold depending on the value of the direction knob.

Of course the age-threshold should not be allowed to go below zero, and it is probably also a good idea to set some upper bound above which the age-threshold is not allowed to go. Under a workload for which GCU as a function of age-threshold has only one local minimum (this holds in every case we examined), this learning algorithm will eventually home in on a good age-threshold.

If the dynamic learning algorithm is too difficult to implement, we propose using the following heuristic value for the normalized age-threshold which seems to give reasonable results in many situations. The heuristic is of the form $F \times (1 - \text{ASU} - m)$, where F is a fraction; we suggest taking $F = 0.5$. For example, with $\text{ASU} = .8$ and $m = .05$, the heuristic with $F = 0.5$ gives a normalized age-threshold of .075, close to the optimal age-threshold for the snake trace. Additional simulations using the snake trace were done with $(\text{ASU}, m) = (.7, .05), (.9, .05), (.9, .01)$. In each case examined, this heuristic gave a close-to-optimal age-threshold.

One direction for future research is additional methods for automatically choosing a good age-threshold. A complication is the possibility that multiple applications could be running concurrently with different requirements. There is existing work on adaptive methods for garbage collection in log-structured file systems (Matthews *et al.*, 1997), although of a different type than the methods considered here. Another future direction is to test the age-threshold garbage collection algorithm in a prototype LSA system, including methods for dynamic selection of the age-threshold value.

Acknowledgments

We thank Bruce McNutt for helpful discussions.

References

Blackwell, T., Harris, J., and Seltzer, M. (1995). Heuristic cleaning algorithms in log-structured file systems. In *Proc. USENIX 1995 Winter Conference*, pages 277–288.

de Jonge, W., Kaashoek, M. F., and Hsieh, W. C. (1993). The Logical Disk: A new approach to improving file systems. In *14th ACM Symposium on Operating System Principles*, pages 15–28.

English, R. M. and Stepanov, A. A. (1992). Loge: A self-organizing disk controller. In *USENIX 1992 Winter Conference*, pages 237–251.

Matthews, J. N., Roselli, D., Costello, A. M., Wang, R. Y., and Anderson, T. E. (1997). Improving the performance of log-structured file systems with adaptive methods. In *16th ACM Symposium on Operating System Principles*.

McNutt, B. (1994). Background data movement in a log-structured disk subsystem. *IBM Journal of Research and Development*, 38(1):47–58.

Menon, J. (1995). A performance comparison of RAID-5 and log-structured arrays. In *4th IEEE Symposium on High-Performance Distributed Computing*.

Menon, J. and Stockmeyer, L. (1998). An age-threshold algorithm for garbage collection in log-structured arrays and file systems. IBM Research Report, IBM Research Division, San Jose, CA.

Patterson, D. A., Gibson, G., and Katz, R. H. (1988). A case for redundant arrays of inexpensive disks (RAID). In *ACM SIGMOD International Conference on Management of Data*, pages 109–116.

Rosenblum, M. and Ousterhout, J. K. (1992). The design and implementation of a log-structured file system. *ACM Trans. Computer Systems*, 10(1):26–52.

Ruemmler, C. and Wilkes, J. (1993). UNIX disk access patterns. In *USENIX 1993 Winter Conference*, pages 405–420.

14 THINKING BIG IN A SMALL WORLD — EFFICIENT QUERY EXECUTION ON SMALL-SCALE SMPs

Stefan Manegold and Florian Waas

Centrum voor Wiskunde en Informatica, The Netherlands

stefan.manegold@@cwi.nl
florian.waas@cwi.nl

Abstract: Many techniques developed for parallel database systems were focused on large-scale, often prototypical, hardware platforms. Therefore, most results cannot easily be transferred to widely available workstation clusters such as multiprocessor workstations. In this paper we address the exploitation of pipelining parallelism in query processing on small multiprocessor environments. We present DTE/R, a strategy for executing pipelining segments of arbitrary length by replicating the segment's operator. Therefore, DTE/R avoids static processor-to-operator assignment of conventional processing techniques. Consequently, DTE/R achieves automatic load-balancing and skew-handling. DTE/R outperforms conventional pipelining execution techniques substantially.

Keywords: parallel and distributed databases, parallel query processing, load balancing.

14.1 INTRODUCTION

Parallel database research is driven by the need for higher performance demanded by new application areas of relational database technology, *e.g.* data warehousing. Much work has been devoted to basic research and produced notable insights to many aspects of parallel query processing. The most prominent representatives are the Gamma Database Machine on a Vax-Cluster and on an Intel Hypercube (DeWitt *et al.*, 1986), Volcano, also running on an Intel Hypercube (Graefe, 1990), EDS/DBS3 (Bergsten *et al.*, 1991) on an Encore MULTIMAX and on a KSR1 in a later version, and PRISMA/DB on top of a Philips/Motorola 128 processor experimental prototype (Apers *et al.*, 1992). The first commercial database machine, Teradata's DBC/1012, was also implemented on non-standard hardware, where processors are connected via a tree-shaped network (Cariño and Kostamaa, 1992).

Figure 14.1   Two-phase optimization approach.

An important aspect that these research prototypes have in common is that they are not affordable to the average user who just wants to reduce response time of his query execution engine. On the other hand, small SMPs (Symmetric Multi-Processors) with about 4 to 8 processors have become quite popular in recent years. These machines provide surprisingly high computing power, efficient shared memory access (fast communication) and are inexpensive, as compared to the previously mentioned research equipment.

In this paper we address the question of how to improve query execution techniques on small-scale shared-everything platforms. We present an execution strategy named DTE/R that exploits pipelining parallelism aggressively. DTE/R replicates the pipelining segment for each processor and avoids any static processor-to-operator assignments. Therefore, DTE/R circumvents the drawbacks of conventional pipelining execution, as given in (Hasan and Motwani, 1994). The basic algorithm distributes the input data over all available processors where the entire pipelining segment is processed. However, in typical skew situations, this may overload some processors while others are idle. DTE/R overcomes this problem by a redistribution of the intermediate processing results. This redistribution adds little overhead, but pays back significantly in almost every case of skew. The quantitative assessment of our implementation, carried out on a 4 processor SMP, shows substantial savings over the conventional pipelining execution technique. Furthermore, DTE/R achieves near-linear speedup and a scaleup ratio close to 1.

14.2   PARALLEL QUERY PROCESSING

Query processing consists of the two components: *query optimization* and *query execution*. The declarative query submitted by the user is transformed into a *query evaluation plan*, which is the actual program the database system has to execute. Following a common approach (Stonebraker, 1993; Chekuri *et al.*, 1995), query optimization in parallel environments can be split into two phases: in the first step, transforming the declarative query into a procedural but sequential one, and, in the second step, parallelizing this sequential plan (Figure 14.1). This way of optimization delivers a parallel plan, consisting of the structural information of a sequential plan and an assignment of processors to sub-plans.

14.2.1   Query Optimization

We assume a conventional sequential optimizer to perform the first optimization step. In Figure 14.2, the possible result of such an optimization is depicted as a tree, consisting of hash-joins only. For convenience we focus on the processing of join trees,

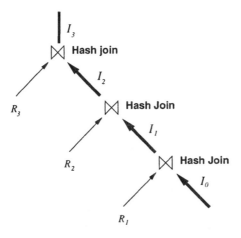

Figure 14.2 Pipelining segment.

although our algorithms are not restricted to a certain set of operators. Joins are considered to be the most expensive operators in relational query processing, involving disk and network access. Thus, most related work focuses on join processing as well (Schneider and DeWitt, 1990; Wilschut and Apers, 1991; Chen *et al.*, 1992; Stonebraker, 1993; Hasan and Motwani, 1995).

A query evaluation plan can be interpreted as a dataflow graph describing *producer-consumer* dependencies between operators. The data to be processed is forwarded from one operator to the next according to the dependencies. Processing a single join involves comparing all data from one input with every data item from the second input. If both tuples fulfill the join's predicate (match on a set of attributes), a new tuple is constructed. If no matching partner is found further processing of this particular tuple is canceled. The most efficient way to perform a join is to store one input in a hash table and do a hash lookup for every tuple of the second input (for more details we refer to (Graefe, 1993)). We call an edge in the dataflow graph *blocking* if it describes input of data that has to be read entirely before any data of the second input can be processed, otherwise we call it *non-blocking* edge. In Figure 14.2, all non-blocking edges are emphasized by thick lines. Furthermore, we call operators along a chain of non-blocking edges *non-blocking operators*.

14.2.2 Exploiting Pipeline Parallelism

The data dependencies, given by the evaluation plan, suggest different feasible kinds of parallelism. *Pipelining parallelism* is of particular interest for the following reasons:

1. Pipelining parallelism is much easier to control than *independent parallelism* where operators without data dependencies are executed in parallel. The resource allocation for a *pipelining segment*—which is a series of operators along non-blocking edges—can be determined more precisely (Srivastava and Elsesser, 1993). Furthermore, intermediate results do not have to be stored in

main memory or on secondary storage; they are immediately processed by the succeeding operator. Only the final output of a pipelining segment has to be stored.

2. For certain classes of queries, all evaluation plans are linear trees. Thus pipelining parallelism is the *only* possibility to achieve parallel processing (Hasan and Motwani, 1994).

3. Every evaluation plan can be decomposed into linear subtrees. Consequently, pipelining parallelism is an option for every evaluation plan.

Parts of the first point also apply to sequential query execution. Thus, we assume that the optimization step already delivers the appropriate pipelining segments. We further assume that (1) all hash tables of a pipelining segment fit into main memory and (2) every hash-look up gives a definite answer (no further comparison is needed).

14.2.3   Query Execution

Once an evaluation plan is generated and decomposed into pipelining segments, the plan is executed on the target platform. Its processing demands a loading phase, where all data belonging to blocking edges is preprocessed in an appropriate way—*e.g.* all hash tables are built. We refer to this phase as the *build-phase*. Hereafter, the actual hash-join processing starts, called the *probe-phase*.

The build-phase is common to all execution strategies and is usually I/O-bound. Therefore, it is not interesting for purposes of this paper.

14.3   EVALUATING PIPELINING SEGMENTS

Before presenting different strategies to evaluate pipelining segments, we informally introduce some definitions and notations used in the remainder of this paper.

14.3.1   Definitions and Notations

Figure 14.2 depicts a sample pipelining segment consisting of three joins. The corresponding notation convention is given in Table 14.1.

We define the *weight w_i* of a single operator op_i as its *sequential execution time T_i*, the time a single processor needs to process all the respective input data through that operator. The weight w of a whole pipelining segment using sequential execution then amounts to the sum of the weight of all the operators within the pipelining segment: $w = \sum_{i=1}^{n} w_i$. The *ideal* parallel execution time of an operator executed on p_i processors is $T_i(p_i) = w_i/p_i$. Thus, the ideal execution time of the whole segment is $T = w/p$.

14.3.2   Pipelining Execution

The conventional *Pipelining Execution (PE)* technique evaluates pipelining segments as follows. Each operator is executed on a distinct processor. To avoid context switching overhead, each processor executes only one operator at a time. Each output tuple

Table 14.1 Notation.

Name	Description
n	number of joins
I_0, R_i	base relations, $i \in \{1, \ldots, n\}$
I_i	intermediate results: $I_i = R_i \bowtie I_{i-1}, i \in \{1, \ldots, n\}$
$\|R\|$	number of tuples in relation R
sf_i	selectivity factor of i-th join: $sf_i = \dfrac{\|R_i \bowtie I_{i-1}\|}{\|R_i\| \cdot \|I_{i-1}\|}$
af_i	augmentation factor of i-th join: $af_i = sf_i \cdot \|R_i\|$
w_i	weight (*i.e.* sequential execution time) of the i-th join
w	weight of the whole pipelining segment: $w = \sum_{i=1}^{n} w_i$
p	number of processors
p_i	number of processors assigned to the i-th stage: $p = \sum_{i=1}^{n} p_i$
$T_i(p)$	parallel execution time of the i-th stage using p processors

produced by op_i is immediately forwarded to its consumer op_{i+1}, and op_i can then process the next input tuple.

Obviously, PE demands $n \leq p$, *i.e.* the number of operators has to be less than the number of processors. We distinguish two cases:

Case 1: $n = p$

Due to the data dependencies between the operators, the total execution time is dominated by the slowest operator, *i.e.* $T_{\text{PE}} = \max_{i=1}^{n}\{T_i\}$. If all operators have the same weight ($w_1 = w_2 = \ldots w_n = w'$), then PE provides optimal parallel performance:

$$T_{\text{PE}} = \max_{i=1}^{n}\{w_i\} = \frac{nw'}{n} = \frac{w}{p}.$$

Otherwise, PE performs suboptimally as it is not able to balance the different loads:

$$T_{\text{PE}} = \max_{i=1}^{n}\{w_i\} = \frac{n \max_{i=1}^{n}\{w_i\}}{n} > \frac{w}{p}.$$

Actually, the performance of PE is even worse. At the beginning, only one processor is active, executing the first operator. The other processors are idle and wait for the output produced by the preceding operator(s). Thus, every processor starts working with a certain *startup delay*:

$$t_{0,1} = 0 ; \qquad t_{0,i+1} = t_{0,i} + \frac{w_i}{\|I_{i-1}\|}\frac{1}{af_i} = \frac{1}{\|I_0\|}\sum_{j=1}^{i}\frac{w_j}{\prod_{k=1}^{j} af_k} .$$

Similarly, as soon as the first processor has finished, the subsequent processors are still active processing the latest output of the first one. Thus, in addition to the aforementioned startup delay, there also is a *shutdown delay* that further decreases the parallel performance of PE. PE provides no means to balance the load variances at run-time.

Case 2: $n < p$

When using more processors than operators, we can assign more than one processor to the same operator. In addition to pipelining parallelism, intra-operator parallelism becomes possible—at least for some operators. To do so, the *processor allocation problem (PAP)*, to assign the appropriate number p_i^* of processors to each operator, has to be solved. We look for a configuration where

$$\frac{w_1}{p_1^*} = \cdots = \frac{w_n}{p_n^*} = \frac{w}{p}$$

holds. Unfortunately, this equation system does not have discrete solutions $(p_1^*, \ldots, p_n^*) \in \mathbb{N}^n$, in general. Thus, integer approximations $(p_1, \ldots, p_n) \in \mathbb{N}^n$ must be found, which provide minimal execution time and fulfill

$$\sum_{i=1}^{n} p_i = p.$$

In this case, the minimal achievable execution time is greater than w/p. This effect is known as *discretization error* (Srivastava and Elsesser, 1993; Wilschut *et al.*, 1995). In (Manegold *et al.*, 1998), we present an algorithm which computes such approximations.

14.3.3   Data Threaded Execution

With *Data Threaded Execution (DTE)* all operators of a pipelining segment are gathered into one stage that is replicated for each processor. DTE creates one thread per processor to perform all operators within the active pipelining segment. A global input queue, accessible by all threads, provides the input tuples for the pipelining segment. Each thread takes one tuple at a time and processes it through all the operators of the pipelining segment. A tuple does not leave the thread—and thus the processor—during its way through the pipelining segment, until it has successfully passed the last operator or it failed to find a join partner. Whenever a tuple leaves a thread, this thread immediately starts processing the next input tuple from the global queue, unless the queue is already empty. In case an operator produces more than one output tuple from one input tuple (a tuple finds more than one partner in a join), the originator thread has to process all these additional tuples first, before it can proceed with the next input tuple from the queue.

Contrary to PE, DTE dynamically assigns processors to the data instead of statically assigning processors to operators according to their weight. This way, DTE avoids the PAP and, hence, cannot suffer from discretization error. Further, as there are no data dependencies between the threads, DTE is not affected by startup delay.

In (Manegold *et al.*, 1997), we presented DTE in detail and showed that in the case of non-skewed data DTE achieves optimal load balancing and, thus, minimal execution time $T_{\text{DTE}} = w/p$.

14.4 SKEW HANDLING

Now we will have a more detailed look at both strategies and discuss how each performs in the presence of highly skewed data. Skew in general means that due to the attribute value distributions not every input tuple finds exactly one output tuple in each join. Some of these skew situations are of little interest; for example, if—per join— only each k-th ($k > 1$) input tuple finds (exactly) one partner or (nearly) each input tuple finds (approximately) $l > 1$ partners. This just involves operators with different weights. While PE suffers from such situations due to discretization error, DTE still achieves optimal load balancing. For a detailed discussion of these kinds of *uniform* skew, we refer the interested reader to (Manegold *et al.*, 1997). We will focus on more extreme kinds of skew in the remainder of this paper.

14.4.1 Execution Skew

PE uses cost estimates to find an appropriate assignment of processors to operators. The errors in cost estimates propagate and intensify while processing a pipelining segment. The best processor assignment according to the (erroneous) cost estimation is, apart from some rare cases, no longer the optimal assignment, according to the actual execution costs. This *execution skew* in turn leads to a non-optimal execution behavior of PE.

In contrast, DTE does not need any cost estimates for processor assignment, but rather assigns processors to the data automatically during execution. Hence, DTE does not suffer from any execution skew.

14.4.2 Join Product Skew

In the following, we assume an ideal setup for PE where neither discretization error nor execution skew occur. We will show that despite these assumptions there are several situation where PE does not perform very well. But we will also show that even DTE may perform badly in some of these situations.

Consider that case of a skewed attribute value distribution. In one of the joins each k-th input tuple ($k \gg 1$) finds k partners while all the other input tuples do not find any partners at all. This scenario, called *join product skew* (Walton *et al.*, 1991), is typical for foreign key joins. In order to trigger just one effect at a time (in this case join product skew), we chose this scenario so that no discretization error occurs. Each join produces as many output tuples as it receives input tuples.

With PE, whenever several subsequent input tuples of an operator do not find any join partner, the next operators are idle, as they get no input data. Thus, join product skew increases the startup delay. In the same way, join product skew increases the shutdown delay, if the last input tuples processed by an operator produce lots of output tuples.

As there are no data dependencies between the threads, DTE does not suffer from startup delay at all. Thus, join product skew obviously cannot increase startup delay with DTE.

The only situation where join product skew can affect the performance of DTE arises when in one operator b input tuples ($b < p$) generate so many additional output tuples that the b threads processing these tuples are occupied significantly longer than the other $p - b$ threads need to process all the remaining input tuples from the global queue. In this case, shutdown delay also occurs with DTE: $p - b$ processors become idle as soon as the global queue is empty, while b processor are still busy. Thus, load balancing is no longer optimal.

14.4.3 DTE/R

To overcome DTE's shutdown delay, we refine the strategy in the following way. Whenever a tuple generates more than one output tuple in an operator, all but one of these output tuples are put back into the global queue. Only one tuple stays on the thread for further precessing. With this *re-distribution*, the additional output tuples can be processed by any thread that becomes idle. Thus, the refined strategy *DTE/R* achieves optimal load balancing even in the case of extreme skew.

To make this re-distribution work, in DTE/R each tuple that is put into the global queue is tagged with the id of the operator that puts it there. Original input tuples are tagged with 0. Thus, each thread knows at which operator it has to start/continue the processing of a tuple it receives from the queue.

As an extension, we changed the global queue into a preemptive queue that uses the tag as the priority of each tuple. Tuples from the queue are then processed according to their priority (highest priority first). This technique achieves two things: the queue does not become unnecessarily long and the output is not unduly deferred. For performance reasons, we implemented the preemptive queue internally as a set of simple queues (one per priority) with one common interface.

14.5   QUANTITATIVE ASSESSMENT

In order to analyze and compare the execution behavior of PE, DTE, and DTE/R, we implemented all three strategies in a prototype query engine to evaluate the probe phase of pipelining segments. We used a 4-processor SGI PowerChallenge shared-memory machine to run our experiments on. Our experiments cover arbitrary randomly generated queries as well as queries particularly designed to examine extreme skew situations. Due to space limitations, we mainly present the results for situations with extreme skew in this section. Further experiments covering non-skew, uniform skew, and execution skew situations can be found in (Manegold *et al.*, 1998).

14.5.1   Query Design and Benching Strategy

In our experiments, queries are marked by the parameters given in Table 14.2. More details about the queries will be given for each set of experiments.

For each configuration, we first generated the base relations according to the query specifications. Then we built all the hash tables and the global input queue in main memory. We chose our query configurations so that the global input queue always fit in main memory. Thus, we could examine the pure performance of the probe phase without any I/O-influence. In case the input queue does not fit completely in main memory, the I/O to read it from disk during the probe phase affects all strategies. After

Table 14.2   Query parameters.

Name	Description	Value
n	number of joins	1 to 16
$\lvert R_i \rvert$	cardinality of base relations	5k to 200k
r	range of join attribute values	$1 \leq r \leq \lvert R_i \rvert$
d	attribute value distribution of join attributes	round-robin, uniform, normal1 (mean=$\frac{r}{2}$, deviation=$\frac{r}{10}$), normal2 (mean=$\frac{r}{2}$, deviation=$\frac{r}{5}$), exponential (mean=$\frac{r}{2}$)

that, we executed the probe phase using PE, DTE, and DTE/R. We only measured the execution time for the probe phase. We ran each distinct configuration 10 times and computed—per strategy—the median of all 10 runs as final execution time T_{PE}, T_{DTE}, and $T_{\text{DTE/R}}$, respectively.

14.5.2   The Average Case

The first series of experiments gives an overall estimate for the average case. The base relation sizes were chosen randomly from our portfolio and one of the five distribution types was used to generate the actual attribute value distribution. For each query, all distributions were of the same type; the particular parameters were chosen as given in Table 14.2. All experiments were carried out on 4 processors.

In Figure 14.3, the response times for uniform attribute value distribution are depicted. The values are scaled to the execution time of DTE. PE is limited by the number of processors and therefore only values for 2, 3 and 4 joins are available. DTE and DTE/R do not differ much as only few skew situations are encountered. Both provide savings of up to 120% compared to PE.

In Figure 14.4, the results for the *normal2* distributed attribute values are plotted. The savings are similar to the previous case. Other distributions showed very similar results—thus we omit them here. Results can be found in (Manegold *et al.*, 1998).

Besides this overall performance comparison, we also ran experiments to measure the speedup and scaleup of the different strategies. Figure 14.5 shows the speedup behavior of PE, DTE, and DTE/R for a two-join-query with $af_1 = 0.5$ and $af_2 = 1$. DTE and DTE/R both provide near-linear speedup, whereas PE obviously suffers from discretization error. Similarly, Figure 14.6 shows the scaleup behavior of PE, DTE, and DTE/R for a two-join-query. We increased the weight of the pipelining segment with the number of processors by increasing af_1, appropriately, while leaving $af_2 = 1$. DTE and DTE/R show a slight performance decrease of 6-7% when moving from one to two processors, but then their scaleup is constant. PE shows a significantly worse scaleup behavior. Experiments with other kinds of queries show the same tendencies for both speedup and scaleup.

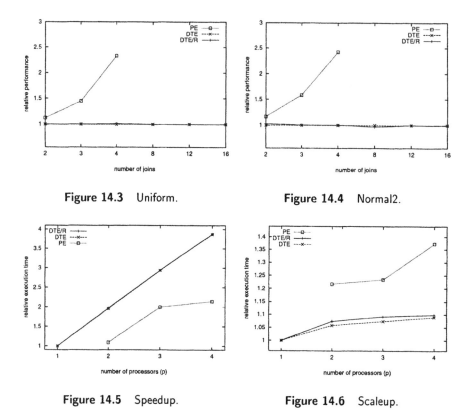

Figure 14.3   Uniform.

Figure 14.4   Normal2.

Figure 14.5   Speedup.

Figure 14.6   Scaleup.

14.5.3   Extreme Skew

As mentioned in the previous section, DTE cannot achieve optimal load balancing once $b < p$ input tuples generate too many additional output tuples. When this happens, the b threads processing these tuples are occupied significantly longer than the other $p - b$ threads need to process all the remaining input tuples from the global queue.

To examine such situations, we used a round-robin distribution on base relations of equal size (24k tuples) and chose the ranges of the join attributes so that:

- In the first join, m input tuples hit and find $k = \dfrac{24,000}{m}$ tuples each. Between two subsequent hitting tuples, $k - 1$ tuples find no partner. In other words, every k-th input tuple finds k partners. For instance, with $m = 2$ the 12,000th and the 24,000th (last) tuple hit and generate 12,000 output tuples each.

- Each of the other joins produces exactly one output tuple from each input tuple.

The parameter m provides a kind of metric for the amount of skew: the smaller m is, the greater the skew is. In our experiments, we varied m from 1 to 8. Figures 14.7 and 14.8 show the relative execution times $\dfrac{T_s}{T_{\text{DTE/R}}}$, $s \in \{\text{PE, DTE, DTE/R}\}$ for queries

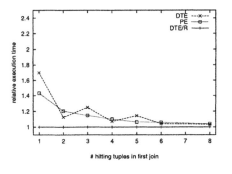

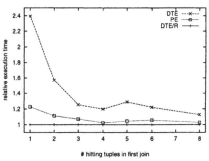

Figure 14.7 Skew on 2 processors.   **Figure 14.8** Skew on 4 processors.

consisting of two joins executed on two and four processors, respectively.

With $m = 1$, DTE provides the worst performance of the three strategies. First, all processors are involved in executing the first join, *i.e.* just probing the input tuples against the first hash table without finding any partner. Only the last input tuple finds partners, which then have to be processed through the second join by one only thread. PE performs better than DTE, as the processors assigned to the second join start working as soon as the first output tuple of the first join is produced. Here, processing tuples through the second join partly overlaps with producing output tuples of the first join. DTE/R performs better than PE, as already the first join is executed on twice as many processors as with PE.

As m increases, the skew decreases. Now the differences between the strategies becomes smaller as the load balancing is easier for PE and DTE. With two processors, DTE is better than PE whenever m is even. With four processors, the performance DTE nearly reaches that of PE whenever m is a multiple of 4, but DTE is never better than PE.

To get better insight in what happens during execution with the different strategies, we also measured the rates at which each thread of each strategy consumes and produces tuples. Figures 14.9 through 14.11 show the respective curves for PE, DTE, and DTE/R executing the skew query with $m = 1$ on two processors. The elapsed time is normalized to the slowest execution time. The different phases during execution are denoted by numbers as follows:

(1) consuming all tuples from the input queue and probing them in the first join without finding any join partner for all but the last tuple,

(2) producing 24k output tuples from the last input tuple of the first join,

(3) consuming 24k tuples from the first join, probing them in the second join, and producing one output tuple from each one, and

(4) idle time.

In our implementation, both probing an input tuple against a hash table as well as producing an output tuple take approximately the same time. Thus, performing only

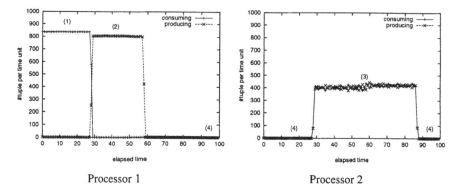

Figure 14.9   Rates of consuming and producing tuples (PE).

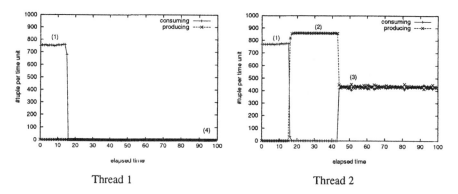

Figure 14.10   Rates of consuming and producing tuples (DTE).

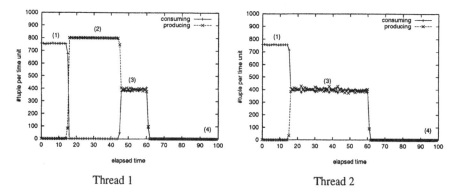

Figure 14.11   Rates of consuming and producing tuples (DTE/R).

one of these actions (1,2) can be done at a rate that is approximately twice as high as that of performing both actions (3).

Figure 14.9 shows the PE results. Only the first processor executes phase (1). As soon as it starts producing output tuples from the last input tuple (2), the second processor starts consuming these tuples and processes them through the second join (3).

Figure 14.10 shows the DTE results. Both threads participate in executing phase (1). Then, only thread 1 executes phases (2) and (3).

Figure 14.11 shows the DTE/R results. Both threads share the execution of phase (1). After that, thread 1 puts each tuple produced in phase (2) immediately into the global queue, so that thread 2 can consume these and process them through phase (3). As soon as thread 1 has finished phase (2), it joins phase (3).

For queries involving up to 16 joins, DTE and DTE/R show the same tendencies as for short queries.

14.6 CONCLUSION

In this paper, we addressed the problem of utilizing the computing power of small off-the-shelf SMP workstations in query execution. We presented the key ideas of DTE/R, a novel strategy to exploit pipelining parallelism in query processing. The principles of operator replication and data threading are building blocks for the desired improvement.

In (Manegold *et al.*, 1997) we reported on DTE, the basic algorithm and gave performance assessments for situations not involving skew. However, in extreme skew situations DTE's performance varies enormously. Here, we show that the potential problems resulting from skew can be solved by a re-distribution mechanism, which adds some extra overhead but makes the algorithm resistant to any skew.

The final contribution of this paper is the assessment of a real implementation, not merely a simulation. We experimented with a broad class of queries and examined both average case and extreme situations. The main characteristics of DTE/R are a nearly linear speedup and a scaleup ratio close to 1. DTE/R outperforms conventional pipelining technique substantially.

Our experiments are promising and show that highly efficient parallel database techniques can be beneficial for small parallel configurations.

References

Apers, P., van den Berg, C., Flokstra, J., Grefen, P., Kersten, M., and Wilschut, A. (1992). PRISMA/DB: A parallel main memory relational DBMS. *IEEE Trans. on Knowledge and Data Eng.*, 4(6):541–554.

Bergsten, B., Couprie, M., and Valduriez, P. (1991). Prototyping DBS3, a shared-memory parallel database system. In *International Conference on Parallel and Distributed Information Systems*, pages 226–235.

Cariño, F. and Kostamaa, P. (1992). Exegesis of DBC/1012 and P-90 – industrial supercomputer database machines. In *International Conference on Parallel Architectures and Languages in Europe*, pages 877–892.

Chekuri, C., Hasan, W., and Motwani, R. (1995). Scheduling problems in parallel query optimization. In *ACM SIGACT-SIGMOD-SIGART Symposium on Principles of Database Systems*, pages 255–265.

Chen, M.-S., Lo, M., Yu, P., and Young, H. (1992). Using segmented right-deep trees for the execution of pipelined hash joins. In *International Conference on Very Large Data Bases*, pages 15–26.

DeWitt, D., Gerber, R., Graefe, G., Heytens, M., Kumar, K., and Muralikrishna, M. (1986). GAMMA — a high performance dataflow database machine. In *International Conference on Very Large Data Bases*, pages 228–237.

Graefe, G. (1990). Encapsulation of parallelism in the Volcano query processing system. In *ACM SIGMOD International Conference*, pages 749–764.

Graefe, G. (1993). Query evaluation techniques for large databases. *ACM Computing Surveys*, 25(2):73–170.

Hasan, W. and Motwani, R. (1994). Optimization algorithms for exploiting the parallelism-communication tradeoff in pipelining parallelism. In *International Conference on Very Large Data Bases*, pages 36–47.

Hasan, W. and Motwani, R. (1995). Coloring away communication in parallel query optimization. In *International Conference on Very Large Data Bases*, pages 239–250.

Manegold, S., Obermaier, J., and Waas, F. (1997). Load balanced query evaluation in shared-everything environments. In *European Conference on Parallel Processing*, pages 1117–1124.

Manegold, S., Waas, F., and Kersten, M. (1998). On optimal pipeline processing in parallel query optimization. Technical Report INS-R9805, CWI, Amsterdam, The Netherlands.

Schneider, D. and DeWitt, D. (1990). Tradeoffs in processing complex join queries via hashing in multiprocessor database machines. In *International Conference on Very Large Data Bases*, pages 469–480.

Srivastava, J. and Elsesser, G. (1993). Optimizing multi-join queries in parallel relational databases. In *International Conference on Parallel and Distributed Information Systems*, pages 84–92.

Stonebraker, W. H. M. (1993). Optimization of parallel query execution plans in XPRS. *Distributed and Parallel Databases*, 1(1):9–32.

Walton, C., Dale, A., and Jenevein, R. (1991). A taxonomy and performance model of data skew effects in parallel joins. In *International Conference on Very Large Data Bases*, pages 537–548.

Wilschut, A. and Apers, P. (1991). Dataflow query execution in a parallel main-memory environment. In *International Conference on Parallel and Distributed Information Systems*, pages 68–77.

Wilschut, A., Flokstra, J., and Apers, P. (1995). Parallel evaluation of multi-join queries. In *ACM SIGMOD International Conference*, pages 115–126.

IV Parallel I/O Issues

15 SPEEDING UP AUTOMATIC PARALLEL I/O PERFORMANCE OPTIMIZATION IN PANDA

Ying Chen and Marianne Winslett

Department of Computer Science,
University of Illinois at Urbana-Champaign, U.S.A.

ying@cs.uiuc.edu
winslett@cs.uiuc.edu

Abstract: The large number of system components and their complex interactions in a parallel I/O system, together with dynamically changing I/O patterns in scientific applications, impose a great challenge in selecting high-quality I/O plans for an anticipated I/O workload in a target execution environment. Previous research has shown that a model-based approach that uses a performance model of the parallel I/O system to predict the performance of the parallel I/O system for a given I/O plan, coupled with an effective search algorithm (simulated annealing) can identify a high quality I/O plan automatically. However, to be truly successful, such automatic strategies must not only be capable of *selecting* high-quality I/O plans for a parallel system, but also be able to select them *quickly*. In this paper, we study the cost of optimization when using the model-based approach. We identify the major performance factors that affect the optimization time, and present techniques used to speed up the process. Our performance results obtained from an IBM SP show that with these techniques, our prototype optimization engine can select high-quality I/O plans quickly.

Keywords: parallel I/O, collective I/O, automatic performance optimization, simulated annealing, performance tuning, constraints, classification, genetic algorithm.

15.1 INTRODUCTION

I/O systems on parallel machines are significantly more complex than those on traditional sequential machines, where only a small number of disks are attached to a host processor. The large number and variety of system components, including I/O processors, communication networks, and disks, calls for sophistication in choosing the proper I/O system configuration and I/O execution strategy, especially if the system has to cope with changing I/O patterns and environments. Often careful tuning of the performance knobs of parallel I/O systems (called "the parallel I/O performance parameters" in this paper) for a target workload in a target execution environment is required. This task is nontrivial for humans considering the complex interaction and tradeoffs among different parameter settings.

Automatic performance optimization is a potential cure for the parameter setting selection problem, in which the details of selecting high-quality I/O parameters for a particular situation are handled internally by the parallel I/O system without human intervention. (Chen et al., 1998a; Chen, 1998) presented such an approach, *a model-based approach*. This approach includes two phases. First, for a given I/O workload, the optimization engine is provided with two types of information: a high-level description of the I/O workload characteristics, and a specification of the system software and hardware configuration. These descriptions tell the optimization engine *what* needs to be optimized (not *how* to optimize it), including the expected number and types of I/O requests issued and the sizes of the requests. An application profiling approach is used to extract the application I/O request characteristics and a system micro-benchmarking approach is used to extract the system characteristics automatically.

In the second phase, the optimization engine digests the description and selects high-quality parameter settings for the given I/O workload using a performance model of the parallel I/O system and a set of optimization algorithms, *i.e.* a rule-based approach and a randomized search algorithm, simulated annealing. The performance model predicts the parallel I/O system performance for different combinations of parameter settings. The optimization algorithms identifies high-quality parameter settings based on the model-predicted costs. The rule-based strategy is used when the high-quality parameter settings for a given situation are known in advance, and simulated annealing is used otherwise. This approach has been validated using Panda, a parallel I/O library for collective I/O operations on multidimensional arrays. The performance results obtained from two IBM SPs show that this approach can identify high-quality I/O parameter settings for a wide range of system conditions.

The work presented in (Chen et al., 1998a) focuses on *the quality of the parameter setting selection* rather than the *cost of the optimization*. However, the true success of such an approach must ensure low cost in the optimization process itself. In this paper, we study the performance issues of the model-based approach. We focus on the optimizations using a randomized search algorithm, simulated annealing, since the rule-based approach is fast. We show that the cost of model evaluation, the size of the parameter search space, and the search algorithm used are the major factors that affect the optimization performance. We present the techniques used to reduce the optimization time.

The remainder of the paper is organized as follows. Section 15.2 presents the Panda basics. Section 15.3 identifies the key performance factors of the model-based approach. Section 15.4 presents the techniques used to speed up the model-based approach. In Section 15.5, we present the performance of the model-based approach obtained from an IBM SP. We discuss related work in Section 15.6, and draw conclusions and outline the future work in Section 15.7.

15.2 PANDA BASICS

Panda is designed for large-scale scientific applications performing collective array I/O operations on distributed-memory multiprocessors, a commonly observed I/O pattern in scientific applications (Poole, 1994). In this paper, we base our discussion on Panda 2.1, where each node is either dedicated to application computation or to I/O. The dedicated I/O nodes are called "Panda servers" and the compute nodes are called "Panda clients". During each I/O request, Panda servers read or write array data from or to disks and move data to and from clients using message passing facilities such as the MPI library.

Arrays in Panda are broken up into rectangular chunks using HPF-style BLOCK and * distributions in memory and on disk. Panda can store arrays in flexible chunked formats on disk as well as a fixed traditional row-major or column-major order. An array layout $(x_0, ..., x_{r-1})$ with distribution (BLOCK, ..., BLOCK) for an r dimensional array indicates that each array dimension i is broken into x_i roughly equally-sized blocks. The "*" distribution along dimension j is equivalent to using the BLOCK distribution along dimension j with $x_j = 1$. Thus, we do not consider it separately. The layout chosen by the users to distribute arrays in memory is called the "array in-memory layout" and the layout used to store arrays on disk is called the "array disk layout".

Figure 15.1 illustrates the array layout concept in Panda using a 2D 8 × 8 array as an example. In this example, the thin grid lines indicate the boundary of each array element. Four compute nodes are arranged in a 2 × 2 in-memory layout, and the array is divided across them using a (BLOCK,BLOCK) in-memory distribution. Each compute node's in-memory array chunk is indicated by a heavy border around its array elements (*e.g.* the first compute node is assigned with array elements between element (1,1) and (4,4)). The array disk layout is 3 × 1, and the distribution is (BLOCK,*), *i.e.* row major order. By default, Panda stripes the array disk chunks across all the I/O nodes in a round-robin fashion. Assuming that three I/O nodes are used, then each I/O node is assigned with one array disk chunk. Each I/O node's disk chunk is indicated by a uniform shading of its array elements. For example, the third I/O node is assigned array elements between element (7,1) and (8,8). Panda servers work on one disk chunk at a time until all the assigned disk chunks are processed.

Since arrays can be stored differently in memory than on disk, gathering or scattering one disk chunk may require multiple data message transfers, each data message corresponding to a logical subarray in an in-memory chunk on one compute node. In Figure 15.1, to gather/scatter the second disk chunk, four data messages are required: the first consists of array elements from (4,1) to (4, 4), the second from (4, 5) to (4, 8), the third from (5, 1) to (6, 4), and the last from (5, 4) to (6, 8). The dotted lines demarcate each subarray that is transferred in a data message between a compute node and an I/O node during a collective I/O call to Panda. In this example, each data message corresponds to a contiguous region of a compute node's memory as well as an I/O node's memory.

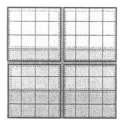

Figure 15.1   Panda's in-memory and on-disk distributions and layouts for a 2D 8×8 array.

15.3 PERFORMANCE FACTORS OF THE MODEL-BASED APPROACH

There are three key factors that affect the performance of the model-based approach:

1. the time required to evaluate the cost of a particular I/O plan using the parallel I/O performance model (called "model evaluation time"),

2. the size of the space in which the search takes place, and

3. the search algorithm used to identify high-quality parameter settings.

Since identifying high-quality I/O parameter settings involves searching in a potentially large parameter setting space ("search space"), the performance model of the parallel I/O system must predict the cost for all the points sampled by the search algorithm. If individual model evaluations incur high cost, this cost will be compounded during the search. We present a classification-based approach to reducing the cost of model evaluation in Section 15.4.1. Since a search algorithm samples a search space to identify high-quality solutions, the larger the search space is, the longer the search may take. We present a constraint-based approach that reduces the size of search space in Section 15.4.2.

Given the complexity of the parallel I/O performance model, and a large search space, exhaustive search is too expensive to be considered. Devising specialized heuristic optimization algorithms for a large complex system with nonlinearly interdependent parameters can be also extremely difficult. Numerical algorithms, such as Newton's method, or direct search methods (Hooke and Jeeves, 1961) are inappropriate since the cost model may not be continuous or differentiable. We believe that randomized search algorithms, such as simulated annealing (SA) (Kirkpatrick *et al.*, 1983) and genetic algorithms (GAs) (Goldberg, 1989), are good candidates to solve this problem. In this paper, we show how simulated annealing (SA) can be used to identify high-quality parameter settings efficiently.

15.4 SPEEDING UP THE MODEL-BASED APPROACH

In this section, we present three techniques used in Panda to reduce the optimization cost:

1. a classification-based approach to reducing the model evaluation costs,

2. a constraint-based approach to reducing the size of the search space, and

3. an adaptive approach, *incremental tuning based on performance goals*.

15.4.1 A classification-based approach to reducing the model evaluation cost

We briefly show how the Panda performance model is derived; details can be found in (Chen *et al.*, 1996). Conceptually, the I/O cost on a particular Panda I/O server i can be simply modeled as follows: $T_i = \sum_{1 \leq j \leq N} T_{ij}$, where N is the number of disk chunks assigned to that I/O node, and T_{ij} is the time spent on disk chunk j. Ideally, if $T_{i1} = ... = T_{iN}$, then $T_i = N \times T_{ij}$. Unfortunately, T_{ij} depends on the number of compute and I/O nodes involved in gathering/scattering disk chunk j, the disk chunk size, the array memory, the disk layout and distribution, and the file system state at the time when disk chunk j is written to/read from disk (*e.g.* the size of the available file system cache). Thus, T_{ij} can vary from one chunk to another. A naïve approach to computing T_i would be to compute T_{ij} for all j, where $1 \leq j \leq N$. This can be expensive when N is large. One way to reduce the time to calculate T_i is to divide disk chunks into a small number of classes such that the costs of processing a disk chunk is the same for all the chunks in the same class.

T_{ij} consists of two cost components: T_{sg}, the cost of gathering or scattering disk chunk j, and T_{fs}, the file system cost for reading or writing disk chunk j. In this paper, we show how we classify disk chunks such that T_{sg} is the same for all the disk chunks in the same class. To classify disk chunks, we examine *the shape of a disk chunk*, which is defined as the number of array elements along each array dimension of the logical subarray that the disk chunk corresponds to. Two disk chunks are said to have *the same shape* iff the logical subarrays that they correspond to have the same size along each array dimension. In Figure 15.1, the first and the second disk chunks are the same shape. T_{sg} for a particular disk chunk is mainly determined by the number, sizes and shapes of the data messages needed to gather/scatter that disk chunk, since these messages may not reside in contiguous memory regions and they must be copied into contiguous memory locations before they are sent/received. Data messages that are of the same size but different shapes could incur different memory-to-memory copying overheads and, hence, they should not be classified into the same class. Given these observations, we devise the following heuristic to classify disk chunks in Panda: *If two disk chunks are of the same shape, and they require the same number of data messages, and the data messages are of the same shape and size, then the two disk chunks belong to the same class.*

15.4.2 A constraint-based approach to reducing the search space

To reduce the size of the search space, we devise a set of constraints based on the target execution environment and I/O workload characteristics. In this section, we present a constraint devised for the array disk layout parameter on the IBM SP as an example.

We use $v = \{v_1 \times ...... \times v_n\}$ to represent an array disk layout, and A to represent an array of rank n, where $A = (a_1 \times ...... \times a_n)$. Here, a_i represents the number of array elements along dimension i. π is used to denote the search space for array disk layout. e is the number of bytes per array element, and S is the array size in bytes. To form π, we use the following *disk layout constraint*:

$$\pi \subseteq \{\{v_1 \times ...... \times v_n\} \mid 1 \leq v_i \leq min(a_i, \lceil \frac{S}{B} \rceil), \text{ and } \frac{S}{\prod_{1 \leq i \leq n} v_i} \geq B\}.$$

On a particular platform, Panda determines the smallest disk chunk size, B. Thus any array disk layout that breaks arrays into chunks that are smaller than B need not be considered, since disk chunks that are too small will not utilize the file system or communication system well.

15.4.3 Adaptive simulated annealing

Previous sections explained why a randomized search algorithm should be used to identify high-quality parameter settings. In Panda, we have experimented with two search algorithms: adaptive simulated annealing (Ingber, 1996) and a genetic algorithm (Goldberg, 1989). (Chen *et al.*, 1998a) and (Chen *et al.*, 1998b) discuss these two algorithms with Panda, respectively. The performance results gathered from an IBM SP indicate that adaptive simulated annealing generally performs better than the genetic algorithm. In this paper, we focus on the optimization techniques used to speed up adaptive simulated annealing, a special simulated annealing described later in this section.

SA is a probabilistic hill-climbing optimization algorithm that resembles the annealing process in statistical mechanics to identify high-quality solutions for a given cost function. In SA, the behavior of a system with many degrees of freedom is highly unstable when the temperature is high. As the temperature cools down, the energy of the system is lowered and the system gradually becomes stable. Finally the system is "frozen". The analogy between SA and the physical annealing process is as follows. The number of parameters of the cost function correspond to the degrees of freedom of the physical system. A particular set of parameter settings, *i.e.* a particular point in the search space, corresponds to a particular system state in the physical annealing process. The value of the cost function at a certain point in the search space corresponds to the energy of the physical system in a particular state.

The efficiency of a SA algorithm is mainly determined by the rate of annealing, which is commonly known as the *annealing schedule*. High-quality annealing schedules largely depend on the properties and solution distributions of the cost function. It can be extremely difficult to find a generally efficient annealing schedule for many problems. As a step towards providing easy-to-use and adapt SA facilities to users to solve their optimization problems, (Ingber, 1996) developed the Adaptive Simulated Annealing package (ASA). ASA attempts to provide an adaptive environment in which users can easily tune a special simulate annealing algorithm that takes into consideration the finite ranges and sensitivities of different parameters and uses importance-sampling techniques to generate efficient annealing schedules, to efficiently solve their own problems. The ASA package provides users with control over many of its parameter settings, such as the annealing rate. Panda uses ASA to identify high-quality parallel I/O parameter settings. In this paper, we study the performance impact of the key ASA parameters, the annealing scales. We also present an adaptive approach, *incremental tuning based on performance goals*, to selecting proper annealing scales in Panda.

The annealing schedule is controlled by three annealing scale parameters in ASA. *Temperature_Ratio_Scale* controls the ratio of the initial and final temperatures. *Temperature_Anneal_Scale* indirectly controls the number of solutions that must be generated to reach the expected final temperature defined by *Temperature_Ratio_Scale*. *Cost_Parameter_Scale_Ratio* controls the temperature cooling rate. Unnecessarily conservative values for these parameters can slow down the search, while over-optimistic values can lead to poor search results. The detailed explanations of these parameters and their mathematical developments can be found in (ASA, 1993; Ingber, 1996).

In Panda, selecting proper annealing scales is I/O workload-specific. For I/O workloads with different I/O characteristics, the best annealing scales for the Panda optimizer can vary. Unless an adaptive approach is used to identify proper annealing scales for a given I/O workload, conservative annealing scales must be used to guarantee the quality of the selected solutions, which can be slow. In Panda, an I/O workload refers to *the sequence of Panda I/O calls generated by a single application execution*.

We present an incremental annealing scale tuning approach based on performance goals to help identify efficient annealing scales for different I/O workloads. We assume that for a particular application on a particular platform, the I/O workloads generated by the same application tend to be similar, and hence these workloads often can use the same or similar annealing scales to identify high-quality Panda parameter settings. This assumption is reasonable since for the same application, often different executions of the same application only change the application's data contents but not its I/O access patterns. Thus, the Panda optimizer can reuse optimization results, and the cost of optimization will be amortized over multiple executions. If the application changes its experiment configurations slightly, such as using a different number of I/O nodes or compute nodes while holding other parameters constant, optimization must be performed again but many of the previously determined annealing scales can be reused.

To determine whether new annealing scales are needed, we use a performance goal as a guard to control the quality of the optimizer-selected solutions. We use *the utilization of the peak file system throughput* as the performance goal, since the file system is the ultimate performance bottleneck on the IBM SP, the platform where this research is carried out. On a platform with different system characteristics, other types of performance goals can be used. If reuse of the initial annealing scales cause ASA to terminate at a solution that does not meet the performance goal, incremental tuning is invoked to adjust the annealing scale settings.

Under the incremental tuning approach, for a given application, the Panda optimizer is initially provided with a set of annealing scales and a performance goal. The initial annealing scales and performance goals are determined based on a calibration of the target platform for a set of common I/O patterns. For instance, on the IBM SP, Panda is often able to achieve more than 90% of the peak, so we can use 90% as the initial performance goal. For annealing scale settings, we select somewhat optimistic settings initially to avoid the use of unnecessarily conservative annealing scale settings at the very beginning of the tuning process.

For the first I/O workload generated by that application, Panda invokes ASA to select high-quality parameter settings using the annealing scales provided. The optimizer compares the performance goal to the predicted performance of the selected I/O plan. If the goal is met, the plan is accepted. Otherwise, the initial annealing scale settings are made more conservative, and the new annealing scale settings are fed into the Panda optimizer to select the high-quality settings again. Currently, we tune the annealing scale settings using a common "one parameter at a time" strategy. Being more precise, we tune the settings for *Temperature_Ratio_Scale* first, and then *Temperature_Anneal_Scale*, since the former has a more significant impact on the annealing time than the latter. In our performance study, we found that the optimization time is not very sensitive to the different values for *Cost_Parameter_Scale_Ratio* in Panda. Hence, we do not tune this annealing scale parameter. This tuning strategy can be improved in the future by using more general experiment design methodologies, such as *the fraction of factorial experiments* (Davies, 1978).

The tuning process is repeated until either the goal is reached or a predefined maximum number of iterations is reached. If the performance goal cannot be met, the best Panda-predicted performance is used to determine a new goal, x, for that application, and a backtrack step is taken to find the first iteration that resulted in a performance level, y, that satisfies $y \geq x$. The annealing scale settings selected in that iteration are accepted and can be reused for the subsequently generated I/O workloads.

15.5   EXPERIMENTAL RESULTS

In this section, we validate the approaches presented in section 15.4 using two synthetic benchmarks. In the first benchmark, we compare ASA with the exhaustive search method and a genetic algorithm. We show that the use of an appropriate search algorithm is crucial to reducing the cost of optimization. The second benchmark validates the incremental tuning strategy used to select proper annealing scales. The two benchmarks we used share the same set of experiment configurations except otherwise specified. The benchmark experiments read and write three-dimensional arrays from and to disks. Arrays are redistributed across compute nodes using different array in-memory layouts at different times.

The array in-memory layout dimensions used in these benchmarks include $n \times 1 \times 1$ and $1 \times 1 \times n$, where n is the number of compute nodes used. We varied the number of compute and I/O nodes used in these benchmarks. The number of I/O nodes used is 4, 5, 6, or 8. The number of compute nodes used are 8, 16, 24, or 32. With 4, 5, or 6 I/O nodes, the array is $256 \times 512 \times 256$ (256 MB), and with 8 I/O nodes, it is $512 \times 512 \times 256$ (512 MB). There are total of 40 requests in the request sequence. Half of the requests are reads and the other half are writes. The total data size for all the requests in each experiment is 5.12 GB for 4, 5 and 6 I/O nodes and 10.24 GB for 8 I/O nodes. The optimization is carried out on the I/O nodes. Hence, there are 16 different combinations of I/O and compute nodes.

Table 15.1   Platform characteristics.

Characteristic	ANL SP
Number of nodes	80
Each node	RS6000/590 or 390, 120 MHz
Node OS	AIX 4.2
Memory per node	256 MB
Interconnect Type and Speed	TB3 switch, 150 MB/s
MPI Latency and Peak throughput	29 microsec, 90 MB/s
Peak FS read and write throughputs:	7.1 MB/s, 6.7 MB/s

The experiments were carried out on the IBM SP at Argonne National Laboratory (ANL). Table 15.1 describes the system characteristics of this platform. More details can be found at http://www.mcs.anl.gov/CCST /computing/quad/. We report the optimization time for each I/O workload in the benchmarks. The Panda optimizer identifies the high-quality settings for the three most important Panda performance parameters: the I/O buffer size used on each I/O node, the array disk layout, and the communication strategy used to move data between clients and servers.

For the ASA and GA results, we report the maximum optimization time over five runs, since different runs have a relatively large variance (the variance can be as high as 30% of the average of five runs). We suspect that this is due to the random sampling property of the algorithms. For the exhaustive search method, the variance is extremely small; we report the average of three runs.

All runs employ disk chunk classification to reduce the time of model evaluation. Constraints are chosen to reduce the search space for different Panda parameters, since the optimization times are too high otherwise (more than 12 hours for one of the tests in the benchmark presented below using ASA).

15.5.1   The effects of optimization algorithms

In this benchmark, we compare the optimization costs when using the exhaustive search, ASA, and GA. Due to the long running time with the exhaustive search method, we only carried out a small set of experiments using 8 compute nodes. However, the performance results obtained are sufficient evidence to demonstrate the effectiveness of these methods.

For ASA, we did not use the incremental tuning approach to select proper annealing scale. Instead, we selected a rather conservative set of ASA parameter settings such that all the experiments in this benchmark can find high-quality settings using the same set of ASA settings. We also compared ASA with GA, an optimization algorithm described in (Chen *et al.*, 1998b). With this set of ASA parameter settings, both the exhaustive search and ASA methods selected the same high-quality Panda parameter settings, but GA stopped at solution points that are about 5% worse than the ASA-selected settings.

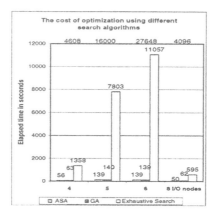

Figure 15.2 The performance comparison among exhaustive search, ASA, and GA. The numbers on the top of the inner box are the sizes of the search space for different I/O node configurations.

Figure 15.2 compares the costs of optimization when using these three methods. It is clear that ASA is at least 11 times faster than the exhaustive search method. Although the ASA and GA performance are comparable, ASA can identify better parameter settings than GA. Using 4 and 8 I/O nodes take much less time than other configurations, since the sizes of the search space for those two cases are much smaller than the other two cases. This is mainly due to the use of the constraints devised for different Panda parameters.

15.5.2 The effects of annealing scales

We run an experiment to validate our incremental tuning approach to selecting proper annealing scales and to demonstrate the effects of different annealing scales in the cost of optimization. We assume that the application generates I/O workloads with different numbers of I/O and compute nodes and different in-memory layouts. In particular, the number of compute nodes is 8, 16, 24, or 32 and the number of I/O nodes is 4, 5, 6, or 8.

In these experiments, we set the initial performance goal to be 90% of the peak underlying file system throughput on each I/O node, and the initial annealing scales are shown in Table 15.2.

The first generated workload uses 8 compute nodes and 4 I/O nodes. With the initial annealing scale settings and the performance goal, the Panda optimizer is able to select Panda parameter settings that utilize 97% of the peak underlying file system throughput in the very first iteration. Thus, there is no need to tune the annealing scale settings. We used the same annealing scale settings for the subsequently generated workloads which varied the number of I/O and compute nodes used. The Panda optimizer is able to select the Panda parameter settings that achieve the goal performance

Table 15.2   The annealing scales.

	Temperature_Ratio_Scale	Temperature_Anneal_Scale
Initial	1.0E-5	10
Incremental	1.0E-4	50
Conservative	1.0E-2	100

using the initial annealing scales for all 16 test cases except for three: when 5 I/O nodes and 24 compute nodes are used, and when 5 and 6 I/O nodes and 32 compute nodes are used. In those three cases, the initial annealing scales led to 86.3%, 78.4%, and 84.2% of the peak file system throughput respectively, which are below the 90% performance goal. The incremental annealing scale tuning is invoked to search for more proper annealing scales in those cases.

We break up the settings for *Temperature_Ratio_Scale* and *Temperature_Anneal_-Scale* into multiple levels based on the conservativeness of the settings. The levels for *Temperature_Ratio_Scale* are 1.0E-5, 1.0E-4, 1.0E-3, 1.0E-2, and 1.0E-1, and the levels for *Temperature_Anneal_Scale* are 10, 50, and 100. We tune these parameters one at a time. Incremental tuning is first invoked for the 24 compute node and 5 I/O node case. The selected annealing scales are listed in Table 15.2 as indicated by "Incremental". The best model-predicted performance using those annealing scales is 87.3%, which is still less than the initial performance goal. Thus, we lowered the performance goal by 5% to 82.3% of the peak based on a performance goal adjustment strategy. This new performance goal and the new annealing scales selected are appropriate for the 5 and 6 I/O node with 32 compute node cases. The model-predicted performance for those two cases is 86.5% and 87.1% respectively when using these new annealing scales.

We also verified the quality of the Panda-selected parameter settings with incremental tuning by comparing them with the results from a set of more conservative annealing scales as listed in Table 15.2. Both methods led to the same set of Panda parameter settings for all the experiments except for one case when 16 compute nodes and 5 I/O nodes were used. In this case, the conservative annealing scales resulted in a 93.4% utilization of the underlying file system, where the optimistic annealing scale settings resulted in a 92.3% utilization. The performance difference is very small.

Figure 15.3 shows the cost of optimization using the annealing scales selected by the incremental tuning approach and the set of conservative annealing scale settings for each of the 16 workloads tested. Clearly, the incremental tuning method is more than 6 times faster than the conservative approach. The choice of annealing scales can significantly impact the optimization time.

15.6   RELATED WORK

Considerable effort has been put into building parallel I/O libraries to provide high-performance support for large-scale scientific applications. However, little work has

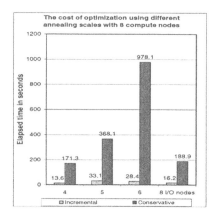

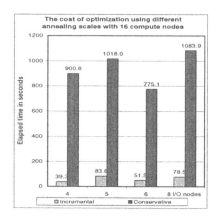

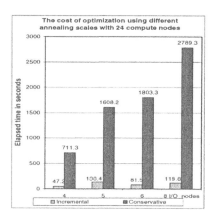

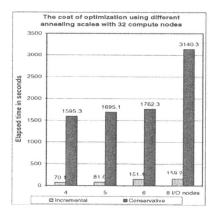

Figure 15.3   The cost of optimization using different annealing scales for the target request sequence with different numbers of compute and I/O nodes.

been done in providing automatic support for performance tuning in parallel I/O systems.

A recent effort has focused on automatically selecting efficient file system caching and prefetching policies in PPFS using two different I/O access pattern classification approaches. In (Madhyasta et al., 1996) and (Madhyasta and Reed, 1996), a trained neural network is used to recognize the application I/O access patterns based on a predefined classification of patterns. A Hidden-Markov model is used to detect the application I/O access patterns automatically. The performance study on the Intel PFS shows that such automatic techniques can select proper file access modes and improve overall application I/O performance. However, no details on the cost of these automatic methods are available.

Despite the lack of automatic optimization work in the parallel I/O world, automatic performance optimization is not uncommon in many other research areas, such as database and operating system research. (Graefe, 1993) provides a fairly complete discussion of many query processing and optimization techniques used for database systems and the design of query optimizers in database systems. (Swami and Gupta, 1988) compared several algorithms used to optimize large join queries, including using iterative improvement, simulated annealing, perturbation walk, and quasi-random sampling. Their performance results showed that simple iterative improvement is superior to all other methods when the amount of time allowed to perform optimization is small. However, as the optimization time increases, the simulated annealing approach becomes the winner. (Ioannidis and Wong, 1987) also showed how to use simulated annealing to optimize recursive queries. Their performance results suggest that with carefully selected parameter settings, simulated annealing can identify high-quality solutions in a relatively short time.

Simulated annealing has been widely used in many different disciplines. (Kirkpatrick *et al.*, 1983) presented examples of using SA to find optimal wiring for computer chips. The early simulated annealing methods typically use Boltzmann annealing (BA) (Szu and Hartley, 1987) that samples an infinite parameter search space and ignores the sensitivities of different parameters. Hence, BA can be inefficient. ASA, however, takes into account such considerations and generates more efficient annealing schedules for a given problem. ASA has been successfully used in many different disciplines, such as combat analysis (Ingber *et al.*, 1991) and neuroscience (Ingber, 1991).

15.7 CONCLUSIONS

In this paper, we identified the major factors that affect the cost optimization using the model-based approach. We discussed how the cost of individual model evaluations, the size of the search space, and the search algorithms have profound implications for the cost of optimization. We presented a classification-based approach to reducing the cost of individual model evaluations, a constraint-based approach to reducing the size of search space, and an incremental tuning approach to selecting proper parameter settings to speed up the annealing process. In the future, we would like to examine the impact of enlarging the number of parameters on the cost of optimization.

Acknowledgments

This research was supported by NASA under grants NAGW 4244 and NCC5 106, Intel Foundation Graduate Fellowship, and Accelerated Strategic Computing Initiative. Computing facilities were provided by Argonne National Laboratory. We thank Ian Foster for his support in providing access to the ANL computing facilities.

References

ASA (1993). Adaptive Simulated Annealing (ASA). ftp.alumni.caltech.edu:/pub-/ingber/ASA-shar.

Chen, Y. (1998). *Automatic parallel I/O performance optimization in Panda*. PhD thesis, Department of Computer Science, University of Illinois.

Chen, Y., Winslett, M., Cho, Y., and Kuo, S. (1998a). Automatic parallel I/O performance optimization in Panda. In *10th Annual ACM Symposium on Parallel Algorithms and Architectures*. ACM Press. To appear.

Chen, Y., Winslett, M., Cho, Y., and Kuo, S. (1998b). Automatic parallel I/O performance optimization using genetic algorithms. In *7th IEEE International Symposium on High Performance Distributed Computing*. To appear.

Chen, Y., Winslett, M., Kuo, S., Cho, Y., Subramaniam, M., and Seamons, K. (1996). Performance modeling for the Panda array I/O library. In *Supercomputing'96*. On CD-ROM.

Davies, O. (1978). *The Design and Analysis of Industrial Experiments*. Longman Group Limited, 2nd edition.

Goldberg, D. (1989). *Genetic Algorithms in Search, Optimization and Machine Learning*. Addison-Wesley, Reading, MA.

Graefe, G. (1993). Query evaluation techniques for large databases. *Computing Surveys*, 25(2):73–170.

Hooke, R. and Jeeves, T. (1961). "Direct Search" solution of numerical and statistical problems. *Journal of the Association for Computing Machinery*, 8:212–229.

Ingber, L. (1991). Statistical mechanics of neocortical interactions: A scaling paradigm applied to electroencephalography. *Physics Review*, A 44(6):4017–4060.

Ingber, L. (1996). Adaptive simulated annealing (ASA): Lessons learned. *Control and Cybernetics*, 25:33–54.

Ingber, L., Fujio, H., and Wehner, M. (1991). Mathematical comparison of combat computer models to exercise data. *Mathl. Comput. Modelling*, 15(1):65–90.

Ioannidis, Y. and Wong, E. (1987). Query optimization by simulated annealing. In *SIGMOD International Conference on Management of Data*, pages 9–22.

Kirkpatrick, S., Gelatt Jr., C., and Vecchi, M. (1983). Optimization by simulated annealing. *Science*, 220(4598):671–680.

Madhyasta, T., Elford, C., and Reed, D. (1996). Optimizing input/output using adaptive file system policies. In *5th NASA Goddard Conference on Mass Storage Systems*, pages II:493–514.

Madhyasta, T. and Reed, D. (1996). Intelligent, adaptive file system policy selection. In *6th Symposium on the Frontiers of Massively Parallel Computation*, pages 172–179. IEEE Computer Society Press.

Poole, J. (1994). Preliminary survey of I/O intensive applications. Technical Report CCSF-38, Scalable I/O Initiative, Caltech Concurrent Supercomputing Facilities, Caltech.

Swami, A. and Gupta, A. (1988). Optimization of large join queries. In *ACM-SIGMOD International Conference on Management of Data*, pages 8–17, Chicago, IL.

Szu, H. and Hartley, R. (1987). Fast simulated annealing. *Physics Review*, A122(3-4):157–162.

16 SOFTWARE CACHING IN A PARALLEL I/O RUNTIME SYSTEM TO SUPPORT IRREGULAR APPLICATIONS

Sung-Soon Park[1], Jaechun No[2], Jesús Carretero[3]
and Alok Choudhary[4]

[1] Department of Computer Science and Engineering,
Anyang University, Korea
[2] Department of Engineering and Computer Science,
Syracuse University, U.S.A.
[3] Arquitectura y Tecnología de S. I.,
Universidad Politécnica de Madrid, Spain
[4] Department of Electrical and Computer Engineering,
Northwestern University, U.S.A.

sspark@ece.nwu.edu
jno@ece.nwu.edu
jcarrete@ece.nwu.edu
choudhar@ece.nwu.edu

Abstract: In this paper we present the design and implementation of a runtime system based on collective I/O techniques for irregular applications. The design is motivated by the requirements of a large number of science and engineering applications, including teraflops applications, where the data is required to be reorganized into a canonical form for further processing or restarting. We also propose the design and implementation of a software caching method to improve the performance of the collective I/O techniques. The main idea of the software caching method is that, in an irregular application, the same data may be accessed repeatedly during the execution of subsequent irregular loops. Thus, improved performance can be achieved by storing the data in a processor's local memory and using it in the subsequent loops. We show the performance results for two implementations: *non-caching* and *software caching* methods. The performance results were collected on an Intel Paragon machine located at Caltech.

Keywords: software caching, parallel I/O, irregular applications, collective I/O.

16.1   INTRODUCTION

In parallel computing environments, I/O intensive applications have been increasing since many of the large-scale applications are data intensive rather than compute intensive (Poole, 1994; Rosario and Choudhary, 1994). These applications have very large I/O needs, and are further complicated by having different I/O types (such as check-pointing of large data sets for restarting, or periodically writing snapshots of the computation for subsequent visualization). A large number of these applications are *irregular*, where accesses to data are performed through one or more levels of indirection (Carretero *et al.*, 1998; No *et al.*, 1998). Sparse matrix computations, particle codes, and many CFD applications, where geometries and meshes are described via indirections, exhibit this characteristic. In I/O intensive applications, especially for irregular problems, the I/O cost for reading and writing data is more significant than communication or computational overheads. To reduce the I/O cost and to enhance performance, collective I/O has been used as a general mechanism (Choudhary *et al.*, 1994). In collective I/O, processors cooperate to combine several fine-grained I/O requests into a single, canonically ordered, coarse-grained request.

In previous works (Carretero *et al.*, 1998; No *et al.*, 1998), we proposed some preliminary ideas to provide parallel I/O for irregular applications. In this paper, we present the design and implementation of a software caching method, using collective I/O, to produce high performance I/O. The motivation is that, in irregular applications, the same data may be accessed repeatedly during an execution. Therefore, some I/O operations can be eliminated by storing the reused data in the processor's local memory. The experimental results, obtained on the Intel Paragon at Caltech, show the feasibility of the software caching method proposed in this paper.

The rest of the paper is organized as follows. In Section 16.2, we present an overview of irregular applications and our collective I/O method to support them. Section 16.3 presents the software caching method. Experimental results for the *non-caching* and *software caching* methods are shown in Section 16.4. Finally, some conclusions are presented in Section 16.5.

16.2   I/O FOR IRREGULAR APPLICATION

In irregular applications, the data domain area to be accessed is determined by computing one or more level of indirections. Figure 16.1 illustrates an irregular loop (Das *et al.*, 1994). Arrays x and y are *data arrays* and $a(i)$, $b(i)$, $c(j)$ and $d(j)$, which are used to reference data, are called *indirection arrays*. This irregular problem can be abstracted into the form shown in Figure 16.2. In this abstraction, we assume that data is distributed using some partitioning scheme, which may be application dependent. There is an indirection array on each node that describes the location of the corresponding data elements in a global array. For example, in Figure 16.2, processor 0 accesses $x(a(0))$, $x(a(4))$, $x(a(5))$, $x(a(2))$, and processor 1 accesses $x(a(7))$, $x(a(8))$, $x(a(6))$, $x(a(9))$, etc.

The objective of the I/O library is to provide a high-level interface for the programmer to write code for reading/writing data from/to files, in the order imposed by the global array. There are several characteristics to be considered in developing an

```
do i = lb1, ub1
  x[a[i]] = F(x[a[i]], y[b[i]])
enddo
do j = lb2, ub2
  x[c[j]] = G(x[c[j]], y[d[j]])
enddo
```

Figure 16.1   An example of an irregular loop.

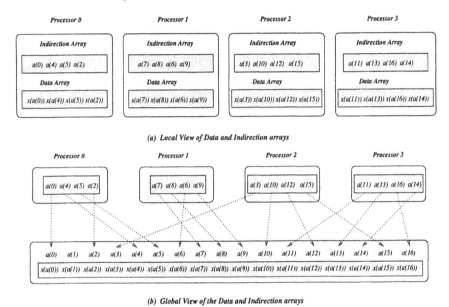

(a) Local View of Data and Indirection arrays

(b) Global View of the Data and Indirection arrays

Figure 16.2   Local and global view of data and indirection arrays.

I/O library for supporting irregular problems. One of them is that irregular problems often generate fine-grained data distributions requiring access to non-contiguous locations in a global array. Therefore, an appropriate collective I/O method is necessary to obtain high I/O performance.

16.2.1   Collective I/O

The design of the collective I/O library relies on the two-phase I/O strategy (Choudhary *et al.*, 1994). The basic idea behind the two-phase collective I/O is to optimize at runtime the access patterns seen by the I/O system. In other words, a large number of small and disjoint I/O requests are converted into a small number of large contiguous requests. This optimization incurs costs in terms of additional communication and buffer space requirements. However, since communication speed is normally several orders of magnitude faster than the I/O speed, the overhead is smaller than the reduction in the I/O cost. Several factors must be considered in the design of a library based on this technique: buffer size used by the library, communication schedule construction and reorganization, the number of processors participating in I/O at any time, and scheduling of I/O requests. In addition, for irregular computations, the number of passes through the data sets for computing an I/O schedule and reorganizing data are also important.

The collective read operations involve three basic steps: schedule construction, reading data from files, and redistributing data into the appropriate locations on each processor. In the write operations, a redistribution step precedes the file write step. Each of these steps consists of several phases, which are described below.

16.2.2   Schedule Construction

The schedule describes the communication and I/O patterns required for each node participating in an I/O operation. For regular multidimensional arrays, the accesses can be described using regular section descriptors. The communication schedule can be built based upon this information. On the other hand, when indirection arrays are involved in referencing the data array, they must be scanned to consider each element individually to determine its place in the global canonical representation as well as its destination processor.

Two factors affect the schedule construction in particular, and the overall I/O library design and performance in general. They are:

1. *Chunk Size*, which is the amount of buffer space available to the runtime library for I/O operations. For example, if the total size of the data per processor to be read/written is 8 MB and the chunk size is 2 MB, then the I/O operation will require four iterations to complete. A schedule must be built for each of these iterations.

2. *Number of processors involved in I/O*, which determines the communication among processors for data redistribution.

The following briefly describes the steps involved in schedule computation:

■   Based upon the chunk size, each processor is assigned a data domain for which it is responsible for reading or writing. For example, if there are four processors and 16 elements to be read/written, with a buffer space of two elements on each processor, the chunk size is 8 elements. Processor 0 will be responsible for elements 0, 1, 8, and 9, processor 1 will be responsible for elements 2, 3, 10, and 11, and so on. For a chunk, each processor computes its part of data to read or write while balancing I/O workload. Next, with each index value in its local memory, each processor first decides from which chunk the appropriate data must be accessed, and then determines which processor is responsible for manipulating the data chunk.

■   Index values in the local memory are rearranged into the *reordered-indirection array* based on the order of the destination processors. Therefore, we can communicate consecutive elements between processors (communication coalescing.)

step1 $\forall i, 0 \leq i \leq C - 1$
step2 $\forall j, 0 \leq j \leq P - 1$
               $p_d = compute\_destination\_processor(p_j), 0 \leq d \leq P - 1$
               $send(p_j, p_d, S_{indx}(i, p_d))$
               $recv(p_j, p_d, R_{indx}(i, p_d))$
               $send(p_j, p_d, S_{data}(i, p_d))$
               $recv(p_j, p_d, R_{data}(i, p_d))$
step3 $write(data\_file, c_i, offset(i)), 0 \leq mynode \leq P - 1$

Figure 16.3   The collective write operation algorithm.

Note that once it is constructed, the schedule information can be used repeatedly in the irregular problems, whose access pattern is not changed during the computation steps, and thereby amortizing its cost.

16.2.3 Parallel Collective Read/Write Operations

A processor involved in the computation is also responsible for reading data from files or writing data into files. Let D bytes be the total size of data and P be the number of processors. If the size of the data chunk is the same as the total size of the data, then each processor reads D/P bytes of data from the file and distributes it among processors based on the schedule information. For the case of writing, each processor collects D/P bytes of data from other processors, and then writes it to the file. By performing I/O this way, the workload can be evenly balanced across processors.

Let C be the number of data chunks. Let $S_{indx}(i, p_j)$ and $S_{data}(i, p_j)$ be the size of the index and the data to be sent to p_j in data chunk i, respectively. Let $R_{indx}(i, p_j)$ and $R_{data}(i, p_j)$ be the size of the index and the data to be received from p_j in data chunk i, respectively. Then, `send(o,d,data)` sends *data* from processor o to d, `recv(d,o,data)` receives *data* from processor o to d, and `write(fid,data, offset)` writes *data* in position *offset* of file *fid*. The steps involved in the collective write operations are shown in Figure 16.3.

16.3 SOFTWARE CACHING

Since irregular applications spend too much time in accessing target data, if a data area can be maintained and reused (Das *et al.*, 1993), we can reduce the I/O cost. Figure 16.4 shows an example of data reuse, where $a(i)$ and $c(j)$ are index values used to reference data in the global array x, with data overlapping between the two loops. *Re-referenced data* is data which has been accessed in a prior loop and that will be accessed in a subsequent loop. So, it will not be read from the file again when the subsequent loop is executed. *Released data* has been accessed in a prior loop and will not be accessed in a subsequent loop. Since this data will no longer be referenced, it should be written back to the file when the execution of the prior loop is finished. *New data* is defined as the data referenced in a loop that has never been accessed in prior loops. Therefore, it is necessary to read this data from the file before executing subsequent loops.

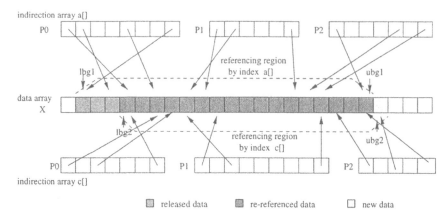

Figure 16.4   Example of data reuse.

In this section, we propose a software caching method that exploits the data reuse property, and also show its implementation details.

16.3.1   Software Caching Method

The motivation for the software caching method is that, in irregular applications, the same data may be accessed repeatedly during the execution of irregular loops. The basic goals for the software caching method are as follows. First and foremost, it is to reduce I/O to the maximum extent possible. To achieve this, an I/O phase, read or write, is divided into two new I/O phases, where the second phase only accesses *new data*. The second goal is to reuse the schedule information constructed in the beginning, and to build only incremental schedule information for *new data*.

To satisfy these aims, we added the following two steps to basic collective read operation: reading data partially from files and redistributing it into appropriate locations of each processor, and performing s/w caching phase to modify schedule information. In the write operation, redistribution step precedes the file write step.

16.3.2   Applying the Software Caching Method

Figure 16.5 represents the execution pattern to perform the irregular problem in Figure 16.1, with and without software caching. The communication and I/O patterns required for each processor are determined in the schedule phase. I/O phases to execute irregular loops are inserted before and after each loop. In Figure 16.5, the read1 phase reads data from a file, and write1 phase writes data to the file after finishing computations, where $x(a(lb1)) \sim x(a(ub1))$ or $x(c(lb2)) \sim x(c(ub2))$ represent data to be accessed in these phases. In this example, the *re-referenced data* need not be written back to the file in the first write1 phase, because it will be used again in the read1 phase.

Figure 16.5(b) illustrates the execution phases when using the software caching method. read1 and write1 are almost the same as for the I/O operations without

	(a) Non-caching Program		(b) S/W Caching Program
schedule	Compute schedule information for *loop*1 and *loop*2	**schedule**	Compute schedule information for *loop*1 and *loop*2
read1	Read all data which will be referred in *loop*1	**read1**	Read all data which will be referred in *loop*1
*loop*1		*loop*1	
write1	Write all data which has been referred in *loop*1	**write2**	Write all data which has been referred in *loop*1 but will not be referred in *loop*2
read1	Read all data which will be referred in *loop*2	**s/w caching**	Modify schedule information to access new data
		read2	Read all data which will be referred in *loop*2 but has never been referred in *loop*1
*loop*2		*loop*2	
write1	Write all data which has been referred in *loop*2	**write1**	Write all data which has been referred in *loop*2

Figure 16.5   Execution format for non-caching and s/w caching programs.

software caching. `read2` and `write2` are the phases for partially reading and writing data, based on the overlap area. In the `write2` phase, some data which will not be used in a subsequent loop is written to the file. Similarly, `read2` reads some data which has never been read from the file, but that will be referenced in a subsequent loop. Since only some part of data is written/read to/from the file, the I/O cost for the `write2` and `read2` is expected to be smaller than for `write1` and `read1`. In the `s/w caching` phase, the schedule information for the new loop is reconstructed. It includes the number of re-referenced data chunks and the modified data domain for each processor in the re-referenced data chunks.

Figure 16.6 shows the data domain of each processor for the irregular loop in Figure 16.1 before and after executing `s/w caching` phase, with 3 processors and 2 data chunks, where *x* is the global array. Before executing the `s/w caching` phase, the *writing back* area is the one referenced only in the prior loop, and the *new reading* area is the one to be referenced only in the later loop. The dark area of *x* represents the data referenced in both loops.

In the first `read1` phase, each processor reads data from the file using 2 data chunks. For example, processor *P0* reads *x(2)* ~ *x(5)* and distributes *x(3)* and *x(4)* to *P1* and *P2* by using a collective I/O method. After finishing the first loop, only data in the *writing back* area is written to the file. The range of the *writing back* area is determined in the `write2` phase. The `s/w caching` phase modifies the schedule information before executing the second loop. Figure 16.6 shows the change on the data domain of each processor after executing the `s/w caching` phase. In the `read2` phase, only data in the *new-reading* area is read from the file. Finally, all data referenced in the second loop are written back to the file in the `write1` phase.

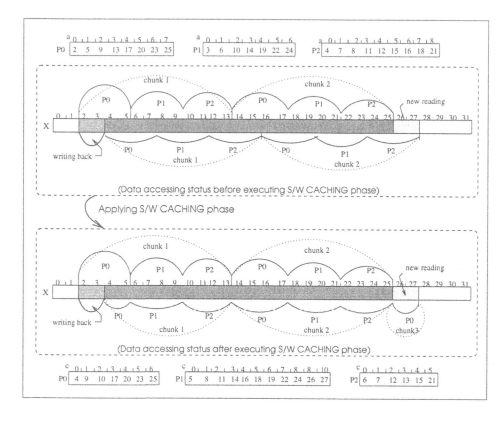

Figure 16.6   Data access pattern before and after executing the s/w caching phase.

16.3.3   Algorithms for the Software Caching Method

We designed and implemented the partial collective I/O operations, read2 and write2, by modifying the basic collective I/O operations. Since the same modifications could be applied for both partial collective I/O operations, we show an algorithm for the partial writing operation, write2. We also show an algorithm for the software caching operation, s/w caching.

To describe the software caching method, let C' be the number of data chunks which will be written in the write2 phase, and k be the number of incomplete chunks. In other words, some part of a chunk may be re-referenced, but the other part may not be, resulting in the writing of a portion of the chunk, called an incomplete chunk. Let $lb1$, $ub1$, $lb2$, and $ub2$ be the lower and upper bounds of two irregular loops. Let P_i' be the number of processors which will execute writing or reading operations for the i-th incomplete chunk. The steps involved in the write2 phase are shown in Figure 16.7.

Let RC be the number of data chunks in a subsequent loop. Also, let $SN_{indx}(i, p_d)$ and $RN_{indx}(i, p_d)$ be the size of the index to be sent and received to/from p_j in the data chunk i. The steps involved in the software caching phase are shown in Figure 16.8.

step0.5 $C' = compute\_data\_chunks\_for\_partial\_writing(lb1, ub1, lb2, ub2)$
               $k = compute\_nbr\_of\_incomplete\_chunks(C', C, lb1, ub1, lb2, ub2)$
step1' $\forall i,\ 0 \leq i \leq C' - 1$
step2 $\forall j,\ 0 \leq j \leq P - 1$
                   $p_d = compute\_destination\_processor(p_j), 0 \leq d \leq P - 1$
                   $send(p_j, p_d, S_{indx}(i, p_d))$
                   $recv(p_j, p_d, R_{indx}(i, p_d))$
                   $send(p_j, p_d, S_{data}(i, p_d))$
                   $recv(p_j, p_d, R_{data}(i, p_d))$
step2.5 $if\ (size_o f(i) \leq complete\_chunk\_size)$
                   $P_i' = compute\_nbr\_of\_partial\_writing\_processor(C', k)$
                   $P = P_i'$
step3' $write(data\_file, c_i', offset(i)), 0 \leq mynode \leq P - 1$

Figure 16.7   The write2 phase algorithm.

step1 $RC = compute\_nbr\_of\_re - referenced\_chunk(lb1, ub1, lb2, ub2)$
step2 $\forall i,\ 0 \leq i \leq RC - 1$
step3 $adjust\_schedule\_information(RC, C', k, P', lb2, ub2)$
step4 $\forall j,\ 0 \leq j \leq P - 1$
             $p_d = compute\_destination\_processor(p_j), 0 \leq d \leq P - 1$
             $send(p_j, p_d, SN_{indx}(i, p_d))$
             $recv(p_j, p_d, RN_{indx}(i, p_d))$

Figure 16.8   The software caching phase algorithm.

16.4   PERFORMANCE EVALUATION

We ran our experiments on the Caltech Intel Paragon machine, called *TREX*. *TREX* is a 550 node Paragon XP/S. We performed our experiments using from 32 to 128 compute nodes on 16 and 64 I/O node partitions, each I/O node having 64 MB memory and a 4 GB Seagate disk. The parameters considered for performance results are:

1. number of processors,

2. size of indirection array in the local memory of each processor,

3. data size, and

4. overlap range, which is the data area overlapped between two irregular loops.

We compared the *software caching method* to the basic collective I/O implementation, called *non-caching method*, which does not use software caching in the I/O library.

To compare the performance results of both implementations, we classified our evaluation environment by changing the size of the indirection and data arrays, the number of processors, and the degree of data reuse. In each case, we analyzed the performance results of both implementations by separating total execution time into three components: *computation time* (t_{comp}), *communication time* (t_{comm}), and *I/O time* (t_{io}). We executed the experiments several times to obtain the average results.

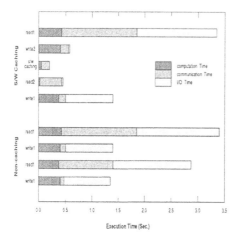

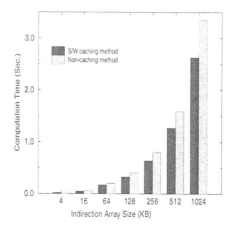

Figure 16.9   Comparing phases.   **Figure 16.10**   Comparing t_{comp}.

16.4.1   Evaluation Based Upon Phases

We ran both implementations with an indirection array of 512 KB and a data array of 32 MB. The number of processors was 32 and the overlap data area between two loops was 7/8, meaning that 7/8 of the data is shared between the loops. Figure 16.9 shows the execution times for each phase of both implementations. In the s/w caching phase, the computation and communication overheads to reconstruct schedule information are added. The communication time for the write2 phase is larger than for the write1, whereas the I/O time for the write2 phase is smaller than for the write1. The reason is that, in the write2 phase, all processors must communicate with all other processors to distribute their computation results, even though the partial writing operation is only performed by a subset of the processors. We observe that the I/O and communication times for the read2 phase are smaller than for the read1 phase. In the read2 phase, a reduced number of processors perform the partial read operation, and then communicate with others to distribute data. Consequently, even though the computation and communication overheads to reconstruct schedule information are added to the software caching method, we can see that its total execution time is smaller than for the non-caching method.

16.4.2   Evaluation Based Upon the Sizes of Indirection and Data Arrays

In this experiment, we used 32 processors and a 7/8 overlap data area between two loops. We increased the size of the indirection array on each processor's local memory from 4 KB to 1 MB, resulting in an increment in the data size from 256 KB to 64 MB. Figure 16.10 shows that the total computation time is reduced when the software caching method is applied to the basic collective I/O. The effect of software caching becomes obvious as the size of the indirection array is increased. Fig-

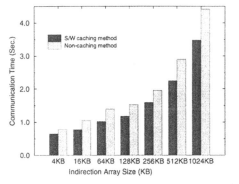

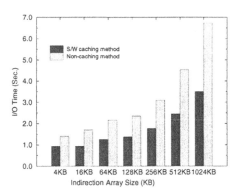

Figure 16.11   Comparing t_{comm}.   **Figure 16.12**   Comparing t_{io}.

ures 16.11 and 16.12 show that total communication and I/O times are also reduced when the software caching method is applied. These two figures also show that the benefits of the software caching method in the I/O time are better than in the communication time. As a result, the overall execution time in the software caching method is smaller than in the non-caching method.

16.4.3   Evaluation Based Upon Number of Processors

In this experiment, we executed both implementations with an indirection array of 4 MB and a data array of 64 MB, while changing the number of processors from 32 to 128. Similarly, we fixed the overlap data area between two loops as 7/8. Figure 16.13 represents the execution time divided into the same three components for both implementations. The results show that each component of the total execution time for the software caching method is smaller than for the non-caching method. In particular, when the number of processors is increased from 32 to 128, the computation time is reduced whereas the communication and I/O times are increased. Since the amount of data to be distributed is determined by dividing the global array into the number of processors, the computation time for packing and unpacking data is reduced.

16.4.4   Evaluation Based Upon Overlap Range

In this experiment, we varied the overlap range from 7/8 to 1/8, and then tried to find out the effect of the software caching method. Figure 16.14 shows that the software caching method provides much better performance when data overlap is increased.

When the software caching method is applied to collective I/O, the I/O and computation times are in general less than for the non-caching implementation. There are two reasons for this behavior: the I/O time is reduced in the partial I/O phases when data overlap exists, and the time to compute the data amount to be distributed is reduced. Therefore, the total execution time for the software caching is smaller than for the non-caching implementation when the overlap range is bigger than 2/8 between

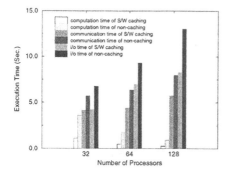

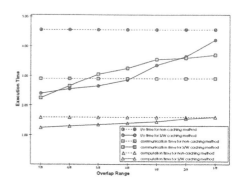

Figure 16.13 Changing the number of processors.

Figure 16.14 Changing the overlap ranges.

two loops. However, when the overlap range is 1/8, the total execution time for the software caching method is greater than the non-caching implementation, because the benefit of the partial I/O operations becomes smaller.

16.5 CONCLUSIONS

In this paper, we presented the design and implementation of the collective I/O with and without the software caching method. The main idea of the software caching method is that the same data may be accessed repeatedly during the execution of subsequent loops in an irregular application. The reused data does not have to be written/read to/from files, thus reducing the amount of I/O data. We also presented the performance results collected on the Caltech Intel Paragon machine. When we change the value of parameters such as the size of indirection (and data) array or the number of processors, the performance results show that, in general, the collective I/O with software caching is better than the collective I/O without software caching.

Acknowledgments

This work was supported in part by Korea Science and Engineering Foundation grants, Sandia National Labs award AV-6193 under the ASCI program, and in part by NSF Young Investigator Award CCR-9357840 and NSF CCR-9509143. This work was developed while all the authors were at Northwestern University.

References

Carretero, J., No, J., Park, S., and Choudhary, A. (1998). COMPASSION: A parallel i/o runtime system including chunking and compression for irregular applications. *International Conference on High-Performance Computing and Networking*, pages 668–677.

Choudhary, A., Bordawekar, R., Harry, M., Krishnaiyer, R., Ponnusamy, R., Singh, T., and Thakur, R. (1994). Passion: Parallel and scalable software for input and output. Technical Report SCCS-636, ECE Department, NPAC and CASE Center, Syracuse University.

Das, R., Saltz, J., and Hanxleden, R. (1993). Slicing analysis and indirect access to distributed arrays. *6th Workshop on Languages and Compilers for Parallel Computing*, pages 152–168.

Das, R., Uysal, M., Saltz, J., and Hwang, Y. (1994). Communication optimizations for irregular scientific computations on distributed memory architectures. *Journal of Parallel and Distributed Computing*, 22(3):462–479.

No, J., Park, S., Carretero, J., and Choudhary, A. (1998). Design and implementation of a parallel i/o runtime system for irregular applications. *12th International Parallel Processing Symposium*. To appear.

Poole, J. (1994). Preliminary survey of i/o intensive application. Technical Report CCSF-38, Scalable I/O Initiative, Caltech Concurrent Supercomputing Facilities, Caltech.

Rosario, J. D. and Choudhary, A. (1994). High performance i/o for parallel computers: Problems and prospects. *IEEE Computer*, 27(3):59–68.

17 HIPPI-6400—DESIGNING FOR SPEED

Don E. Tolmie

Los Alamos National Laboratory,
Los Alamos, U.S.A.

det@lanl.gov

Abstract: The emerging High-Performance Parallel Interface – 6400 Mbit/s Physical Layer (HIPPI-6400-PH) is targeted as a local area network (LAN) or system area network (SAN), supporting data rates of 6400 Mbit/s (800 Mbytes/s). This is eight times the speed of Gigabit Ethernet. The features used and the design choices made for the data link and physical layers of HIPPI-6400 to achieve this unprecedented speed are the subject of this paper. HIPPI-6400 borrowed freely from other successful technologies such as ATM, Ethernet and the original HIPPI, taking the best features of each and melding them with some new features. HIPPI-6400 is a cost-effective reliable interconnect for distances up to 1 kilometer; it intermixes large and small messages efficiently.
Keywords: HIPPI, gigabit, gigabyte, parallel, LAN, deskew.

17.1 BACKGROUND

The increasing complexity of server and cluster computing, and bandwidth-hungry applications such as scientific computing, imaging, and modeling, are demanding unprecedented interconnect speeds. Out of all the available gigabit and gigabyte technologies, Gigabit Ethernet (based on the framing Ethernet) has become the leading choice in meeting demands at a gigabit by offering greater bandwidth and improved client/server response times. Now, however, the emerging use of gigabit connections at the departmental server and desktop is creating a need for even higher-speed network technology at the backbone and in the cluster.

The High-Performance Parallel Interface, 6400 Mbit/s Physical Layer (HIPPI-6400-PH) and 6400 Mbit/s Physical Switch Control (HIPPI-6400-SC), are an answer to this need (ANSI, 1998). They will initially be deployed in a gigabyte system area network interconnecting high performance shared-memory multiprocessors (SMPs), clustered to provide an aggregate computing power, even beyond that achievable with the highest speed SMPs of today or tomorrow.

HIPPI-6400 represents the next generation beyond the current gigabit (and near gigabit) interconnect standards. Operating at 6400 Mbit/s, full-duplex, HIPPI-6400 ensures maximum compatibility with the Ethernet, Gigabit Ethernet, ATM, and HIPPI installed base. The original HIPPI standards, running at 800 and 1600 Mbit/s, developed and first deployed almost 10 years ago, pioneered higher speed interconnect technology. Along with a proposal from Silicon Graphics Inc., the original HIPPI provided the starting point for HIPPI-6400.

HIPPI-6400 is based on the best features of several successful interfaces, drawing from ATM, Ethernet and the original HIPPI specifications. From ATM it borrowed a small 32-byte micropacket (like a 48-byte ATM cell), and four Virtual Circuits (fewer than ATM, but limited for performance reasons). From Ethernet it borrowed the MAC header to allow easy translation to other popular protocols, and to use existing Ethernet-based control and management tools. From the original HIPPI it borrowed the large message size capability, credit-based flow control, encoding scheme for dc-balance, and a cable using multiple twisted-pairs (or optical fibers) for the data path. Features of HIPPI-6400 not found in any of these interfaces include end-to-end as well as link-level checksums, automatic retransmission at the physical layer to correct flawed data, a data rate of 6400 Mbit/s, and very low latency. As in other gigabit technologies, HIPPI-6400 systems will be switched rather than have multiple devices sharing a common bus or medium.

The HIPPI-6400 standards are being developed in ANSI Task Group T11.1 (see the web page at http://www.cic-5.lanl.gov/lanp/ANSI/ for meeting notices, meeting minutes, and draft documents). In relation to the OSI Reference Model, HIPPI-6400-PH (Physical Layer) specifies the physical and data link layers. HIPPI 6400-SC (Physical Switch Control) specifies a network layer for controlling physical layer switches. T11.1 completed their work on these documents in October 1997, and forwarded them for further review and balloting. The HIPPI-6400-PH and -SC documents are expected to complete their processing and become approved ANSI standards in late 1998. In addition, Task Group T11.1 is working on a transport layer standard, initially part of HIPPI-6400-PH, called the Scheduled Transfer Protocol (ST). Scheduled Transfer takes advantage of the high-speed reliable HIPPI-6400 lower layers, and provides additional performance by bypassing parts of the host's operating system. Scheduled Transfer specifies mappings for use on Ethernet, ATM, and Fibre Channel, as well as HIPPI-6400.

17.2   SYSTEM FEATURES

Figure 17.1 shows a system overview with a HIPPI-6400 switch interconnecting four nodes, two of which are translators to other media (*e.g.* to Gigabit Ethernet to talk to Ethernet-based devices in a local environment, and to ATM to connect to other far-flung sites over the telephone network). The networking aspects of HIPPI-6400 are detailed in the HIPPI-6400-SC document.

HIPPI-6400-PH defines a symmetric point-to-point physical link for transferring micropackets. The physical links are bidirectional and capable of the full 6400 Mbit/s

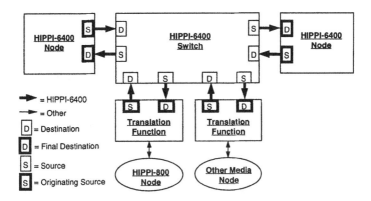

Figure 17.1   System overview.

bandwidth in both direction simultaneously. The logical links are simplex; the data inbound and outbound are completely separate.

A link's control information is carried on separate wires in parallel with the user's data (*i.e.* out-of-band). The control information is not counted in the 6400 Mbit/s bandwidth number (the rate available for the user's data is 99.6% of the 6400 Mbit/s).

17.3   VIRTUAL CHANNELS

Four Virtual Channels (VC0, VC1, VC2, and VC3) are available in each direction on each link. The VCs are assigned to specific message sizes and transfer methods. All of the micropackets of a message are transmitted on a single VC; the VC number does not change as the micropackets travel from the Originating Source to the Final Destination over one or more links. Messages to a Final Destination are delivered in order on a single VC.

Worm-hole routing is used in the HIPPI-6400 switches rather than the virtual connections used in ATM, or the end-to-end connections used in the original HIPPI. Worm-hole routing means that a message is sent into the network without prior knowledge if a free path is currently available to the Final Destination. If the message hits a link (*e.g.* on the output of a switch), that is using the same Virtual Channel, then the new message must wait for the existing message to complete (Tail bit = 1), before progressing further. On the plus side, worm-hole routing does not need time-consuming circuit setup or teardown, or for the links and switches to maintain large amounts of state information.

The VCs provide a multiplexing mechanism which can be used to prevent a large message from blocking a small message until the large message has completed, in contrast to the original HIPPI where a large message blocked any messages queued behind it. The number of Virtual Channels was deliberately limited to four (as opposed to the almost unlimited number in ATM), since the buffering needed to be on-chip for performance reasons. Three message sizes are supported: VC0 $\leq$ 2,176 bytes, VC1 $\leq$ 128 KB, VC2 $\leq$ 128 KB, and VC3 $\leq$ 4 GB. The intent was to separate the small control messages from the larger messages, (*i.e.* as shown by the bi-modal packet sizes in most networks).

17.4   MICROPACKETS

Micropackets are the basic transfer unit from Source to Destination on a link. As shown in Table 17.1, a micropacket is composed of 32 data bytes and 8 bytes of control information. This small transfer unit (the micropacket), results in a low latency for short messages and a component for large transfers. At 6400 Mbit/s, a micropacket is transmitted every 40 nanoseconds, with Null micropackets transmitted when other micropackets are not available. Credit and retransmit operations are performed on a micropacket basis.

Table 17.2 details the different micropacket Types, and the Data byte contents for each Type. In addition, the control fields carrying flow control information are detailed as to whether the field carries a valid value for that micropacket Type. A field with an invalid value is ignored.

17.5   MESSAGES

A message is an ordered sequence of one or more micropackets which have the same VC, Originating Source, and Final Destination. Messages carry the payload data. The first micropacket of a message, the Header micropacket, contains a HIPPI-6400 Header (24 bytes of information used to route through a HIPPI-6400 fabric), and 8 bytes of user data. The last micropacket of the message is marked with the Tail bit (much like an ATM AAL5 packet).

Table 17.1   Micropacket contents.

		Control Information (8 bytes)
	bits	Function
	4	Micropacket type
	2	Virtual channel selector
	8	Transmit sequence number
	8	Receive sequence number (ACK)
User data	1	Tail bit (end of message)
(32 bytes)	1	Error (unrecoverable upstream error)
	6	Credit update value
	2	Virtual channel number for credit update
	16	End-to-end CRC ($x^{16} + x^{12} + x^5 + 1$)
	16	Link level CRC ($x^{16} + x^{12} + x^3 + x + 1$)

Table 17.2 Capabilities of each type of micropacket.

Micropacket Type	Micropacket carries:			
	Data byte contents	Transmit sequence #	Receive sequence #	Credit update
Header	24-byte Header and 8 bytes of user date	Yes	Yes	Yes
Data	32 bytes of user data	Yes	Yes	Yes
Admin	Admin message	Yes	Yes	Yes
Credit-only	0's	Yes	Yes	Yes
Null	0's	Invalid	Yes	Invalid
Reset or Initialize	0's	Invalid	Invalid	Invalid

The contents of a HIPPI-6400 Header are shown in Figure 17.2. The MAC header is the same as the IEEE 802.3 header except that the length field (M_len) is 32 bits in HIPPI-6400 for longer messages, while in IEEE 802.3 it is 16 bits. The D_ULA and S_ULA are the 48-bit IEEE Universal LAN Addresses for the Originating Source and Final Destination. The IEEE 802.2 LLC/SNAP header is used to carry the Ether-Type, which selects the upper-layer protocol. Translating to other common networks is facilitated by using the IEEE network formats.

Table 17.3 shows a message contained in five micropackets. Bytes N–M are the user payload bytes. If a message does not end on a micropacket boundary, the last micropacket is padded with zeroes.

17.6 FLOW CONTROL

Link-level credit-based flow control is used between a Source and Destination to prevent over-running a Destination's buffers. Note that the flow control is between a

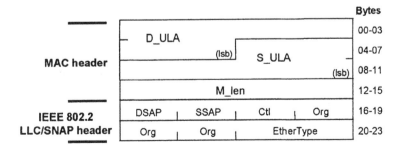

Figure 17.2 HIPPI-6400 header.

Table 17.3   Message contained in five micropackets.

Micropacket number	Data Bytes contents	Tail bit
1	Header, Bytes 0 - 7	0
2	Bytes 8 - 39	0
3	Bytes 40 - 71	0
4	Bytes 72 - 103	0
5	Bytes 104 - 135	1

Source and Destination, not necessarily the Originating Source and Final Destination (see Figure 17.1).

As shown in Figure 17.3, the credits are assigned on a VC basis; VC0's credits are separate from VC1's credits (hence congestion on VC3 will not stall traffic on VC0). The Destination end of a link grants credits to match the number of free receive buffers for a particular VC. The Source end of the link consumes credits as it moves micropackets from the VC Buffers to the Output Buffer. Note that flow control is on a link basis.

If a link has credit information, but no data, to transmit, then "credit-only" micropackets are transmitted. The micropackets containing credit information are checked for delivery, and included in the retransmission if an error occurs. Credit information in the original HIPPI was not as reliable, and in error cases could be lost, possibly leading to credit starvation. This is not possible in HIPPI-6400-PH. We feel that credit-based flow is the optimum method in a local area network environment where the distances are short and the buffering limited, but in a wide-area network environment, rate-based control is preferred.

It was the permissible buffer size that limited the link to one kilometer without speed degradation. For performance reasons the Destination buffers had to be on-chip, and about 10 KB was available for each of the four VCs. At 6400 Mbit/s (800 Mbytes/s), 5 nanoseconds per meter propagation delay, and 10 KB in flight (assuming the worst case with all of the in-flight data directed to a single receive buffer), the distance can be calculated as 2.5 kilometers.

The 2.5 kilometer is a round trip distance (giving time for acknowledgments to get back to the Source), and does not include any processing overhead. Hence, the link distance was specified as one kilometer maximum; the speed may decrease at greater distances. Note that the distance limit, before speed degradation, is dependent on fully loading a single VC with data. Spreading the load over multiple VCs or not trying to send at the full rate gives longer distances.

17.7   RETRANSMISSION

Retransmission is performed to correct flawed micropackets; providing in-order, reliable data delivery. Go-back-N retransmission is used; if an error is detected then the flawed micropacket, and all micropackets transmitted after it, are retransmitted. The

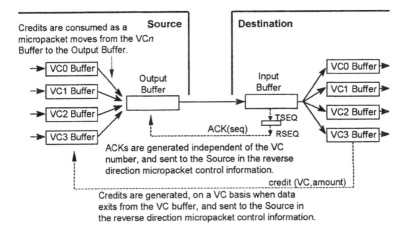

Figure 17.3 Reverse direction control information.

CRCs in each micropacket are checked at the Destination side of a link, at the Input Buffer in Figure 17.3. Correct micropackets are acknowledged, flawed micropackets are discarded. Note that retransmission is independent of the VC used, and also independent of the credit information. That is, retransmission occurs between the Output and Input Buffers in Figure 17.3, while VC and credit information pertains only to the VC Buffers. Retransmission is on a link basis.

Sequence numbers, in a micropacket's control information, are transmitted with all micropackets that contain data or credit information. Other micropackets, such as Type = Null, use hexadecimal sequence number 'FF'. The receiver acknowledges micropackets by returning the highest sequence number of contiguously good micropackets. Hence, if a micropacket is received in error, the receive sequence number sticks on the value of the last correct micropacket. A timeout mechanism at the sender detects that a transmitted micropacket was not acknowledged, and retransmits all micropackets starting with the one in error. Note that only micropackets with transmit sequence numbers (see Table 17.2) are retransmitted. The timeout mechanism was chosen because it was more robust than sending an ACK; if an ACK is dropped the protocol will just wait for the next ACK. A timeout mechanism may not be appropriate for a link with a long delay, but is preferred when the link delay is low (on the order of 10 microseconds for HIPPI-6400), and adequate buffering is available. The 8-bit sequence numbers allow up to 256 unacknowledged micropackets (10 KB, the size of the receive buffer).

17.8 CHECK FUNCTIONS

Two 16-bit cyclic redundancy checks (CRCs), with different polynomials, are used. The LCRC is the link-level checksum; the ECRC is the end-to-end checksum. Table 17.4 shows a 5-micropacket message, and the coverage for each CRC. Bytes x–y

Table 17.4   Checksum coverage for a 5-micropacket message.

Packet number	Data Bytes contents	LCRC checksum coverage	ECRC checksum coverage
1	Header, Bytes 0-7	Header, Bytes 0-7, c00-c47	Header, Bytes 0-7
2	Bytes 8-39	Bytes 8-39, c00-c47	Header, Bytes 0-39
3	Bytes 40-71	Bytes 40-71, c00-c47	Header, Bytes 0-71
4	Bytes 72-103	Bytes 72-103, c00-c47	Header, Bytes 0-103
5	Bytes 104-135	Bytes 104-135, c00-c47	Header, Bytes 0-135

are the user payload; c00-c47 are the first 48 control bits, and c48-c63 contain the ECRC and LCRC (see Table 17.1).

The end-to-end CRC (ECRC) covers the data bytes of all of the micropackets in a message, which includes the Header micropacket and all of the Data micropackets (if any) up to this point in a message. The ECRC does not cover the control bits. The ECRC is unchanged from the Originating Source to the Final Destination, *e.g.* through switches and bridges. The ECRC is accumulated over an entire message; it is not re-initialized for intermediate Data micropackets. Note that in Table 17.4, the second micropacket's ECRC covers the information in the first and second micropacket; the third micropacket's ECRC covers the information in the first, second, and third micropacket, *etc.* The ECRC generator polynomial is:

$$x^{16} + x^{12} + x^3 + x + 1.$$

The link CRC (LCRC) covers all of the data and control bits of a micropacket, with the exception of itself. The LCRC is initialized for each micropacket, and must be calculated fresh for each link since some values change hop-to-hop, *e.g.* Received sequence number and credit information. The LCRC polynomial is:

$$x^{16} + x^{12} + x^5 + 1.$$

Both CRCs are checked at each HIPPI-6400 node, be it a switch or end device. The combination of two 16-bit CRCs provides a stronger check than a single 16-bit CRC for link-level checking of individual micropackets. Analysis has shown that there are no undetected errors unless at least 6 bits in a micropacket are in error (1 in 1.86 billion bits) (Hoffman, 1996). Not only must there be at least 6 bits in error, but the bits must be strategically located and not contiguous.

In addition, the two separate CRCs are easier to calculate than a single 32-bit CRC. While many CRC implementations are done in a serial bit-by-bit fashion, at the speeds of HIPPI-6400 this may not be feasible. As an aid to the designer, example circuits and equations for parallel CRC implementations are included in an informative annex in HIPPI-6400-PH (ANSI, 1998).

The Error bit in Data micropackets is used to inform downstream HIPPI-6400 nodes that an uncorrectable error occurred upstream, for example from a translator to another media that does not provide retransmissions. Received Data micropackets with the Error bit set are passed on and not reported. This helps pinpoint where the error occurred; it would be next to impossible if everyone downstream also reported the error.

A Source also has the capability to abort a micropacket by forcing a specific LCRC value (called a "stomp code"). Downstream HIPPI-6400 nodes receiving a stomped micropacket will discard it as if were a Null micropacket. Other checks are made for out-of-order or missing micropackets (*e.g.* two Header micropackets without an intermediate Tail bit), lack of credit for a timeout period, *etc.* All error events are logged. There are no known error cases that would cause a link to lock up. An upper-layer protocol only needs to retransmit those messages that had unrecoverable errors, and these should be few and far between on a properly installed and maintained HIPPI-6400 system.

17.9 LATENCY

The Silicon Graphics Inc. SuMAC chip, which implements HIPPI-6400-PH, has shown latencies of 90 nanoseconds in one direction, and 120 nanoseconds in the other direction, for a total end-to-end latency of 210 nanoseconds. A switch path between two hosts would most likely contain two SuMAC chips (one for input and one for output). In addition, a switch may service up to 69 micropackets on each of the other three Virtual Channels before getting to your Virtual Channel (giving a worst case to-tal of 207 micropackets). At 40 nanoseconds per micropacket, and a cable delay of about 1.5 nanoseconds per meter, this translates to a worst case latency of about 10 microseconds. Typical latencies should be on the order of 1 microseconds.

17.10 MEDIA INTERFACES

The data is transmitted in parallel over the cable, and strobed with the clock signal. Figure 17.4 shows the signal lines between two end devices. Figure 17.5 shows the signal waveforms during a micropacket time (all of the time except the 40 nanoseconds when retraining the deskew circuitry).

The parallel architecture allowed the use of CMOS circuits and available drivers and receivers, a real cost and time-to-market saving. A serial implementation of HIPPI-6400 would have required a serial rate of about 10 Gbit/s, costly with optics and impossible with copper cable.

A copper cable interface is defined for the 16-bit system, using a total of 23 signals in each direction. Each signal operates at 500 MBaud. The cable assembly (cable and connectors) provides differential paths for 46 signals, 23 in each direction. The cable's characteristic impedance is 150 Ω and the maximum distance supported is 40 meters. The cable to support this speed and distance is not cheap, but is available from several vendors. Some testing has shown that passive equalizers aid signal quality for cables greater than 10 meters. Active equalizers would have given longer distances, but they required power, took considerable room, and added cost.

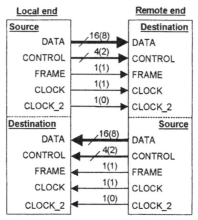

(Numbers in parenthesis are for an 8-bit system.
CLOCK_2 is only used in 16-bit systems.)

Figure 17.4   HIPPI-6400-PH link showing signal lines.

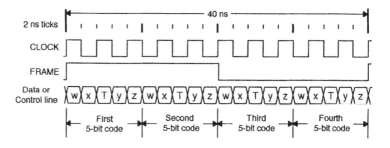

Figure 17.5   16-bit system micropacket waveforms.

A local electrical interface is also defined, with the intent to drive parallel optical transceivers on the same circuit board. The optical interface is defined for an 8-bit system, with a total of 12 signals in each direction (see Figure 17.4). Each signal operates at 1 GBaud. A 12-fiber ribbon cable is used in each direction. The optical interface is not as far along in design and standardization, and has been split out into a separate standards document called the High-Performance Parallel Interface— 6400 Mbit/s Optical Specification (HIPPI-6400-OPT) (ANSI, 1998). Several optical variants are being explored. One uses 850 nanometer laser arrays, 62.5/125 micron multimode fiber, and may use an open-fiber-control system to detect an open fiber and power down the lasers (to avoid potential eye damage). Another variant uses the same lasers and fiber, but decreases the power to avoid eye safety problems. The third variant uses 1300 nanometer lasers and either single-mode or multimode fiber. The human eye is much less susceptible to the 1300 nanometer wavelength, and that system will probably not need an open fiber control safety system. The 850 nanometer variants will probably be limited to 200–300 meters, while the 1300 nanometer variant with single-mode fiber may operate up to 10 kilometers. The HIPPI-6400-OPT specification is being written with the intent that it can also be used for other systems needing high-speed parallel fiber paths.

17.11 AC COUPLING

When driving long cables it is highly desirable to AC couple the signals and to keep them DC balanced. The AC coupling separates the ground paths between the end devices and avoids ground loops. DC balance means that a signal is above the switching threshold as much of the time as it is below the threshold. This considerably improves jitter and signal quality. HIPPI-6400-PH specifies the 4-bit to 5-bit (4B/5B) encoders/decoders, one encoder/decoder on each signal line. The 4B/5B encoding is adapted from the HIPPI-Serial standard (ANSI, 1997) and derived from some U.S. Patents (Crandall *et al.*, 1995; Hornak *et al.*, 1991). The 4B/5B encoding scheme transmits four data bits as a 5-bit code group (w, x, T, y, and z in Figure 17.5). The 4B/5B encoding was chosen for its implementation simplicity since 20 copies are required on the chip.

Figure 17.6 is a simplified schematic. A 4-bit to 5-bit encoder is shown on the left, and a 5-bit to 4-bit decoder is on the right. For each signal line, a running count, called the Disparity Count, is kept of all the ones and zeros transmitted on that line since the link was reset. The Disparity Count is incremented for each "1" transmitted, and decremented for each "0" transmitted. The 5-bit code transmitted (w, x, T, y, and z in Figure 17.6), is based on the current value of the Disparity Count and the input data 4-bit code (a, b, c, and d in Figure 17.6).

For example, if the Disparity Count is negative (more 0's than 1's transmitted), and the incoming 4-bit data also has more 0's than 1's, then the incoming 4-bit code is complemented (to generate more 1's), and the "T" bit is set to 0. At the receive end the incoming bits are passed straight through (un-complemented if T = 1), or complemented (if T = 0).

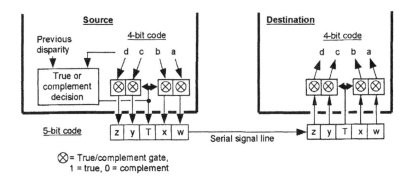

Figure 17.6 4B/5B encoder/decoder.

This algorithm gives a maximum run length of 11 bits, and a maximum disparity of +6 and -7. While the run length and maximum disparity are not as good as the 8B/10B code used in Fibre Channel, developed by (Widmer and Franaszek, 1983) and covered under a U.S. Patent (Franaszek and Widmer, 1983), the 4B/5B algorithm is much simpler to implement, and simplicity is mandatory when you remember that a single link requires 20 copies of the circuit (one for each data and control bit line).

A design goal for the 4B/5B encoding was to minimize the average run length for real data. As a test case, the operating system of a Silicon Graphics workstation was used as the random data input for a 4B/5B simulator. Rather than start in the middle of a 5-bit data pattern, the "T" bit was put on the end. The simulation showed that the operating system had many 4-bit zero patterns (binary '0000'), and these gave long run lengths when the zeros were back-to-back. Moving the "T" bit to the center of the 5-bit code shortened the average run length considerably. Since real user data is more likely to contain '0000' rather than '1111' patterns, this move was also considered useful for the general case.

17.12   DESKEWING THE PARALLEL SIGNALS

The CLOCK signal, used to strobe the other received signals, is carried on a separate line, negating the need for clock recovery circuits on every data line. Up to 10 nanoseconds of differential skew is allowed between the signals lines at the receiver, and the deskew circuits are dynamically adjusted every 10 microseconds. The deskew adjustment eats up one micropacket time (40 nanoseconds) every 10 microseconds, accounting for the missing 0.04% of the 6400 Mbit/s total bandwidth. Figure 17.7 is a block diagram of the deskew circuit on one signal line; there are a total of 20 such circuits on an interface chip. Each received signal drives a tapped delay line (implemented as a series of inverters in the SuMAC chip), and the output is selected from one of the taps. A special signal pattern is used to train the deskew logic.

Figure 17.8 shows four signals being deskewed. They are transmitted edge-aligned, but entering the receiver they are skewed due to differences in wire lengths, propagation delay, etc. Delay Ckt 1 is adjusted to $\Delta 1$, Delay Ckt 2 to $\Delta 2$, etc., so that all of the signals are again edge-aligned as they leave the delay circuits. This implements the deskew function. Not shown is a half-cycle shift of the CLOCK signal so that the CLOCK can sample the other signals in the middle of a bit period.

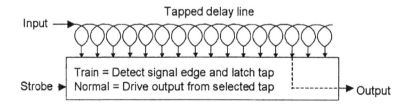

Figure 17.7   Tapped delay line deskew circuit.

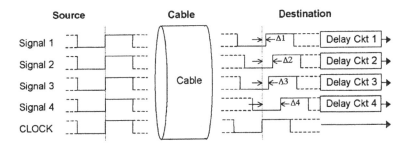

Figure 17.8 Dynamic deskew in operation.

17.13 SUMMARY

HIPPI-6400 is an emerging standard for moving digital data at speeds of up to 6400 Mbit/s (800 Mbytes/s) with very low latency between devices in a LAN-like environment. Many innovative design techniques are employed, resulting in a robust full-duplex link with efficient, reliable, in-order, data delivery. The links use parallel copper or fiber paths so that today's CMOS technology can be used to implement the links.

Acknowledgments

The Los Alamos National Laboratory is operated by the University of California for the United States Department of Energy under contract W-7405-ENG-36. The author's work was performed under the auspices of the U.S. Department of Energy (this paper is LA-UR 97-4906). Silicon Graphics Inc., with Dr. Greg Chesson leading their efforts, has contributed the majority of the HIPPI-6400 technical innovations. The HIPPI standards committee, with participation from many people throughout the industry, has worked tirelessly to document HIPPI-6400 and Scheduled Transfer as ANSI standards.

References

ANSI (1997). ANSI X3.300-1997, High-Performance Parallel Interface – Serial Specification HIPPI-Serial.

ANSI (1998). The following documents are "draft proposed American National Standard" as of March 16, 1998. The x's in the NCITS xxx-199x will be replaced with digits as the documents progress through the ANSI processing.

(1) ANSI NCITS 323-199x, High-Performance Parallel Interface – 6400 Mbit/s Physical Layer (HIPPI-6400-PH).

(2) ANSI NCITS 324-199x, High-Performance Parallel Interface – 6400 Mbit/s Optical Specification (HIPPI-6400-OPT).

(3) ANSI NCITS xxx-199x, High-Performance Parallel Interface – 6400 Mbit/s Physical Switch Control (HIPPI-6400-SC).

Crandall, D., Hessel, S., Hornak, T., Nordby, R., Springer, K., and Walker, R. (1995). Dc-free code for arbitrary data transmission. *U.S. Patent 5438621.*

Franaszek, P. and Widmer, A. (1983). Byte-oriented dc balanced (0,4) 8b/10b partitioned block transmission code. *U.S. Patent 4486739*.

Hoffman, J. (1996). HIPPI-6400: Analysis of a high-throughput network interface. Master's thesis, University of Arizona.

Hornak, T., Lai, B., Petruno, P., Stout, C., Walker, R., Wu, J., and Yen, C. (1991). Dc-free line code and bit and frame synchronization for arbitrary data transmission. *U.S. Patent 5022051*.

Widmer, A. and Franaszek, P. (1983). Dc-balanced, partioned-block, 8b/10b transmission code. *IBM Journal of Research and Development*, 27(5):440–451.

V Performance Issues

18 RSD – RESOURCE AND SERVICE DESCRIPTION

Matthias Brune, Jörn Gehring, Axel Keller and Alexander Reinefeld

Paderborn Center for Parallel Computing
Universität Paderborn, Germany

ar@uni-paderborn.de

Abstract: RSD (Resource and Service Description) is a scheme for specifying resources and services in complex heterogeneous computing systems and metacomputing environments. At the system administrator level, RSD is used to specify the available system components, such as the number of nodes, their interconnection topology, CPU speeds, and available software packages. At the user level, a GUI provides a comfortable, high-level interface for specifying system requests. A textual editor can be used for defining repetitive and recursive structures. This gives service providers the necessary flexibility for fine-grained specification of system topologies, interconnection networks, system and software dependent properties. All these representations are mapped onto a single, coherent internal object-oriented resource representation.

Dynamic aspects (like network performance, availability of compute nodes, and compute node loads) are traced at runtime and included in the resource description to allow for optimal process mapping and dynamic task load balancing at runtime at the metacomputer level. This is done in a self-organizing way, with human system operators becoming only involved when new hardware/software components are installed.

Keywords: distributed computing, metacomputing, resource management, specification language, multi-site applications.

18.1 INTRODUCTION

Metacomputing environments come in many flavors: A wide spectrum of network-based metacomputing services is offered, ranging from traditional client-server applications over distributed Intranet- or Internet-services, virtual organizations, teleworking, conferencing to autonomous Internet agents (*e.g.* search engines). All these services (and many more) are coordinated by a 'metacomputing middleware', which is also geographically distributed. Clearly, this middleware must be generic and versatile to also serve newly emerging technologies.

In many areas of computing science, a general tendency towards an open, distributed computation paradigm can be spotted:

- *User interfaces* have been improved from the early command line interfaces, towards system-specific graphical user interfaces, and on to web-based job submission sheets that are executed under control of standard browsers.

- *Resource management systems* have mutated from a single-system view, via a compute-center view, towards fully distributed metacomputing environments.

- *Parallel programming models and execution environments* have emerged from the initial proprietary libraries (*e.g.* MPL, NX) via vendor-independent programming models (*e.g.* PVM, MPI) towards interconnected environments (*e.g.* PVMPI (Fagg and Dongarra, 1996), PACX (Beisel *et al.*, 1997), PLUS (Brune *et al.*, 1997)).

As it turns out, in all these domains the representation of resources plays a major role. System services and resource requests can be represented by structured, attributed resource descriptions. Just like bids and offers at a public market place, the metacomputing market also deals with two sides: the user's resource requests and the system provider's resource offers. Assuming a set of compatible resource descriptions, the task is now to determine a suitable mapping between the two representations with respect to constraints such as node performance, connectivity, required software packages, *etc.*

This paper deals with resource and service description in metacomputing environments. For a general motivation and introduction to the subject of metacomputing the reader is referred to other publications, *e.g.* (Brune *et al.*, 1998; Smarr and Catlett, 1992). Section 18.2 briefly reviews representation schemes that have been proposed for other metacomputing environments. Section 18.3 presents our *resource and service description* scheme *RSD*, and the last section gives a brief summary.

18.2 OTHER WORK ON RESOURCE DESCRIPTION

18.2.1 *MPP Management*

Several initiatives have been initiated to standardize resource descriptions and command line interfaces of the current batch queuing systems (POSIX standard 1003.2d). Even though, the specific requirements of parallel and distributed computing systems, of graphical interface technologies (*e.g.* JavaSwing), and the facilities of HPC management systems (*e.g.* CCS (Keller and Reinefeld, 1998), or PBS (Bayucan *et al.*, 1996)) are still not recognized. Further divergence is added by hardware vendors promoting proprietary interfaces.

On an IBM SP2 with LoadLeveler, for example, the command "`qsub -l nodes=2:hippi+mem+mem:disk+12`" is used to allocate two processors with HIPPI connections, one processor with extra memory and disk-space, and 12 other processors (LoadLeveler, 1997). An alternative to the command line interface are batch script files, such as the one shown in Figure 18.1.

```
#@ min_processors = 6
#@ max_processors = 6
#@ cpu_limit = 14 000
#@ requirements = (Adapter == "hps_ip")
#@ environment = MP_EUILIB=us
#@ requirements = (Memory >= 64)
#@ requirements = (Feature == "Thin")
...
```

Figure 18.1 Resource request with LoadLeveler.

18.2.2 Management of Distributed Client-Server Architectures

The change from mainframe based computing to a distributed client-server architecture raises the need for distributed system management software. Two popular representatives of this class are the Tivoli Management Environment TME10 (Tivoli, 1998) and the Athena project (MIT, 1998).

The *Tivoli* system was designed for the system management of enterprise-wide distributed computing environments. Supported platforms are MVS mainframes, UNIX, Windows NT, Windows 95, and OS/2. Tivoli aims at four disciplines: deployment (*i.e.* manage software configurations), availability (*i.e.* maximize network utilization), security (*i.e.* control the user access), and operations (*i.e.* perform routine tasks).

The *Athena* project started in 1983 with the goal to explore innovative uses of computing in the MIT curriculum. Currently, the Athena system provides computing resources for roughly one thousand users across the MIT campus through a system of 1,300 computers connected to a campus-wide network. It follows a distributed client/server model for delivering services like: file servers scattered throughout the campus, print job spooling services, reservation of (sub-) clusters, or specialized software for courses. Application servers provide services like electronic mail, access to the Internet (www, news, ftp, *etc.*), electronic databases (English dictionary, encyclopedias, *etc.*), or programming tools (compiler, debugger, editor, *etc.*)

This kind of middleware is designed to simplify the management of large workstation clusters with distributed services. In contrast to our project, Tivoli and Athena do not consider the requirements of distributed high-performance computing (*e.g.* system topology or interconnection networks). They focus on sequential applications and do not support multi-site applications or linked parallel applications.

18.2.3 MDS (Globus)

The Globus project (Foster and Kesselman, 1997) aims at building an adaptive wide area resource environment AWARE with a set of tools that enables applications to *adapt to heterogeneous and dynamically changing metacomputing environments*. In this environment, applications must be able to obtain answers to questions such as: "Which average bandwidth is available from 3pm until 7pm between host A and host B?" and "Which PVM version is running on the MPP xy?" This information is obtained from multiple sources like the Network Information Service NIS, the Simple Network Management Protocol SNMP, or from system specific configuration files.

The *Metacomputing Directory Service MDS* (Fitzgerald *et al.*, 1997) is a tool that addresses the need for efficient and scalable access to diverse, dynamic, and distributed information. Both static information (*e.g.* amount of memory or CPU speed) and dynamic information (*e.g.* network latency or CPU load) are handled by MDS. It has been built on the data representation and API defined by the *lightweight directory access protocol LDAP* (Yeong *et al.*, 1995), which in turn is based on the X.500 standard. This standard defines a directory service that provides access to general data about entities like people, institutions, or computers. The information base in MDS is therefore structured as a set of such entries. Each entry is an attribute/value pair which may be either mandatory or optional. The type of an entry defines which attributes are associated with that entry and what type of values those attributes may contain. All entries are organized in a hierarchical, tree-structured name space, the *directory information tree DIT*. Unique entry names are constructed by specifying the path from the root of the DIT to the entry being named. The DIT is used to organize and manage a collection of entries and to allow the distribution of these entries over multiple sites. MDS servers are responsible for complete DIT subtrees (*e.g.* for a single machine or for all entries of a computing center).

To describe resources for a job request the Globus *Resource Manager RM* provides a *Resource Specification Language*, given by a string of parameter specifications and conditions on MDS entries. As an example, 32 nodes with at least 128 MB and three nodes with ATM interface are specified by:

```
+(&(count=32)  (memory >=128M)  )
 (&(count=3)   (network=ATM)  )
```

The RM matches resource requests against the information obtained from MDS.

18.2.4   RDL

The *Resource Description Language RDL* (Bauer and Ramme, 1991) is a language-based approach for specifying system resources in heterogeneous environments. At the administrator's level it is used for describing the type and topology of the available resources, and at the user's level it is used for specifying the required system components for a given application. RDL originated in the "Transputer world" with massively parallel systems of up to 1024 nodes and was later used for specifying PowerPC-based systems and workstation clusters.

In RDL, system resources consist of nodes that are interconnected by an arbitrary topology. A node may be either active or passive and the links between nodes may be static or dynamic. Active nodes are indicated by the keyword PROC while passive nodes are specified by PORT. Depending on whether RDL is used to describe hardware or software topologies, PROC may be read as "processor" or as "process". A port node may be a physical socket, a process that behaves passively within the parallel program, or a passive hardware entity like a crossbar. In addition to these built-in properties, new attributes can be introduced by the DEF *Identifier* [= (*value,...*)] statement (see Figure 18.2).

```
DECLARATION
BEGIN PROC MPP_with_frontend
    DYNAMIC;              -- system can be configured dynamically
    EXCLUSIVE;            -- allocate resources for exclusive use
    SYSTEM = (GC_Cluster_Size, 4, 4, 1); -- cluster size of MPP
    DECLARATION
    BEGIN PROC Parsytec_GCel
        DECLARATION
        { PROC; CPU=T8; MEMORY=4; SPEED=30; REPEAT=1024;}
        { PORT; REPEAT=4; }
    END PROC

    BEGIN PROC Unix_frontend
        DECLARATION
        { PORT; SBUS; OS=UNIX; REPEAT=4 }
    END PROC

    CONNECTION            -- of the MPP with the frontend
        FOR i=0 TO 3 DO
            Parsytec_GCel LINK i <=> Unix_frontend LINK i;
        OD
END PROC MPP_with_frontend
```

Figure 18.2   RDL specification of a 1024 node Transputer system with a UNIX frontend.

The hierarchical concept of RDL allows nodes to be grouped to build more complex nodes. Such a group is introduced by BEGIN [PROC | NODE] *ComplexNode-Name*. A complex node definition consists of three parts. First, there is a definition section where attributes are assigned to the node. This is followed by the declaration section in which the subnodes are described. The final part defines connections between the various subnodes (*node_a* LINK *n* <=> *node_b* LINK *m*) and to the outside world (ASSIGN *node_a* LINK *n* <=> LINK *m*). Figure 18.2 depicts an example of a parallel machine that is connected to a UNIX frontend via four external links.

18.3 RSD: RESOURCE AND SERVICE DESCRIPTION

18.3.1   *Requirements and Concept*

A versatile resource description facility should meet the following requirements:

- **Simple resource requirements should be easy to generate:** Users do not want to type in long and complex descriptions just for launching a small program. Therefore, a user-friendly graphical editor is needed.

- **Powerful tools for generating complex descriptions:** System administrators need adequate tools for describing complex systems made up of heterogeneous computing nodes and various kinds of interconnection networks. Although most of the system configuration is likely to remain constant over a long time, it might be quite difficult to specify all components, especially for large HPC centers.

Furthermore it is necessary to manage dynamic system data like network traffic and CPU load. Therefore, a simple graphical interface is insufficient. Rather, a combination of a text-based description language and a GUI is needed.

- **Application programming interface (API):** For automatic evaluation and mapping of resource descriptions an API is needed that allows to access the data by function calls and/or method invocations. Furthermore, the API encapsulates the internal structure of the resource description and allows future extensions without the need to update other software.

- **Portable representation:** Resource descriptions are sent across the network and they are exchanged between a vast variety of different hardware architectures and operating systems. The representation should be designed to be understood by each participating system.

- **Recursive structure:** Computing resources are usually organized in a hierarchical way. A metacomputer might comprise several HPC centers, which include several supercomputers that in turn consist of a number of processing elements. The resource description facility should reflect this general approach by supporting recursive constructs.

- **Graph structured:** HPC hardware, software and dynamic (time dependent) data flow are often described by graphs. Thus, the resource description facility should allow to define graph based structures.

- **Attributed components:** For describing properties of processors, network connections, jobs, and communication requirements, it should be possible to assign valued attributes to arbitrary nodes and edges of the resource description.

Figure 18.3 depicts the general concept of the RSD framework. It was designed to fit the needs of both, the user as well as the system administrator. This is achieved by providing three different representations: a graphical interface, a textual interface, and an application programming interface. Users and administrators are expected to use a graphical editor for specifying their resource descriptions (Section 18.3.2). While the GUI editor will usually suffice for the end-user, system administrators may need to describe the more complex components by an additional text-based language. The editor combines the textual parts with the graphical input and creates an internal data representation (Section 18.3.4). The resulting data structure is bundled with the API access methods and sent as an attributed object to the target systems in order to be matched against other hardware or software descriptions.

The internal data description can only be accessed through the API. For later modifications it is re-translated into its original form of graphic primitives and textual components. This is possible, because the internal data representation also contains a description of the component's graphical layout. In the following, we describe the core components of RSD in more detail.

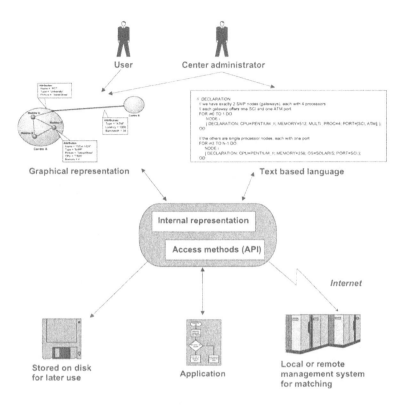

Figure 18.3   General concept of the RSD approach.

18.3.2   Graphical Interface

In this section we present the general layout of a graphical RSD editor. It contains a toolbox of modules that can be edited and linked together to build a complex dependency graph or a system resource description.

At the *administrator level*, the graphical interface is used to describe the basic computing and networking components in a (meta-)center. Figure 18.4 illustrates a typical administrator session. In this example, the components of the center are specified in a top-down manner with the interconnection topology as a starting point. With drag-and-drop, the administrator specifies the available machines, their links and the interconnection to the outside world. It is possible to specify new resources by using predefined objects and attributes via pulldown menus, radio buttons, and check boxes.

In the next step, the machines are specified in more detail. This is done by clicking on a node, whereby the editor opens a window showing detailed information on the machine (if available). The GUI offers a set of standard machine layouts (such as the Cray T3E, IBM SP2, and Parsytec) and some generic topologies like grid or torus. The administrator defines the size of the machine and the general attributes of the whole machine. As described in Section 18.3.3, attributes can also be specified in a textual manner. When the machine has been specified, a window with a graphical

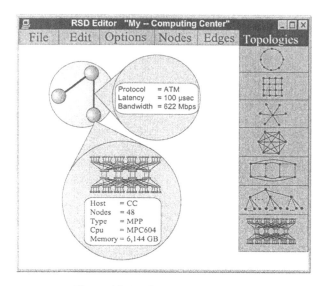

Figure 18.4   Graphical RSD editor.

representation of the machine opens, in which single nodes can be selected. Attributes like network interface cards, main memory, disk capacity, I/O throughput, CPU load, network traffic, disk space, or the automatic start of daemons, *etc.*, can be assigned.

At the *user level*, the interface can be customized. User-generated configuration files contain the most common resource descriptions. Additionally, it is possible to connect to a remote site to load its RSD dialect (*i.e.* the attribute names and their valid values). This allows the user interface to perform online syntax checks, even if the user specifies remote resources. Likewise, it is possible to join multiple sites to a meta-site, using a different RSD dialect without affecting the site-specific RSD dialects.

Analogously to the wide-area metacomputer manager *WAMM* (Baraglia *et al.*, 1996), the user may click on a center and the target machines. Interconnection topologies, node availability, and the current job schedule may be inspected. Partitions can be selected via drag and drop or in a textual manner. For multi-site applications, the user may either specify the intended target machines, or constraints in order to let the RMS choose a suitable set of systems.

18.3.3   Language Interface

For system administrators, graphical user interfaces may not be powerful enough for describing complex metacomputing environments with a large number of services and resources. Administrators need an additional tool for specifying irregularly interconnected, attributed structures. Hence, we devised a language interface that is used to specify arbitrary topologies. The hierarchical concept allows different dependency graphs to be grouped for building even more complex nodes, *i.e.* hypernodes.

Active nodes are indicated by the keyword NODE. Depending on whether RSD is used to describe hardware or software topologies, the keyword NODE is interpreted as a "processor" or a "process". Communication interfaces are declared by the keyword PORT. A PORT may be a socket, a passive hardware entity like a network interface card, a crossbar, or a process that behaves passively within the parallel program.

A NODE definition consists of three parts:

1. In the optional DEFINITION section, identifiers and attributes are introduced by *Identifier* [= (*value,...*)].

2. The DECLARATION section declares all nodes with corresponding attributes. The notion of a 'node' is recursive. They are described by NODE *NodeName* {PORT *PortName*; attribute 1, ...}.

3. The CONNECTION section is again optional. It is used to define attributed edges between the ports of the nodes declared above: EDGE *NameOfEdge* {NODE *w* PORT *x* <=> NODE *y* PORT *z; attribute 1; ...*}. In addition, the notion of a 'virtual edge' is used to provide a link between different levels of the hierarchy in the graph. This allows for the establishment of a link from the described module to the outside world by 'exporting' a physical port to the next higher level. These edges are defined by: ASSIGN *NameOfVirtualEdge* { NODE *w* PORT *x* <=> PORT *a*}. Note, that NODE *w* and PORT *a* are the only entities known to the outside world.

Figures 18.5 illustrates a resource specification for the metacomputer application testbed G-WAAT (Global - Wide Area Application Testbed), a demonstration presented at the Supercomputing conference in San Jose, 1997. In this project, a Cray T3E at Pittsburgh Supercomputing Center was connected via a trans-Atlantic ATM network to a Cray T3E at the High Performance Computing Center in Stuttgart, Germany. On this global metacomputer, the CFD code URANUS was run to simulate the re-entry of a space vehicle into the atmosphere.

The RSD definition of G-WAAT is straightforward (see Figure 18.6). Figure 18.7 shows a more detailed specification of Stuttgarts' Cray T3E containing 512 nodes. For each node, the following attributes are specified: CPU type, the CPU clock rate, memory per node, and the port to the Gigaring. All nodes are interconnected in a

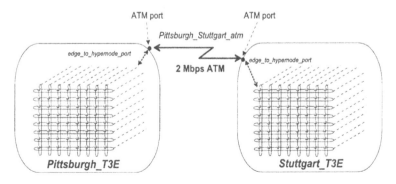

Figure 18.5   RSD example for multi-site application G-WAAT.

```
NODE G-WAAT { // Global Wide Area Application Testbed
   // DEFINITION: define attributes, values and ranges
      BANDWIDTH = (1..1200); // valid range of bandwidth in Mbps
   // DECLARATION: include the two hyper nodes
      INCLUDE "Pittsburgh_Cray_T3E";
      INCLUDE "Stuttgart_Cray_T3E";
   // CONNECTION:
   // dedicated trans-Atlantic ATM net between Pittsburgh and Stuttgart
      EDGE Pittsburgh_Stuttgart_atm {
         NODE Pittsburgh_T3E PORT ATM <=> NODE Stuttgart_T3E PORT ATM;
         BANDWIDTH = 2 Mbps; LATENCY = 75 msecs;
         AVAILABLE_DAY = Thursday; FROM = 3pm UTC; UNTIL = 6pm UTC;};
};
```

Figure 18.6   RSD specification of Figure 18.5.

3D-Torus topology by the 1.0 Gbps Gigaring. In the example, the first node acts as gateway between the Cray T3E system and the outside world. It presents its ATM port to the next higher node level (see ASSIGN statement in Figure 18.7) to allow for remote connections.

18.3.4   *Internal Data Representation*

In this section, we describe the abstract data type that establishes the link between graphical and text-based representations. This RSD data format is also used to store descriptions on disk and to exchange them across networks.

Abstract Data Types.   As stated in Sections 18.3.1, 18.3.2, and 18.3.3, the internal data representation must be capable of describing:

- arbitrary graph structures,

- hierarchical systems or organizations,

- nodes and edges with arbitrary sets of valued attributes.

Furthermore it should be possible to reconstruct the original representation, either graphical or text based. This facilitates the maintenance of large descriptions (*e.g.* a complex HPC center) and allows visualization at remote sites.

Following these considerations we define the data structure RSD_{Graph} as the fundamental concept of the internal data representation:

$$RSD_{Graph} := (V, E) \ \ with \ \ \left\{ \begin{array}{l} V \ being \ a \ set \ of \ RSD_{Node} \ and \\ E \subseteq (V \times V \times RSD_{Attribute}) \end{array} \right. \tag{18.1}$$

and

$$RSD_{Node} := RSD_{Attribute} \mid (RSD_{Graph}, RSD_{Attribute}). \tag{18.2}$$

Thus, an RSD_{Graph} is a graph which may have assigned attributes to its nodes and edges. Since a node may also represent a complete substructure it is possible

```
NODE Stuttgart_T3E {
  // DEFINITION:
  CONST N = 512;                          // number of nodes
  CONST DIMX = 8, DIMY = 8, DIMZ = 8;  // dimensions of machine in nodes
  // DECLARATION: // we have 512 nodes, node 000 is the gateway
  // the gateway provides one GIGARING port and one ATM port
  FOR x=0 TO DIMX-1 DO
      FOR y=0 TO DIMY-1 DO
          FOR z=0 TO DIMZ-1 DO
              NODE $x$y$z {
              PORT GIGARING; IF (x == 0 && y == 0 && z == 0) THEN PORT ATM;
              CPU=Alpha; CLOCKRATE=450 Mhz; MEMORY=128 MB; OS=UNICOS};
          OD
      OD
  OD
  // CONNECTION: build the 1.0 Gbps 3D-Torus
  FOR x=0 TO DIMX-1 DO
      FOR y=0 TO DIMY-1 DO
          FOR z=0 TO DIMZ-1 DO
              neighborX = (x+1) MOD DIMX;
              neighborY = (y+1) MOD DIMY;
              neighborZ = (z+1) MOD DIMZ;
              EDGE edge_$x$y$z_to_$neighborX$y$z {
                      NODE $x$y$z PORT GIGARING =>
                      NODE $neighborX$y$z PORT GIGARING;
                      BANDWIDTH = 1.0 Gbps;};
              EDGE edge_$x$y$z_to_$x$neighborY$z {
                      NODE $x$y$z PORT GIGARING =>
                      NODE $x$neighborY$z PORT GIGARING;
                      BANDWIDTH = 1.0 Gbps;};
              EDGE edge_$x$y$z_to_$x$y$neighborZ {
                      NODE $x$y$z PORT GIGARING =>
                      NODE $x$y$neighborZ PORT GIGARING;
                      BANDWIDTH = 1.0 Gbps;};
          OD
      OD
  OD
  // establish a virtual edge from node 000 to the
  // port of the hyper node Stuttgart_T3E (=outside world)
  ASSIGN edge_to_hypernode_port {
      NODE 000 PORT ATM <=> PORT ATM;};
};
```

Figure 18.7   RSD specification of Stuttgarts' Cray T3E.

to represent hierarchies of any depth. For example, this may be used to describe a computing center that has its own set of attributes but of which the internal hardware description needs further refinement.

For maximum flexibility, we define $RSD_{Attribute}$ to be either a (possibly empty) list of names and values, or a single name for a more complex structure of attributes:

$$RSD_{Attribute} := \emptyset \mid (Name, \ Value), \ldots \mid (Name, \ RSD_{Attribute}).$$

(18.3)

The name of an attribute follows the common naming conventions of programming languages while the value may have one of the following types: *boolean, integer, real, string, bytearray,* and *bitfield.*

Figure 18.8 depicts a small example of this concept. There are two RSD_{Node} structures representing computing centers that are connected by a 34 Mbps ATM link.

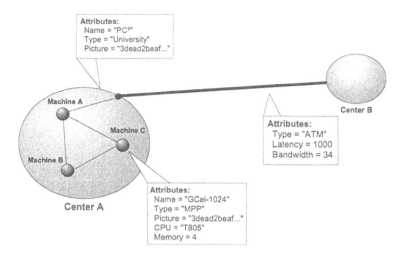

Figure 18.8   Example configuration for RSD.

The substructure contained in Center A indicates that there are three machines within this center, one of them being an MPP system called "GCel-1024".

Methods for Handling Abstract Data Types.   Having described the data structure, we now introduce the basic methods for dealing with the RSD abstract data type. Only those methods are provided that are absolutely necessary. All higher level functions build upon them.

- $Nodes(RSD_{Graph})$ – Returns a list of nodes in the highest level of the graph. Referring to Figure 18.8 these would be *Center A* and *Center B*.

- $Neighbors(RSD_{Node})$ – Gives a list of all nodes connected to RSD_{Node} in the same level. In Figure 18.8, *Center B* is a neighbor of *Center A* but not of *Machine A*.

- $SubNodes(RSD_{Node})$ – These are all nodes contained in RSD_{Node}. As an example the subnodes of *Center A* comprise *Machine A*, *Machine B*, and *Machine C*, but none of their processors.

- $Attributes (RSD_{Node} \mid (RSD_{Node} , RSD_{Node}))$ – Returns a list of all top-level attributes of the specified node or edge.

- $Value(RSD_{Attribute})$ – Returns the value of an $RSD_{Attribute}$. This may also be a set of attributes in the case of hierarchically structured attributes.

In addition, there exists a number of auxiliary methods, *e.g.* for list handling, that build on the basic methods. These are not described in this paper.

Note that the internal data types are also used for reconstructing the original graphical and/or text representation. As an example, a system administrator may have defined a large *Fast Fourier Transformation* FFT network by a few lines of the RSD language. The internal representation of such a node is much more complex and not easy to translate back into its original form. However, it is possible to keep the source code of this special node as one of its attributes. This can then be used by the editor whenever the FFT is to be modified. The same is possible with nodes that were created graphically and require a uniform layout of their components.

18.3.5   Implementation

In order to use RSD in a distributed environment a common format for exchanging RSD data structures is needed. The traditional approach to this would be to define a data stream format. However, this would involve two additional transformation steps whenever RSD data is to be exchanged (internal representation into data stream and back). Since the RSD internal representation has been defined in an object-oriented way, this overhead can be avoided, if the complete object is sent across the net.

Today there exists a variety of standards for transmitting objects over the Internet, *e.g. CORBA, Java*, or *COM+*. Unfortunately, it is currently impossible to tell which of them will survive in the future. Therefore, we only define the interfaces of the RSD object class but not its private implementation. This allows others to choose an implementation that fits best to their own data structures. Interoperability between different implementations can be improved by defining translation constructors, *i.e.* constructors that take an RSD object as an argument and create a copy of it using another internal representation.

This concept is much more flexible than the traditional approach. It allows enhancements of the RSD definition and of the communication paradigm while still maintaining downward compatibility with older implementations.

18.4   CONCLUSION

Due to their distributed nature with a vast number of (possibly recursively structured) components and their dynamic behavior, metacomputing environments are much more difficult to access, organize and maintain than any single high-performance computing system.

With RSD, we presented a graph-oriented scheme for specifying and controlling resources and services in heterogeneous environments. Its graphical user interface allows the user to specify resource requests. Its textual interface gives a service provider enough flexibility for specifying computing nodes, network topology, system properties and software attributes. Its internal object-oriented resource representation is used to link different resource management systems and service tools.

References

Baraglia, R., Faieta, G., Formica, M., and Laforenza, D. (1996). Experiences with a wide area network metacomputing management tool using IBM SP-2 parallel systems. *Concurrency: Practice and Experience*, 8.

Bauer, B. and Ramme, F. (1991). A general purpose resource description language. In Grebe, R. and Baumann, M., editors, *Parallele Datenverarbeitung mit dem Transputer*, pages 68–75, Berlin. Springer-Verlag.

Bayucan, A., Henderson, R., Proett, T., Tweten, D., and Kelly, B. (1996). *Portable Batch System: External Reference Specification. Release 1.1.7.* NASA Ames Research Center.

Beisel, T., Gabriel, E., and Resch, M. (1997). An extension to MPI for distributed computing on MPPs. In Bubak, M., Dongarra, J., and Wasniewski, J., editors, *Recent Advances in Parallel Virtual Machine and Message Passing Interface*, pages 25–33. Springer-Verlag LNCS.

Brune, M., Gehring, J., Keller, A., Monien, B., Ramme, F., and Reinefeld, A. (1998). Specifying resources and services in metacomputing environments. *Parallel Computing*. To appear.

Brune, M., Gehring, J., and Reinefeld, A. (1997). Heterogeneous message passing and a link to resource management. *Journal of Supercomputing*, 11:355–369.

Fagg, G. and Dongarra, J. (1996). PVMPI: An integration of the PVM and MPI systems. *Calculateurs Parallèles*, 8(2):151–166.

Fitzgerald, S., Foster, I., Kesselman, C., Laszewski, G. V., Smith, W., and Tuecke, S. (1997). A directory service for configuring high-performance distributed computations. Preprint. Mathematics and Computer Science Division, Argonne National Laboratory, Argonne, IL.

Foster, I. and Kesselman, C. (1997). Globus: A metacomputing infrastructure toolkit. *Journal of Supercomputer Applications*, pages 115–128.

Keller, A. and Reinefeld, A. (1998). CCS resource management in networked HPC systems. In *Heterogeneous Computing Workshop HCW'98*, Orlando.

LoadLeveler (1997). *SP Parallel Programming Workshop: LoadLeveler.* http://www.mhpcc.edu/training/workshop/html/loadleveler/LoadLeveler.html.

MIT (1998). *The Athena Project.* Massachusetts Institute of Technology. http://web.mit.edu/olh/Welcome/index.html.

Smarr, L. and Catlett, C. (1992). Metacomputing. *Communications of the ACM*, 35(6): 45–52.

Tivoli (1998). *The Tivoli Management Environment.* Tivoli Systems Inc. http://www.tivoli.com.

Yeong, W., Howes, T., and Kille, S. (1995). Lightweight directory access protocol. RFC 1777, 03/2895, Draft Standard.

19 AN EXAMINATION OF THE PERFORMANCE OF TWO ELECTROMAGNETIC SIMULATIONS ON A BEOWULF-CLASS COMPUTER

Daniel S. Katz, Tom Cwik and Thomas Sterling

Jet Propulsion Laboratory,
California Institute of Technology, U.S.A.

Daniel.S.Katz@jpl.nasa.gov
cwik@jpl.nasa.gov
tron@cacr.caltech.edu

Abstract: This paper uses two electromagnetic simulations to examine some performance and compiler issues on a Beowulf-class computer. This type of computer, built from mass-market, commodity, off-the-shelf components, has limited communications performance and therefore also has a limited regime of codes for which it is suitable. This paper first shows that these codes fall within this regime, and then examines performance data. Comparisons are made between a Beowulf, a Cray T3D, and a Cray T3E, examining execution time, performance scaling, compiler choices, and the use of some hand-tuned optimizations.

Keywords: Beowulf, cluster, pile of PCs, parallel computation, electromagnetics, finite-difference time-domain, physical optics, radiation integral.

19.1 INTRODUCTION

A typical small Beowulf system, such as the machine at the Jet Propulsion Laboratory (JPL) may consist of 16 nodes interconnected by 100Base-T Fast Ethernet. Each node may include a single Intel Pentium Pro 200 MHz microprocessor, 128 MBytes of DRAM, 2.5 GBytes of IDE disk, PCI bus backplane, and an assortment of other devices. At least one node will have a video card, monitor, keyboard, CD-ROM, floppy drive, and so forth. The technology is rapidly evolving, and price-performance and price-feature curves are changing so fast that no two Beowulfs ever look exactly alike. Of course, this is also because the pieces are almost always acquired from a mix of vendors and distributors. The power of *de facto* standards for interoperability of subsystems has generated an open market that provides a wealth of choices for customizing one's own version of a Beowulf, or just maximizing cost advantage as prices fluctuate among sources. Such a system will run the Linux operating system (Husain *et al.*, 1996), freely available over the net or in low-cost and convenient CD-ROM distributions. In addition, publicly available parallel processing libraries such as MPI (Snir *et al.*, 1996) and PVM (Giest *et al.*, 1994) are used to harness the power of parallelism for large application programs. A Beowulf system such as described here with a maximum of 16 processors, taking advantage of appropriate discounts, costs about $20K including all incidental components such as low-cost packaging.

At this time, there is no clearly typical medium to large Beowulf system, since as the number of processors grows, the choice of communications network is no longer as clear[1]. Many choices exist for various topologies of small and large switches and hubs, and combinations thereof.

Naegling, the Beowulf-class system at the California Institute of Technology, which currently has 140 nodes, has evolved through a number of communications networks. The first network was a tree of 8- and 16-port hubs. At the top of the tree was a standard 100 Mbits/s 16-port crossbar, with full backplane bandwidth. Each port of this was connected to a hub, each hub having 100 Mbits/s ports connected to 8 or 16 computers. However, the backplane bandwidth of each hub was also 100 Mbits/s. The second topology used additional 16-port crossbars at the lowest level of the tree, where 15 ports of each crossbar were connected to computers, and the last port was connected to a high-level crossbar. A third network, the current topology, involves 2 80-port switches connected by 4 Gbits links. Each switch is intended to have 100 Mbits ports and full backplane bandwidth.

The Beowulf approach represents a new business model for acquiring computational capabilities. It complements rather than competes with the more conventional vendor-centric systems-supplier approach. Beowulf is not for everyone. Any site that would include a Beowulf cluster should have a systems administrator already involved in supporting the network of workstations and PCs that inhabit the workers' desks. Beowulf is a parallel computer and, as such, the site must be willing to run parallel programs, either developed in-house or acquired from others. Beowulf is a loosely coupled, distributed memory system, running message-passing parallel programs that do not assume a shared memory space across processors. Its long latencies require a favorable balance of computation to communication and code written to balance the workload across processing nodes. Within the constrained regime in which Beowulf is appropriate, it should provide the best performance to cost and often comparable performance per node to vendor offerings (Katz *et al.*, 1998).

This paper will examine two electromagnetic simulation codes which fit within this regime. The applications are described in Sections 19.2 and 19.3, while their performance on three parallel architectures is presented in Sections 19.4 and 19.5, respectively. Section 19.6 summarizes our experiences with Beowulf-class computing.

---

[1]If a crossbar switch can support the entire machine, then that is the desired choice.

19.2 PHYSICAL OPTICS SIMULATION

The first code is used to design and analyze reflector antennas and telescope systems (Imbriale and Cwik, 1994). It is based simply on a discrete approximation of the radiation integral (Imbriale and Hodges, 1991). This calculation replaces the actual reflector surface with a triangularly faceted representation so that the reflector resembles a geodesic dome. The Physical Optics (PO) current is assumed to be constant in magnitude and phase over each facet so the radiation integral is reduced to a simple summation. This program has proven to be surprisingly robust and useful for the analysis of arbitrary reflectors, particularly when the near-field is desired and the surface derivatives are not known.

Because of its simplicity, the algorithm has proven to be extremely easy to adapt to the parallel computing architecture of a modest number of large-grain computing elements. The code was initially parallelized on the Intel Paragon, and has since been ported to the Cray T3D, T3E, and Beowulf architectures.

For generality, the code considers a dual-reflector calculation, as illustrated in Figure 19.1, which can be thought of as three sequential operations:

1. computing the currents on the first (sub-)reflector using the standard PO approximation;

2. computing the currents on the second (main) reflector by utilizing the currents on the first (sub-)reflector as the field generator; and

3. computing the required observed field values by summing the fields from the currents on the second (main) reflector.

The most time-consuming part of the calculation is the computation of currents on the second reflector due to the currents on the first, since for N triangles on the first reflector, each of the M triangles on the second reflector require an N-element sum over the first. At this time, the code has been parallelized by distributing the M triangles on the second reflector, and having all processors store all the currents on the N triangles of the first reflector (though the computation of the currents of the first reflector is done in parallel). The calculation of observed field data has been parallelized. So, the three steps listed above are all performed in parallel. There are also sequential operations involved, such as I/O and the triangulation of the reflector surfaces. Some of these could be performed in parallel, but this would require a serious effort that has not been done at this time.

This code is written in Fortran, and has been parallelized using MPI. Communication is required in two locations of the code. At the end of the first step, after each processor has computed a portion of the currents on the first reflector, the currents must be broadcast to all the processors. While this may be done in many ways, a call to *MPI_Allgatherv* is currently used. During the third step, each processor calculates a partial value for each final observed field, by integrating over the main reflector currents local to that processor. A global sum (an *MPI_Reduce* call) is required to compute the complete result for each observed field value. Since there are normally a number of observed fields computed, currently there are that number of global sums. These could be combined into a single global sum of larger length, but this has not

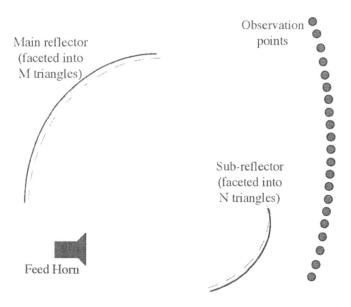

Figure 19.1 The dual reflector physical optics problem, showing the source, the two reflectors, and the observation points.

been done at this time, since the communication takes up such a small portion of the overall run time.

19.3 FDTD SIMULATION

The Finite-Difference Time-Domain (FDTD) code studied here is a fairly simple example of a time-stepping partial differential equation (PDE) solution over a physical problem domain which is distributed over the memory of a mesh of processors (Taflove, 1995). In traditional FDTD electromagnetic codes, there are generally six field unknowns which are staggered in time and space. For the purposes of this paper, they can be thought of as residing in a spatial cell, where each cell is updated at each time step. This particular code adds two specific features. First, each field component is split into two sub-components which are stored in memory and updated separately. This is done to implement a boundary condition (Berenger, 1994). Second, the parallelization that was done for this code tried to reduce the required communication, and therefore redundantly updates some of the sub-components on the face of each processor's domain, and communicates only four of the sub-components on each face.

This code is written in Fortran using MPI. The decomposition performed is two-dimensional (in x and y), while the spatial region modeled is three-dimensional. The processors are mapped into a two-dimensional Cartesian mesh using MPI's facilities for Cartesian communicators, and each processor models a physical domain that contains a subset of the entire physical domain in x and y, and the entire domain in z. Because of this, at each time each processor swaps one face (a complete $y - z$ plane) of four sub-components in the $\pm x$ direction, and one face (a complete $x - z$ plane) of four other sub-components in the $\pm y$ direction. The communication is done as follows: each processor issues an *MPI_IRecv* call to each neighboring processor (usually 4, except on the edges of the processor mesh); each processor fills and sends a buffer in each appropriate direction, suing an *MPI_SSend* call; and finally, each processor does a number of *MPI_Wait* operations, followed by unpacking the received data. This combination of calls should produce no additional buffering of data, since the program is already doing some in an effort to reduce the number of messages.

19.4 PO RESULTS

Timing results for the PO code in this paper are presented by breaking down the overall timing into three parts. Part I is input I/O and triangulation of the main reflector surface, some of which is done in parallel. No communication occurs in part I. Part II is triangulation of the sub-reflector surface (sequential), evaluation of the currents on the sub-reflector (parallel), and evaluation of the currents on the main reflector (parallel). A single *MPI_Allgatherv* occurs in part II. Part III is an evaluation of the observed fields (parallel) and I/O (on only one processor). A number of 3-word global sums occur in part III, one for each observation point. In the test cases used here, there are 122 observation points.

Two different compilers were compared (Gnu g77 and Absoft f77) on the Beowulf system. One set of indicative results from these runs are shown in Table 19.1. For this code, the Absoft compiler produced code that was approximately 30% faster, and this compiler was used hereafter.

It should be mentioned that the computation of the radiation integral in two places (in parts II and III) originally had code of the form:

```
CEJK = CDEXP(-AJ*AKR),
```

where AJ = (0.d0,1.d0). This can be changed to:

```
CEJK = DCMPLX(DCOS(AKR),-DSIN(AKR)).
```

On the T3D, these two changes led to improved results (the run-times were reduced by 35 to 40%,) which are shown in this paper. When these changes were applied to the Beowulf code using the Absoft f77 compiler, no significant performance change

Table 19.1 The effect of two Beowulf compilers (Gnu g77 and Absoft f77), shown by timing results (in minutes) for PO code, for M=40,000, N=4,900.

Number of Processors	Gnu g77			Absoft f77		
	I	II	III	I	II	III
1	0.085	64.3	1.64	0.0482	46.4	0.932
4	0.051	16.2	0.43	0.0303	11.6	0.237
16	0.047	4.2	0.11	0.0308	2.9	0.065

Table 19.2 Timing results (in seconds) for PO code, for M=40,000, N=400.

Number of Processors	Beowulf			T3D		
	I	II	III	I	II	III
1	3.19	230.0	56.0	14.50	249.0	56.4
4	1.85	57.7	14.2	8.94	62.5	14.7
16	1.52	14.6	3.9	8.97	16.6	4.1

Table 19.3 Timing results (in minutes) for PO code, for M=40,000, N=4,900.

Number of Processors	Beowulf			T3D		
	I	II	III	I	II	III
1	0.0482	46.4	0.932	0.254	48.7	0.941
4	0.0303	11.6	0.237	0.149	12.2	0.240
16	0.0308	2.9	0.065	0.138	3.1	0.074

Table 19.4 Timing results (in minutes) for PO code, for M=160,000, N=10,000.

Number of Processors	Beowulf			T3D		
	I	II	III	I	II	III
4	0.095	94.6	0.845	0.547	101.0	0.965
16	0.099	23.9	0.794	0.463	25.6	0.355
64	0.095	6.4	0.541	0.520	6.9	0.116

was observed, leading to the conclusion that one of the optimizations performed by this compiler was similar to this hand-optimization.

Tables 19.2, 19.3 and 19.4 compare the Beowuld and Cray T3D using three different problem sizes. Table 19.4's results were obtained from Naegling, and the other two table's data used Hyglac, a 6-node Beowulf at JPL. It may be observed from the tables that the Beowulf code performs slightly better than the T3D code, both in terms of absolute performance as well as scaling from 1 to 64 processors. This performance difference can be explained by the faster CPU on the Beowulf versus the T3D, and the simple, limited communication that negates the benefits of the T3D's faster network. The scaling difference is more a function of I/O, which is both more direct and more simple on the Beowulf, and thus faster. By reducing this part of the sequential time, scaling performance is improved. Another way to look at this is to compare the results in the three tables. Clearly, scaling is better in the larger test case, in which I/O is a smaller percentage of overall time. It is also clear that the communications network used on Naegling is behaving as designed for the PO code running on 4, 16, or 64 processors. Since the majority of communication is a single-word global sums, this basically demonstrates that the network has reasonable latency.

Table 19.5   Timing results (in seconds) for complete PO code, for M=40,000, N=400.

Number of Processors	Beowulf (Hyglac)	Cray T3D	Cray T3E-600
1	289	320	107.0
4	74	86	29.6
16	20	29	8.4

Table 19.6   Timing results (in minutes) for complete PO code, for M=40,000, N=4,900.

Number of Processors	Beowulf (Hyglac)	Cray T3D	Cray T3E-600
1	47.4	49.4	18.4
4	11.9	12.6	4.4
16	3.0	3.3	1.1

Table 19.7   Timing results (in minutes) for complete PO code, for M=160,000, N=10,000.

Number of Processors	Beowulf (Naegling)	Cray T3D	Cray T3E-600
4	95.5	102.0	35.1
16	24.8	26.4	8.8
64	7.0	7.6	2.3

Tables 19.5, 19.6, and 19.7 show comparisons of the complete execution time for the 3 test problems sizes, for the Beowulf, T3D, and T3E-600 systems. These demonstrate good performance on the two Beowulf-class machines when compared with the T3D in terms of overall performance, as well as when compared with the T3E-600 in terms of price-performance. For all three test cases, the Beowulf scaling is better than the T3D scaling, but the results are fairly close for the largest test case (using Naegling). This can be explained in large part by I/O requirements and timings on the various machines. The I/O is close to constant for all test cases over all machine sizes, so in some way it acts as serial code that hurts scaling performance. The I/O is the fastest on Hyglac, and slowest on the T3D. This is due to the number of nodes being used on the Beowulf machines, since disks are NFS-mounted, and the more nodes there are, the slower the performance is using NFS. The T3D forces all I/O to travel through its Y-MP front end, which causes it to be very slow. Scaling on the T3D is generally as good as the small Beowulf, and faster than the large Beowulf, again due mostly to I/O. It may be observed that the speedup of the second test case on the T3E

Table 19.8 Timing results (in computation–communication CPU seconds per time step) for FDTD code, for fixed problem size per processor of $69 \times 69 \times 76$ cells.

Number of Processors	Beowulf	Cray T3D	Cray T3E-600
1	2.44 - 0.0	2.71 - 0.0	0.851 - 0.0
4	2.46 - 0.01	2.79 - 0.026	0.859 - 0.019
16	2.46 - 0.21	2.79 - 0.024	0.859 - 0.051
64	2.46 - 0.32	2.74 - 0.076	0.859 - 0.052

is superlinear in going from 1 to 4 processors. This is likely caused by a change in the ratio of the size of some arrays to the cache size.

A hardware monitoring tool was used on the T3E to measure the number of floating point operations in the M=40,000, N=4,900 test case. The result was 1.32×10^{11} floating point operations. This gives a rate of 46, 44, and 120 MFLOP/second on one processor of the Beowulf, T3D, and T3E-600, respectively. These are fairly good (23, 29, and 20% of peak, respectively) for RISC processors running Fortran code.

19.5   FDTD RESULTS

All FDTD results that are shown in this section use a fixed size local (per processor) grid of $69 \times 69 \times 76$ cells. The overall grid sizes therefore range from $69 \times 69 \times 76$ to $552 \times 552 \times 76$ (on 1 to 64 processors). This is the largest local problem size that may be solved on the T3D, and while the other machines have more local memory and could solve larger problems, it seems more fair to use the same amount of local work for these comparisons. In general, the FDTD method requires 10 to 20 points per wavelength for accurate solutions, and a boundary region of 10 to 20 cells in each direction is also needed. These grid sizes therefore correspond to scattering targets ranging in size from $5 \times 5 \times 5$ to $53 \times 53 \times 5$ wavelengths.

Both available compilers were used on the Beowulf version of the FDTD code. While the results are not tabulated in this paper, the Gnu g77 compiler produced code which ran faster than the code produced by the Absoft f77 compiler. However, the results were just a few percent different, rather than on the scale of the differences shown by the PO code.

Table 19.8 shows results on various machines and various numbers of processors in units of CPU seconds per simulated time step. All Beowulf results are from Naegling, and all compiles were done using Gnu g77. Complete simulations might require hundreds to hundreds of thousands time steps, and the results can be scaled accordingly, if complete simulation times are desired. Results are shown broken into computation and communication times, where communication includes send, receive, and buffer copy times.

It is clear that the Beowulf and T3D computation times are comparable, while the T3E times are about 3 times faster. This is reasonable, given the relative clock rates

(200, 150, and 300 MHz) and peak performances (200, 150, 600 MFLOP/s) of the CPUs. As with the PO code, the T3D attains the highest fraction of peak performance, the higher clock rate of the Beowulf gives it a slightly better performance than the T3D, and the T3E obtains about the same fraction of peak performance as the Beowulf. As this code has much more communication that the PO code, there is a clear difference of an order of magnitude between the communication times on the Beowulf and the T3D and T3E. However, since this is still a relatively small amount of communication as compared with the amount of computation, it does not really affect the overall results.

19.6 CONCLUSIONS

This paper has shown that for both parallel calculation of the radiation integral and parallel finite-difference time-domain calculations, a Beowulf-class computer provides slightly better performance that a Cray T3D, at a much lower cost. The limited amount of communication in the physical optics code defines it as being in the heart of the regime in which Beowulf-class computing is appropriate, and thus it makes a good test code for an examination of code performance and scaling, as well as an examination of compiler options and other optimizations. The FDTD code contains more communication, but the amount is still fairly small when compared with the amount of computation, and this code is a good example of domain decomposition PDE solvers. The timing results from this code show trends that are very similar to the results of other domain decomposition PDE solvers that have been examined at JPL.

An interesting observation is that for Beowulf-class computing, using commodity hardware, the user also must be concerned with commodity software, including compilers. As compared with the T3D, where Cray supplies and updates the best compiler it has available, the Beowulf system has many compilers available from various vendors, and it is not clear that any one always produces better code than the others. In addition to the compilers used in this paper, at least one other exists (to which the authors did not have good access). The compilers also accept various extensions to Fortran, which may make compilation of any given code difficult or impossible without re-writing some of it, unless the code was written strictly in standard Fortran 77 (or Fortran 90), which seems to be extremely uncommon.

It is also interesting to notice that the use of hand-optimizations produces indeterminate results in the final run times, again depending on which compiler and which machine is used. Specific compiler optimization flags have not been discussed in this paper, but the set of flags that was used in each case were those that produced the fastest running code. In most, but not all, cases, various compiler flag options produced greater variation in run times than any hand optimizations. The implication of this is that the user should try to be certain there are no gross inefficiencies in the code to be compiled, and that it is more important to choose the correct compiler and compiler flags. This is not a good situation.

The choice of communication network for a large Beowulf is certainly not obvious. Current products in the marketplace have demonstrated scalable latencies, but not scalable bandwidths. However, this may be changing, as seems to be demonstrated by the new portions of Naegling's network.

Overall, this paper has validated the choice of a Beowulf-class computer for both the physical optics application (and other similar low-communication applications) as well as for the finite-difference time-domain application (and other domain decomposition PDE solvers). It has examined performance of these codes in terms of comparison with the Cray T3D and T3E, scaling, and compiler issues, and pointed out some "features" of which users of Beowulf systems should be aware.

Acknowledgments

The authors would like to acknowledge helpful conversations with John Salmon and Jan Lindheim at Caltech, as well as the contribution of Bill Imbriale at JPL in developing the original POPO code studied here. The FDTD code studied here includes large contributions made by Allen Taflove at Northwestern University. The work described was performed at the Jet Propulsion Laboratory, California Institute of Technology under contract with the National Aeronautics and Space Administration. The Cray T3D supercomputer used in this investigation was provided by funding from the NASA Offices of Earth Science, Aeronautics, and Space Science. Part of the research reported here was performed using the Beowulf system operated by the Center for Advanced Computing Research at Caltech; access to this facility was provided by Caltech. Access to the Cray T3E-600 was provided by the Earth and Space Science (ESS) component of the NASA High Performance Computing and Communication (HPCC) program.

References

Berenger, J.-P. (1994). A perfectly matched layer for the absorption of electromagnetic waves. *J. Comp. Physics*, 114:185–200.

Giest, A., Beguelin, A., Dongarra, J., Jiang, W., Manchek, R., and Sunderam, V. (1994). *PVM: A Users's Guide and Tutorial for Networked and Parallel Computing*. The MIT Press, Cambridge, Mass.

Husain, K., Parker, T., McKinnon, P., McMullin, R., Treijs, E., and Pfister, R. (1996). *Red Hat Linux Unleashed*. Sams Publishing, Indianapolis, Ind.

Imbriale, W. and Cwik, T. (1994). A simple physical optics algorithm perfect for parallel computing architecture. *10th Annual Review of Progress in Appl. Comp. Electromag.*, pages 434–441.

Imbriale, W. and Hodges, R. (1991). Linear phase approximation in the triangular facet near-field physical optics computer program. *Appl. Comp. Electromag. Society Journal*, 6:74–85.

Katz, D., Cwik, T., Kwan, B., Lou, J., Springer, P., Sterling, T., and Wang, P. (1998). An assessment of a Beowulf system for a wide class of analysis and design software. *Advances in Engineering Software*, 29. To appear.

Snir, M., Otto, S., Huss-Lederman, S., Walker, D., and Dongarra, J. (1996). *MPI: The Complete Reference*. The MIT Press, Cambridge, Mass.

Taflove, A. (1995). *Computational Electrodynamics: The Finite-Difference Time-Domain Method*. Artech House, Norwood, Mass.

20 MANAGING DIVISIBLE LOAD ON PARTITIONABLE NETWORKS

Keqin Li

Department of Mathematics and Computer Science,
State University of New York, U.S.A.

li@mcs.newpaltz.edu

Abstract: We analyze the parallel time and speedup for processing divisible load on (1) a linear array with a corner initial processor, (2) a linear array with an interior initial processor, (3) a mesh with a corner initial processor, (4) a mesh with an interior initial processor, (5) a b-ary complete tree with the root as the initial processor, and (6) a pyramid with the apex as the initial processor. Due to communication overhead, and limited network connectivity, the speedup of parallel processing for divisible load is bounded from above by a quantity independent of the network size. It is shown that for the above six cases, as the network size becomes large, the asymptotic speedup is approximately $\sqrt{\beta}$, $2\sqrt{\beta}$, $\beta^{3/4}$, $4\beta^{3/4}$, $(b-1)\beta$, and 3β respectively, where β is the ratio of the time for computing a unit load to the time for communicating a unit load.

Keywords: Divisible load, load distribution, recurrence relation, speedup.

20.1 INTRODUCTION

A divisible load has the property that it can be arbitrarily divided into small pieces (load fractions) which are assigned to parallel processors for execution. Example applications include large-scale data file processing, signal and image processing, engineering computations, and many real-time computing such as target identification, searching, and data processing in distributed sensor networks (Bharadwaj *et al.*, 1996). The problem was first proposed in (Cheng and Robertazzi, 1988). Since then, scheduling divisible load has been investigated by a number of researchers in recent years for the bus (Bataineh *et al.*, 1994; Sohn and Robertazzi, 1996), linear array (Mani and Ghose, 1994), tree (Bataineh *et al.*, 1994; Cheng and Robertazzi, 1990), mesh (Blazewicz and Drozdowski, 1996), and hypercube (Blazewicz and Drozdowski, 1995; Li *et al.*, 1998) networks.

The well-known Amdahl's Law (Amdahl, 1967) states that if a fraction f of a computation cannot be parallelized, the speedup is bounded from above by $1/f$, no matter how many processors are used. For a divisible load, there is no inherently sequential part, *i.e.* $f = 0$. However, this does not imply that unbounded speedup can be achieved. The reason is that Amdahl's Law has no restriction on a parallel system, where processors can communicate at no cost. When a divisible load is processed on a multicomputer with a static network, there is communication overhead. Also, the network topology, that determines the speed at which a load is distributed over a network, has a strong impact on performance.

In this paper, we analyze the parallel time and speedup for processing a divisible load on partitionable networks. The following six cases are considered:

1. a linear array with a corner initial processor,

2. a linear array with an interior initial processor,

3. a mesh with a corner initial processor,

4. a mesh with an interior initial processor,

5. a b-ary complete tree with the root as the initial processor, and

6. a pyramid with the apex as the initial processor.

We provide recurrence relations to calculate the total processing time. It is discovered that due to communication overhead and limited network connectivity, the speedup of parallel processing for divisible load is bounded from above by a quantity independent of the network size. It is shown that for the above six cases, as network size becomes large, the asymptotic speedup is approximately $\sqrt{\beta}$, $2\sqrt{\beta}$, $\beta^{3/4}$, $4\beta^{3/4}$, $(b - 1)\beta$, and 3β respectively, where β is the ratio of the time for computing a unit load to the time for communicating a unit load. It turns out that the size of a network may have less impact since the parallel processing time and speedup converge very fast as the network size increases.

20.2   THE MODEL

We consider divisible load distribution on a static network with N processors P_1, P_2, ..., P_N. Each processor P_i has n_i neighbors. It is assumed that P_i has n_i separate ports for communication with each of the n_i neighbors. That is, processor P_i can send messages to all its n_i neighbors simultaneously. Let T_{cm} be the time to transmit a unit load along a link. The time to send a load to a neighbor is proportional to the size of the load, with a negligible communication start-up time.

Let T_{cp} be the time to process a unit load on a processor. Again, the computation time is proportional to the size of a load. We use $\beta = T_{cp}/T_{cm}$ to denote the granularity, which is a parameter indicating the nature of a parallel computation and a parallel architecture. A large (small) β gives a small (large) communication overhead. A computation-intensive load has a large β, and a communication-intensive load has a small β. An infinite β implies that the communication cost is negligible.

It is noticed that once a processor sends part of a load (load fraction) to a neighbor, it can proceed with other computation and communication activities. This provides the capability to overlap computation and communication, and enhances the system performance. However, the neighbor (receiver) must wait until the load fraction arrives, and then starts the processing of the load fraction. It is this waiting time that limits the overall system performance. (One can also imagine that each processor is equipped with a communication coprocessor that handles load distribution. However, based on the above discussion, this seems unnecessary.)

We will consider a divisible load distribution on partitionable networks. Let $\mathcal{G} = (\mathcal{P}, \mathcal{E})$ be a static network, where $\mathcal{P} = \{P_1, P_2, ..., P_N\}$ is a set of N processors, and $\mathcal{E}$ is a set of interprocessor connections. Consider a processor P_i which has n_i neighbors $P_{j_1}, P_{j_2}, ..., P_{j_{n_i}}$. We say that $\mathcal{G}$ is partitionable at processor P_i, if

1. $\mathcal{P} - \{P_i\}$ can be partitioned into n_i disjoint subsets $\mathcal{P}_1, \mathcal{P}_2, ..., \mathcal{P}_{n_i}$, such that $P_{j_k} \in \mathcal{P}_k$, for all $1 \leq k \leq n_i$; and

2. the subgraphs induced by $\mathcal{P}_1, \mathcal{P}_2, ..., \mathcal{P}_{n_i}$ have the same topology (or, belong to the same family of graphs) as the original graph $\mathcal{G}$, and each $\mathcal{P}_k$ is partitionable at $P_{j_k}, 1 \leq k \leq n_i$.

Example partitionable networks are the linear array and mesh at any processor, tree at the root, and pyramid at the apex. Another example, the hypercube, has been studied in (Li *et al.*, 1998), where it is shown that due to the unlimited node degree in a hypercube, the speedup is unbounded.

20.3   LINEAR ARRAY

A linear array with N processors $P_1, P_2, ..., P_N$ is a one-dimensional mesh $M_1 = (P_1, P_2, ..., P_N)$. There is a connection between P_j and P_{j+1}, for all $1 \leq j \leq N - 1$. Assume that processor P_i is the initial processor (it holds an initial load x) Our strategies for divisible load distribution on M_1 as well as their analysis depend on the position of the initial processor.

20.3.1   Corner Initial Processor

In this case, i can be 1 or N; the initial processor is a corner processor. Without loss of generality, let us assume that $i = N$. Assume that there are N numbers $\alpha_N, \alpha_{N-1}, ..., \alpha_1$ in the interval $(0, 1)$.

Algorithm C_N

(1) If $N = 1$, processor P_1 handles the load itself. (This is essentially algorithm C_1.)

(2) In general, when $N > 1$, processor P_N sends a faction α_N of the load x to P_{N-1}, and then processes the remaining load $(1 - \alpha_N)x$.

(3) Upon receiving the load $\alpha_N x$ by processor P_{N-1} (which is regarded as a corner initial processor of the linear array $M_1' = (P_1, P_2, ..., P_{N-1})$), the linear array M_1' processes the load $\alpha_N x$ using the load distribution algorithm C_{N-1}.

We now examine the performance of algorithm C_N. Let T_N denote the total processing time for one unit of load on an M_1 with N processors when the initial processor is on the boundary. Since both computation and communication times are linearly proportional to the amount of load, the time to process x units of load on an M_1 with N processors is xT_N, where $x \geq 0$.

The value T_N can be obtained recursively as follows. First, by (1), we have $T_1 = T_{cp}$. In general, when $N > 1$, processor P_N can proceed without waiting after sending the load fraction α_N in (2). Hence, the time spent by P_N is $(1 - \alpha_N)T_{cp}$. In (3), it takes $\alpha_N T_{cm}$ time for the load fraction α_N to reach P_{N-1}. Then the linear array M_1' requires $\alpha_N T_{N-1}$ time to process the load fraction α_N. To minimize the total processing time T_N, we need to make sure that both P_N and M_1' spend the same amount of time, that is,

$$T_N = (1 - \alpha_N)T_{cp} = \alpha_N(T_{cm} + T_{N-1}).$$

It is not difficult to see that α_N satisfies the following recurrence relation,

$$\alpha_1 = 0;$$
$$\alpha_N = \frac{\beta}{\beta(2 - \alpha_{N-1}) + 1}, \quad N > 1;$$

and T_N satisfies the following recurrence relation,

$$T_1 = T_{cp};$$
$$T_N = \left(\frac{T_{N-1} + T_{cm}}{T_{N-1} + T_{cp} + T_{cm}}\right)T_{cp}, \quad N > 1. \tag{20.1}$$

The speedup is defined as the ratio of the sequential processing time to the parallel processing time, namely, $S_N = T_1/T_N = T_{cp}/T_N$. Let $T_\infty = \lim_{N\to\infty} T_N$, and $S_\infty = \lim_{N\to\infty} S_N$. We notice that $T_N > 0$ is a decreasing function of N. Hence, T_∞ exists. Taking the limit of both side of Equation (20.1), we obtain

$$T_\infty^2 + T_{cm}T_\infty - T_{cp}T_{cm} = 0.$$

Solving this quadratic equation, we get T_∞ and S_∞ in the following theorem.

Theorem 1. For a linear array with a corner initial processor, we have

$$T_\infty = \frac{\sqrt{T_{cm}^2 + 4T_{cp}T_{cm}} - T_{cm}}{2}.$$

Furthermore,

$$S_\infty = \frac{2\beta}{\sqrt{4\beta + 1} - 1} = \frac{\sqrt{4\beta + 1} + 1}{2} \approx \sqrt{\beta},$$

as $\beta \to \infty$. ∎

20.3.2 Interior Initial Processor

In this case, the initial processor P_i is an interior processor, where $1 \leq i \leq N$. We need to have $\alpha_N(i, i-1)$ and $\alpha_N(i, i+1)$ in the interval $(0,1)$, where $2 \leq i \leq N-1$.

Algorithm $I_N(i)$

(1) If $i = 1, N$, we use algorithm C_N to process the load. (Hence, algorithm C_N is a special case of I_N.)

(2) If $2 \leq i \leq N-1$, processor P_i sends a faction $\alpha_N(i, i-1)$ of the load x to P_{i-1}, and also sends a faction $\alpha_N(i, i+1)$ of the load x to P_{i+1}. Processor P_i then processes the remaining load $(1 - \alpha_N(i, i-1) - \alpha_N(i, i+1))x$.

(3) Upon receiving the load $\alpha_N(i, i-1)x$ by processor P_{i-1} (which is regarded as a corner initial processor of the linear array $M_1' = (P_1, P_2, ..., P_{i-1})$), the linear array M_1' processes the load $\alpha_N(i, i-1)x$ using the load distribution algorithm C_{i-1}.

(4) Upon receiving the load $\alpha_N(i, i+1)x$ by processor P_{i+1} (which is regarded as a corner initial processor of the linear array $M_1'' = (P_{i+1}, P_{i+2}, ..., P_N)$), the linear array M_1'' processes the load $\alpha_N(i, i+1)x$ using the load distribution algorithm C_{N-i}.

Let $T_N(i)$ denote the total processing time for one unit of load on an M_1 with N processors when P_i is the initial processor. To make the total processing time minimal, we need to choose $\alpha_N(i, i-1)$ and $\alpha_N(i, i+1)$ such that P_i, M_1', and M_1'' take the same time, $i.e.$

$$
\begin{aligned}
T_N(i) &= (1 - \alpha_N(i, i-1) - \alpha_N(i, i+1))T_{cp} \\
&= \alpha_N(i, i-1)(T_{cm} + T_{i-1}) \\
&= \alpha_N(i, i+1)(T_{cm} + T_{N-i}).
\end{aligned}
$$

Hence, $T_N(i)$ can be calculated based on the T_N's:

$$
T_N(i) = T_N, \quad i = 1, N;
$$

$$
T_N(i) = \left(\frac{1}{T_{cp}} + \frac{1}{T_{i-1} + T_{cm}} + \frac{1}{T_{N-i} + T_{cm}} \right)^{-1}, \tag{20.2}
$$
$$
2 \leq i \leq N-1.
$$

Using $T_N(i)$, we get

$$
\alpha_N(i, i-1) = \frac{T_N(i)}{T_{cm} + T_{i-1}},
$$

and

$$
\alpha_N(i, i+1) = \frac{T_N(i)}{T_{cm} + T_{N-i}},
$$

for all $2 \leq i \leq N - 1$.

It can be shown that $T_N(i)$ is minimized when $i = \lceil N/2 \rceil$. Let $T_N^* = T_N(\lceil N/2 \rceil)$, and $T_\infty^* = \lim_{N \to \infty} T_N^*$. Taking the limit of both sides of Equation (20.2), we obtain

$$T_\infty^* = \left(\frac{1}{T_{cp}} + \frac{2}{T_\infty + T_{cm}} \right)^{-1}.$$

Theorem 2. For a linear array with an interior initial processor, we have

$$T_\infty^* = \frac{T_{cp}(\sqrt{T_{cm}^2 + 4T_{cp}T_{cm}} + T_{cm})}{\sqrt{T_{cm}^2 + 4T_{cp}T_{cm}} + 4T_{cp} + T_{cm}}.$$

Furthermore,

$$S_\infty^* = \frac{4\beta + 1 + \sqrt{4\beta + 1}}{\sqrt{4\beta + 1} + 1} \approx 2\sqrt{\beta},$$

as $\beta \to \infty$.   ■

20.4   MESH

We now consider divisible load distribution on a mesh with $N = N_1 \times N_2$ processors $M_2 = \{P_{j_1,j_2} \mid 1 \leq j_1 \leq N_1, \ 1 \leq j_2 \leq N_2\}$. Processor P_{j_1,j_2} is connected to four neighbors $P_{j_1 \pm 1, j_2 \pm 1}$, if the neighbors exist.

20.4.1   Corner Initial Processor

Let us look at the case where a corner processor is the initial processor. Without loss of generality, assume that P_{N_1,N_2} is the initial processor, and $N_1 \geq N_2$.

Algorithm C_{N_1,N_2}

(1) If $N_1 = 1$, we have a linear array of size N_2, hence algorithm C_{N_2} is invoked. Or, if $N_2 = 1$, we have a linear array of size N_1, hence algorithm C_{N_1} is invoked.

(2) If $N_1 \geq 2$ and $N_2 \geq 2$, processor P_{N_1,N_2} sends a fraction $\alpha_{N_1,N_2}^{(1)}$ of the load x to processor P_{N_1-1,N_2}, and a fraction $\alpha_{N_1,N_2}^{(2)}$ of the load x to processor P_{N_1,N_2-1}. Processor P_{N_1,N_2} then processes the remaining load $(1 - \alpha_{N_1,N_2}^{(1)} - \alpha_{N_1,N_2}^{(2)})x$.

(3) Upon receiving the load $\alpha_{N_1,N_2}^{(1)} x$ by processor P_{N_1-1,N_2} (which is regarded as a corner initial processor of the linear array $M_1 = (P_{1,N_2}, P_{2,N_2}, ..., P_{N_1-1,N_2})$), the linear array M_1 processes the load $\alpha_{N_1,N_2}^{(1)} x$ using algorithm C_{N_1-1}.

(4) Upon receiving the load $\alpha_{N_1,N_2}^{(2)} x$ by processor P_{N_1,N_2-1} (which is regarded as a corner initial processor of a 2-dimensional submesh M_2' of size $N_1 \times (N_2 - 1)$), M_2' processes the load $\alpha_{N_1,N_2}^{(2)} x$ using algorithm C_{N_1,N_2-1}.

Let T_{N_1,N_2} denote the total processing time for one unit of load on an M_2 with $N = N_1 \times N_2$ processors when the initial processor is at the corner P_{N_1,N_2}. Then, we need to have

$$
\begin{aligned}
T_{N_1,N_2} &= (1 - \alpha^{(1)}_{N_1,N_2} - \alpha^{(2)}_{N_1,N_2})T_{cp} \\
&= \alpha^{(1)}_{N_1,N_2}(T_{cm} + T_{N_1-1}) \\
&= \alpha^{(2)}_{N_1,N_2}(T_{cm} + T_{N_1,N_2-1}).
\end{aligned}
$$

Thus, we obtain the following recurrence relation for T_{N_1,N_2},

$$
\begin{aligned}
T_{N_1,N_2} &= T_{N_2}, \quad N_1 = 1; \\
T_{N_1,N_2} &= T_{N_1}, \quad N_2 = 1; \\
T_{N_1,N_2} &= T_{N_2,N_1}, \quad N_2 > N_1 \geq 2; \\
T_{N_1,N_2} &= \left(\frac{1}{T_{cp}} + \frac{1}{T_{N_1-1} + T_{cm}} + \frac{1}{T_{N_1,N_2-1} + T_{cm}} \right)^{-1}, \quad (20.3) \\
& \hspace{7cm} N_1 \geq N_2 \geq 2.
\end{aligned}
$$

The fractions $\alpha^{(1)}_{N_1,N_2}$ and $\alpha^{(2)}_{N_1,N_2}$ can be obtained based on T_{N_1,N_2}.

Let $T_{\infty,\infty} = \lim_{N_1 \to \infty, N_2 \to \infty} T_{N_1,N_2}$. Taking the limit in both side of Equation (20.3), we have

$$
T_{\infty,\infty} = \left(\frac{1}{T_{cp}} + \frac{1}{T_\infty + T_{cm}} + \frac{1}{T_{\infty,\infty} + T_{cm}} \right)^{-1},
$$

that is,

$$
(T_\infty + T_{cp} + T_{cm})T_{\infty,\infty}^2 + T_{cm}(T_\infty + T_{cp} + T_{cm})T_{\infty,\infty} - T_{cp}T_{cm}(T_\infty + T_{cm}) = 0.
$$

The above quadratic equation yields

Theorem 3. For a mesh with a corner initial processor, we have

$$
T_{\infty,\infty} = \frac{1}{2}\left(\sqrt{T_{cm}^2 + 4T_{cm}T_\infty} - T_{cm} \right).
$$

Furthermore,

$$
\begin{aligned}
S_{\infty,\infty} &= \frac{2\beta}{\sqrt{2\sqrt{4\beta + 1} - 1} - 1} \\
&= \frac{(\sqrt{2\sqrt{4\beta + 1} - 1} + 1)(\sqrt{4\beta + 1} + 1)}{4} \\
&\approx \beta^{3/4},
\end{aligned}
$$

as $\beta \to \infty$. ∎

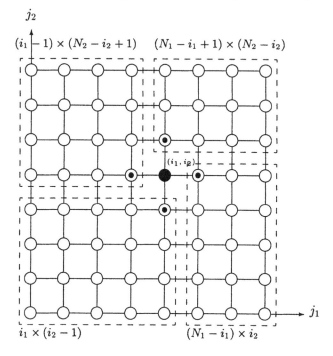

Figure 20.1   The partitioning of a mesh network with $N = N_1 \times N_2$ processors.

20.4.2 Interior Initial Processor

Assume that P_{i_1,i_2} is the initial processor. The algorithm in this case is called $I_{N_1,N_2}(i_1, i_2)$. Without loss of generality, we assume that $\lfloor N_1/2 \rfloor + 1 \leq i_1 \leq N_1$, and $\lfloor N_2/2 \rfloor + 1 \leq i_2 \leq N_2$, since a mesh is symmetric with respect to its center.

The case for interior initial processor can be reduced to the case of corner initial processor by mesh partitioning. When $i_1 \leq N_1 - 1$ and $i_2 \leq N_2 - 1$, P_{i_1,i_2} has four neighbors, *i.e.* P_{i_1,i_2-1}, P_{i_1-1,i_2}, P_{i_1,i_2+1}, and P_{i_1+1,i_2}. As illustrated in Figure 20.1, these neighbors are corner processors of four submeshes of sizes $i_1 \times (i_2 - 1)$, $(i_1 - 1) \times (N_2 - i_2 + 1)$, $(N_1 - i_1 + 1) \times (N_2 - i_2)$, and $(N_1 - i_1) \times i_2$.

Let us use $\alpha_{N_1,N_2}^{(1)(2-)}(i_1, i_2)$, $\alpha_{N_1,N_2}^{(1-)(2)}(i_1, i_2)$, $\alpha_{N_1,N_2}^{(1)(2+)}(i_1, i_2)$, and $\alpha_{N_1,N_2}^{(1+)(2)}(i_1, i_2)$ to denote the load fraction sent to the four neighbors of P_{i_1,i_2} respectively. Let $T_{N_1,N_2}(i_1, i_2)$ denote the total processing time for one unit of load on an M_2 with $N = N_1 \times N_2$ processors, when the initial processor is P_{i_1,i_2}. Then, we have

$$\begin{aligned}
T_{N_1,N_2}(i_1,i_2) &= (1 - \alpha_{N_1,N_2}^{(1)(2-)}(i_1,i_2) - \alpha_{N_1,N_2}^{(1-)(2)}(i_1,i_2) \\
&\quad - \alpha_{N_1,N_2}^{(1)(2+)}(i_1,i_2) - \alpha_{N_1,N_2}^{(1+)(2)}(i_1,i_2))T_{cp} \\
&= \alpha_{N_1,N_2}^{(1)(2-)}(i_1,i_2)(T_{cm} + T_{i_1,i_2-1}) \\
&= \alpha_{N_1,N_2}^{(1-)(2)}(i_1,i_2)(T_{cm} + T_{i_1-1,N_2-i_2+1}) \\
&= \alpha_{N_1,N_2}^{(1)(2+)}(i_1,i_2)(T_{cm} + T_{N_1-i_1+1,N_2-i_2}) \\
&= \alpha_{N_1,N_2}^{(1+)(2)}(i_1,i_2)(T_{cm} + T_{N_1-i_1,i_2}).
\end{aligned}$$

Special consideration should be given to P_{i_1,i_2} which is on a boundary or at a corner, since it has only three or two neighbors. We omit the detailed explanations, and give the following recurrence relation of $T_{N_1,N_2}(i_1,i_2)$, where $N_1, N_2 \geq 2$:

- For $i_1 = N_1$ and $i_2 = N_2$, $T_{N_1,N_2}(i_1,i_2) = T_{N_1,N_2}$.
- For $i_1 = N_1$ and $i_2 < N_2$, $T_{N_1,N_2}(i_1,i_2) =$

$$\left(\frac{1}{T_{cp}} + \frac{1}{T_{i_1,i_2-1}+T_{cm}} + \frac{1}{T_{i_1-1,N_2-i_2+1}+T_{cm}} + \frac{1}{T_{N_1-i_1+1,N_2-i_2}+T_{cm}} \right)^{-1}.$$

- For $i_1 < N_1$ and $i_2 = N_2$, $T_{N_1,N_2}(i_1,i_2) =$

$$\left(\frac{1}{T_{cp}} + \frac{1}{T_{i_1,i_2-1}+T_{cm}} + \frac{1}{T_{i_1-1,N_2-i_2+1}+T_{cm}} + \frac{1}{T_{N_1-i_1,i_2}+T_{cm}} \right)^{-1}.$$

- For $i_1 < N_1$ and $i_2 < N_2$, $T_{N_1,N_2}(i_1,i_2) =$

$$\left(\frac{1}{T_{cp}} + \frac{1}{T_{i_1,i_2-1}+T_{cm}} + \frac{1}{T_{i_1-1,N_2-i_2+1}+T_{cm}} \right.$$

$$\left. + \frac{1}{T_{N_1-i_1+1,N_2-i_2}+T_{cm}} + \frac{1}{T_{N_1-i_1,i_2}+T_{cm}} \right)^{-1}. \tag{20.4}$$

Let $T_{N_1,N_2}^* = T_{N_1,N_2}(\lceil N_1/2 \rceil, \lceil N_2/2 \rceil)$, and $T_{\infty,\infty}^* = \lim_{N_1 \to \infty, N_2 \to \infty} T_{N_1,N_2}^*$. Then, by Equation (20.4),

$$T_{\infty,\infty}^* = \left(\frac{1}{T_{cp}} + \frac{4}{T_{\infty,\infty} + T_{cm}} \right)^{-1}.$$

Theorem 4. For a mesh with an interior initial processor, we have

$$T_{\infty,\infty}^* \quad \frac{T_{cp}(\sqrt{T_{cm}^2 + 4T_{cm}T_\infty} + T_{cm})}{\sqrt{T_{cm}^2 + 4T_{cm}T_\infty + 8T_{cp}} + T_{cm}}.$$

Furthermore,

$$S_{\infty,\infty}^* = \frac{8\beta + 1 + \sqrt{2\sqrt{4\beta+1}-1}}{\sqrt{2\sqrt{4\beta+1}-1}+1} \approx 4\beta^{3/4},$$

as $\beta \to \infty$. ∎

20.5   TREE AND PYRAMID

A complete b-ary tree network of height $h > 0$ is partitionable at the root. The b children of the root are roots of b subtrees of height $h - 1$. The root sends a fraction α_h of the load x to each of the children, and continues to process the remaining load $(1 - b\alpha_h)x$. Upon receiving the load $\alpha_h x$, a subtree processes the load using the same strategy.

Let T_h denote the total processing time of a unit load in a b-ary complete tree network of height h. Then, when $h > 0$, we have

$$T_h = (1 - b\alpha_h)T_{cp} = \alpha_h(T_{cm} + T_{h-1}),$$

which implies that

$$\alpha_h = \frac{T_{cp}}{T_{h-1} + bT_{cp} + T_{cm}}.$$

Hence, T_h satisfies the following recurrence relation,

$$T_0 = T_{cp};$$
$$T_h = \left(\frac{T_{h-1} + T_{cm}}{T_{h-1} + bT_{cp} + T_{cm}} \right) T_{cp}, \quad h \geq 1. \tag{20.5}$$

Taking the limit in Equation (20.5), we have

$$T_\infty^2 + ((b - 1)T_{cp} + T_{cm})T_\infty - T_{cp}T_{cm} = 0.$$

Solving the above quadratic equation, we obtain

Theorem 5. For a b-ary complete tree network with the root as the initial processor, we have

$$T_\infty = \frac{1}{2}\left(\sqrt{(b - 1)^2 T_{cp}^2 + 2(b + 1)T_{cp}T_{cm} + T_{cm}^2} - ((b - 1)T_{cp} + T_{cm}) \right).$$

Furthermore,

$$S_\infty = \frac{\sqrt{(b - 1)^2\beta^2 + 2(b + 1)\beta + 1} + ((b - 1)\beta + 1)}{2} \approx (b - 1)\beta,$$

as $\beta \to \infty$.   ∎

The pyramid network is partitionable at the apex. The analysis is similar to that of complete b-ary tree network with $b = 4$. Let T_h denote the total processing time of a unit load in a pyramid network of height h, and $T_\infty = \lim_{h \to \infty} T_h$.

Theorem 6. For a pyramid with the root at the apex, we have

$$T_\infty = \frac{1}{2}\left(\sqrt{9T_{cp}^2 + 10T_{cp}T_{cm} + T_{cm}^2} - (3T_{cp} + T_{cm}) \right).$$

Furthermore,

$$S_\infty = \frac{\sqrt{9\beta^2 + 10\beta + 1} + (3\beta + 1)}{2} \approx 3\beta,$$

as $\beta \to \infty$.   ∎

20.6 NOTES ON RELATED RESEARCH

The limits of parallel processing performance for divisible loads in static networks have been reported in the literature. Our Theorems 1 and 2 are essentially similar to those in (Bataineh and Robertazzi, 1992), which are included here for the purpose of deriving Theorems 3 and 4. An infinite mesh structure with the initial processor in the center was considered in (Blazewicz and Drozdowski, 1996). However, our Theorems 3 and 4 deal with finite meshes. Asymptotic performance analysis for linear arrays and single-level trees were conducted in (Ghose and Mani, 1994). Our Theorem 5 considers more general trees.

20.7 CLOSING REMARKS

We have analyzed the parallel time and speedup for processing divisible load on partitionable networks. We provided recurrence relations to calculate the total processing time. It is discovered that due to communication overhead and limited network connectivity, the speedup of parallel processing for a divisible load is bounded from above by a quantity independent of the network size.

Acknowledgments

This research was partially supported by the National Aeronautics and Space Administration and the Research Foundation of State University of New York through NASA/University Joint Venture in Space Science Program under Grant NAG8-1313.

References

Amdahl, G. (1967). Validity of the single processor approach to achieving large scale computing capabilities. In *AFIPS Spring Joint Computer Conference*, volume 30, pages 483–485.

Bataineh, S., Hsiung, T., and Robertazzi, T. (1994). Closed form solutions for bus and tree networks of processors load sharing a divisible job. *IEEE Trans. on Computers*, 43(10):1184–1196.

Bataineh, S. and Robertazzi, T. (1992). Ultimate performance limits for networks of load sharing processors. In *Conference on Information Sciences and Systems*, pages 794–799.

Bharadwaj, V., Ghose, D., Mani, V., and Robertazzi, T. (1996). *Scheduling Divisible Loads in Parallel and Distributed Systems*. IEEE Computer Society Press.

Blazewicz, J. and Drozdowski, M. (1995). Scheduling divisible jobs on hypercubes. *Parallel Computing*, 21:1945–1956.

Blazewicz, J. and Drozdowski, M. (1996). The performance limits of a two-dimensional network of load sharing processors. *Foundations of Computing and Decision Sciences*, 21(1):3–15.

Cheng, Y. and Robertazzi, T. (1988). Distributed computation with communication delays. *IEEE Trans. on Aerospace and Electronic Systems*, 24(6):700–712.

Cheng, Y. and Robertazzi, T. (1990). Distributed computation for a tree network with communication delays. *IEEE Trans. on Aerospace and Electronic Systems*, 26(3): 511–516.

Ghose, D. and Mani, V. (1994). Distributed computation with communication delays: asymptotic performance analysis. *Journal of Parallel and Distributed Computing*, 23(3):293–305.

Li, K., Lin, X., Sun, Y., and Cheung, P. (1998). Divisible load distribution on hypercubes. In *10th International Conference on Parallel and Distributed Computing and Systems*, Las Vegas, Nevada.

Mani, V. and Ghose, D. (1994). Distributed computation in linear networks: Closed-form solutions. *IEEE Trans. on Aerospace and Electronic Systems*, 30(2):471–483.

Sohn, J. and Robertazzi, T. (1996). Optimal divisible job load sharing for bus networks. *IEEE Trans. on Aerospace and Electronic Systems*, 32(1):34–40.

VI Computational Chemistry

21 NWCHEM MOLECULAR DYNAMICS SIMULATION

T.P. Straatsma

Environmental Molecular Sciences Laboratory,
Pacific Northwest National Laboratory, U.S.A.

Straatsma@emsl.pnl.gov

Abstract: NWChem is the computational chemistry software package developed by the High Performance Computational Chemistry group for the Environmental Molecular Sciences Laboratory at the Pacific Northwest National Laboratory. This software includes quantum-mechanical and classical computational chemistry modules. The parallel implementation of the molecular dynamics module is based on domain decomposition, which is an efficient approach for simulations of large molecular systems on massively parallel architectures, despite the additional complexity of periodic reassignment of atoms and the need for dynamic load balancing techniques for heterogeneous systems. The module relies on the Global Array tools for one-sided asynchronous communication. The calculation of the forces, which is the most computation intensive part of a simulation, can be performed without synchronization, allowing for the implementation of efficient load balancing techniques. Results of benchmark simulations show the efficiency of the implementation as a function of both molecular system size and the number of processors.

Keywords: Molecular dynamics simulation, spacial decomposition, Global Array tools, load balancing.

21.1 INTRODUCTION

Theoretical chemistry, from the chemistry of materials to the rational design of pharmaceutical agents and protein engineering, increasingly relies on computer simulation methods to complement analytic theories and experimental studies. Molecular simulation can provide unique, atomic level information about molecular systems that can be fundamentally or technically difficult to obtain with other methods, contributing to a detailed understanding of the function and properties of these systems. The microscopic trajectories generated by molecular dynamics simulations are directly related to measurable macroscopic properties through statistical mechanical averages that can, however, be slow to converge with increasing system size and simulation length. Even with simple molecular interaction models such simulations require considerable computational resources. Parallel computer systems are rapidly developing to be the main architectures for scientific computing, able to deliver extremely high peak performance that, however, in practice is often difficult to realize because of the complexity of the required programming models and the fraction of inherent non-concurrency in many scientific problems.

Molecular dynamics simulations typically propagate the evolution of a molecular system in time by taking successive time steps. In the time domain such simulations are inherently non-parallel. On the other hand, in a single time step all particles in the system move simultaneously, and the calculation of forces and particle displacement is amenable to parallelization. In recent years a number of strategies for the parallelization of molecular dynamics simulation have been suggested and successfully implemented, based on replicated data (Lin *et al.*, 1992), atom decomposition, force decomposition (Plimpton and Hendrickson, 1996) and spacial decomposition (Clark *et al.*, 1994) approaches. The parallel performance of these approaches depends on the type and size of the molecular system, the type of interaction functions used, the computer architecture, and the availability of efficient communication protocols. For large molecular systems simulated with short-range interactions on distributed memory massively parallel computing systems, memory requirement, computation and communication theoretically scale as indicated in Table 21.1.

This article describes the implementation of molecular dynamics simulation in NWChem, the computational chemistry software developed for the Environmental Molecular Sciences Laboratory at the Pacific Northwest National Laboratory, which is based on a spacial decomposition of the molecular volume. Section 21.2 describes the spacial decomposition. The parallelization uses the Global Array tools (Nieplocha *et al.*, 1996), and the communication issues that arise are discussed in Section 21.3. Section 21.4 describes the load balancing that is essential for high parallel efficiency. Section 21.5 presents some performance results.

Table 21.1 Theoretical scaling of molecular dynamics simulation parallelization approaches for large molecular systems on distributed memory massively parallel computer architectures. For the spacial decomposition the assumption is made that interactions extend into neighboring nodes only. The scaling for computation assumes use of Newton's third law. N is the number of atoms in the system, and p is the number of processors.

	Memory	Computation	Communication
Replicated data	N	$\frac{1}{2}\frac{N}{p}$	$2N$
Atom decomposition	$\frac{2N}{p}$	$\frac{1}{2}\frac{N}{p}$	N
Force decomposition	$\frac{2N}{\sqrt{p}}$	$\frac{1}{2}\frac{N}{p}$	$\frac{4N}{\sqrt{p}}$
Spacial decomposition	$\frac{2N}{p}$	$\frac{1}{2}\frac{N}{p}$	$\frac{26N}{p}$

21.2   SPACIAL DECOMPOSITION

The spacial decomposition approach to parallelizing molecular simulations is based on a decomposition of the physical molecular simulation volume over processing elements available to the calculation. A two-dimensional representation is shown in Figure 21.1, illustrating the two main advantages of spacial decomposition. First, by distributing the atomic data over the available nodes, memory requirements are significantly reduced compared to replication of all the data on all nodes. Secondly, the locality of short-range interactions significantly reduces the required communication between nodes to evaluate inter-atomic forces and energies.

This type of decomposition also has disadvantages. In molecular dynamics simulations, particles continuously move, including between regions assigned to different processors. Consequently periodic redistribution of atoms will be required. A second complication arises from an unequal distribution of computational work over nodes caused, for example, by an inhomogeneous distribution of atoms. These factors make

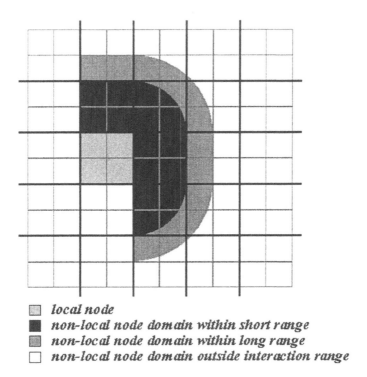

☐ *local node*
■ *non-local node domain within short range*
▨ *non-local node domain within long range*
☐ *non-local node domain outside interaction range*

Figure 21.1 Two-dimensional representation of spacial decomposition. To evaluate forces and energies a node communicates only with immediate neighbors. For long-range interactions beyond the physical space assigned to nearest neighbors the communication is still with a limited set of non-nearest neighboring nodes.

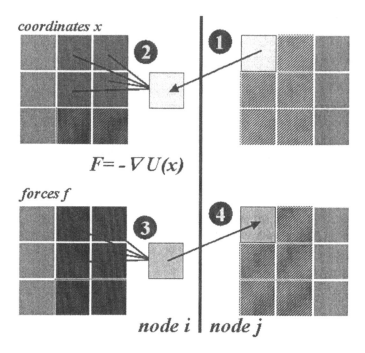

coordinates x

$$F = -\nabla U(x)$$

forces f

node i | *node j*

Figure 21.2 Force evaluation using global array communication consists of (1) asynchronous retrieval of coordinates by node i from node j, (2) calculation of all forces for which the retrieved coordinates are required, (3) fast accumulation of the local forces, and (4) asynchronous accumulation by node i of the forces on the remote node j.

it necessary to periodically and dynamically balance the computational load in order to reduce excessive synchronization times and increase parallel efficiency.

21.3 COMMUNICATION

The communication in the molecular dynamics module of NWChem is implemented using the Global Array tools (Nieplocha *et al.*, 1996). This public domain library allows physically distributed memory to be used as a single logical data object, using logical, topology independent array addressing for simple data communication as well as linear algebra operations. This library has been ported to a variety of architectures and is implemented using the appropriate message passing or shared memory primitives. An important aspect of communicating using global arrays is that remote memory access is one-sided and asynchronous.

Because communication is one-sided and asynchronous, data needed on one node can be retrieved by that node when it is needed, without the node on which the data exists being involved in the transfer. Figure 21.2 illustrates the steps involved in the retrieval of remote coordinates, the calculation of the forces, and the remote accumu-

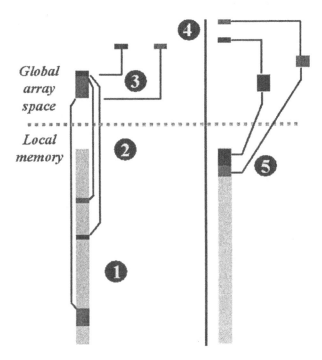

Figure 21.3 Redistribution of atoms consists of (1) re-determination of node ownership for each local atom, (2) copy of atomic data changing ownership into the local portion of a global array, (3) one-sided communication to the global array space of the receiving nodes of pointers to the atomic data that need to be moved, (4) global synchronization to ensure that all nodes copied data to be moved to the global array space, and communicated the pointers to that data to the receiving nodes, and (5) one-sided retrieval of atomic data from other nodes for which pointers were obtained.

lation of forces. All of these steps are done without synchronization or remote node involvement in initiating the data transfer.

Atoms change node ownership when they move from the domain assigned to one node to the domain assigned to another node. This requires a periodic redistribution of the atoms. This is implemented in NWChem such that point-to-point communication between nodes is only needed when atoms are changing ownership. As illustrated in Figure 21.3, the implementation always requires one global synchronization.

21.4 DYNAMIC LOAD BALANCING

The parallel efficiency of a spacial decomposition molecular dynamics algorithm can be increased by dynamic load balancing. In NWChem two complementary methods have been implemented. The first method resizes the physical space assigned to each node such that the domain of the busiest node is decreased, thereby increasing the domain size on all other nodes. This collective load balancing technique requires ad-

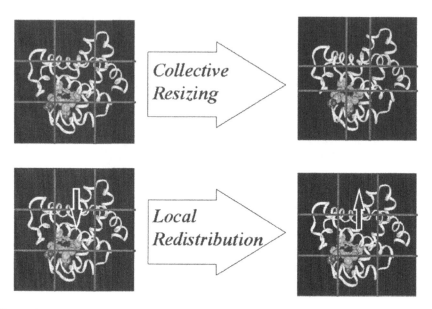

Figure 21.4   Dynamic load balancing techniques for spacial decomposition parallel molecular dynamics simulations: node domain resizing (top) and inter-node computation redistribution (bottom).

ditional communication of atom data. The second procedure transfers the calculation of forces between atoms on different nodes from the busiest node to the least busy neighboring node. This second method is local and requires only minimal additional communication. The two methods are illustrated in Figure 21.4 and can be used simultaneously in molecular simulations.

The basic components of a molecular dynamics step is illustrated in Figure 21.5. The calculation of the forces usually constitutes the bulk of the computation and can, with the global array tools, be performed with one-sided asynchronous communication only. By placing time stamps around the synchronization immediately following the calculation of the forces timing information is obtained that can be used in the load balancing procedure.

Particle-mesh Ewald summation (pme) is a method that allows the calculation of electrostatic forces and energies to be separated into explicitly calculated short-range local interactions and long-range interactions. The long-range interactions are approximated by a discrete convolution on an interpolating grid, using 3-dimensional fast Fourier transforms to perform the convolution efficiently. Molecular dynamics simulations using the pme method for long-range electrostatic interactions can be made more efficient by performing the calculation of energies and forces in reciprocal space on a subset of the available nodes (Essmann *et al.*, 1995). As illustrated in Figure 21.5, all nodes are involved in setting up the charge grid, but only a subset of the nodes perform the fast Fourier transforms and the computations in reciprocal space. By splitting

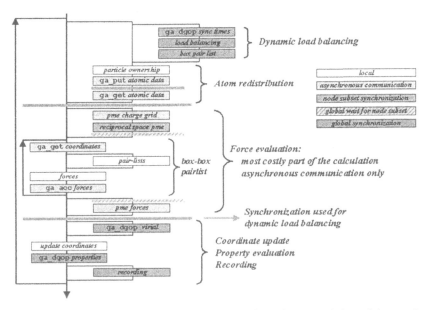

Figure 21.5   Molecular Dynamics Simulation basic flow chart, consisting of (top to bottom) periodic dynamics load balancing, atom redistribution, synchronization to ensure all coordinates are updated, force evaluation including particle-mesh Ewald long-range interactions and local and non-local forces, synchronization to ensure all forces are updated with time stamps used for load balancing, and coordinate and property updates.

this work from the calculation of the pme atomic forces, the nodes that are not involved in the reciprocal work can immediately continue with the real space forces.

21.5   PARALLEL EFFICIENCY

The parallel efficiency of the molecular dynamics module is illustrated by the timing and speedup as a function of the number of atoms in the chemical system and the number of nodes used for a molecular dynamics simulation of liquid water as shown in Figure 21.6. The speedup curves show that the spacial decomposition approach becomes increasingly efficient with increasing system size. In practice, for the IBM SP on which these benchmark calculation were performed, simulations will only be efficient with at least a few hundred atoms per node.

Timing and speedup data are given in Figure 21.7 for a simulation of an aqueous solution of haloalkane-dehalogenase, using the particle-mesh Ewald method to account for long-range energies and forces. This type of enzyme plays an important role in the detoxification of aliphatic halogenated hydrocarbons through replacement of the halogen substituent by a hydroxyl group under the release of halide ions (Pries *et al.*, 1995). The system consists of 41,259 atoms and was simulated using a cutoff radius of 1.0 nanometers on the IBM-SP.

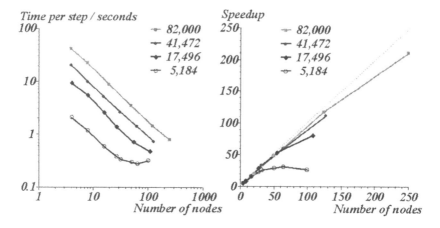

Figure 21.6   Timing and speedup information from benchmark simulations of liquid water with different system size as a function of the number of nodes used on an IBM SP.

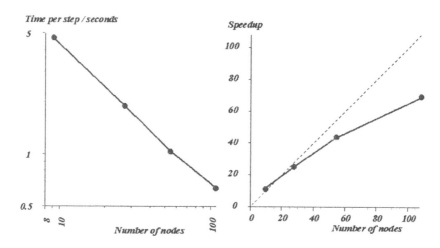

Figure 21.7   Timing and speedup information from benchmark simulations of the solvated enzyme haloalkane-dehalogenase.

Figure 21.8 gives the wall clock times and scaling for a molecular dynamics simulation of a 216,000 atom system of liquid octanol. A cutoff radius of 2.4 nanometers was used. The timings were done on a Cray T3E-900.

Acknowledgments

This work was performed under the auspices of the High Performance Computing and Communications Initiative of the Division of Mathematical, Information and Computational Sciences, Office of Computational and Technology Research, U.S. Department of Energy (U.S.DOE),

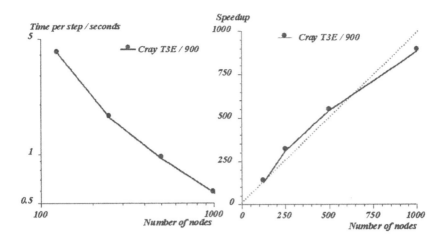

Figure 21.8   Timing and speedup information from benchmark simulations of liquid octanol.

and the U.S.DOE Office of Biological and Environmental Research (OBER), which funds the Environmental Molecular Sciences Laboratory (EMSL). This work was performed subject to contract DE-AC-6-76RLO 1830 with Battelle Memorial Institute, which operates the Pacific Northwest National Laboratory.

Development of NWChem and benchmark calculations were performed on the IBM-SP in the EMSL Molecular Science Computing Facility (MSCF), supported by OBER. The author thanks Cray Research, the supercomputer subsidiary of Silicon Graphics, for the opportunity to run benchmark calculations on a 1328 processor Cray T3E-900 at the Eagan, MN, research facilities.

References

Clark, T., Hanxleden, R., McCammon, J., and Scott, L. (1994). Parallelizing molecular dynamics using spatial decomposition. *Scalable High Performance Computing Conference Proceedings*, pages 95–102.

Essmann, U., Perera, L., Berkowitz, M., Dorden, T., Lee, H., and Pedersen, L. (1995). A smooth particle-mesh ewald method. *J. Chem. Phys.*, 103:8577–8593.

Lin, S., Mellor-Crummey, J., Pettitt, B., and Phillips, G. (1992). Molecular dynamics on a distributed-memory multiprocessor. *J. Comp. Chem.*, 13:1022–1035.

Nieplocha, J., Harrison, R., and Littlefield, R. (1996). Global arrays: A nonuniform memory access programming model for high-performance computers. *Journal of Supercomputing*, 10:197–220.

Plimpton, S. and Hendrickson, B. (1996). A new parallel method for molecular dynamics simulation of macromolecular systems. *J. Comp. Chem.*, 17:326–337.

Pries, F., Kingma, J., Krooshof, G., Jeronimus-Stratingh, C., Bruins, A., and D.B, J. (1995). Histidine 289 id essential for hydrolusis of the alkyl-enzyme intermediate of haloalkane dehalogenase. *J. Biol. Chem.*, 270:10405–10411.

22 USING GAUSSIAN GEMINAL FUNCTIONS IN MOLECULAR CALCULATIONS: INTEGRAL EVALUATION

Andrew Komornicki

Polyatomics Research Institute, U.S.A.

komornic@crl.com

Abstract: It has been known since the early days of quantum chemistry that a good representation of the cusp condition is a necessary condition for the very accurate evaluation of wavefunctions and the properties contained in them. An efficient way of introducing such character into a wavefunction was proposed almost 50 years ago by Boys who advocated the use of gaussian geminal functions, rather than the usual one electron functions. These functions explicitly contain dependence on the coordinates of two electrons. Cast in this form, there are many aspects to the correlation problem. We will focus on the mathematics, physics and computational implementation of very efficient algorithms for the evaluation of all the necessary integrals required in such calculations. Recursion relations will be discussed for two, three, and four electron integrals and extensions shown which enable one to compute arbitrary n-electron integrals. The computational consequences of such an approach will be discussed in detail given the rapidly changing complexion of high performance computing.

23 THE DEVELOPMENT OF PARALLEL GAMESS

Michael W. Schmidt

Iowa State University, U.S.A.

mike@si.fi.ameslab.gov

Abstract: This talk will describe the development of parallel capabilities with the quantum chemistry program GAMESS (General Atomic and Molecular Electronic Structure System) (Fletcher *et al.*, 1998). The conversion of a large sequential Fortran package to parallel computation involves an interplay between four factors, the driving force of which is the need to run larger chemical applications and the availability of personnel interested in parallel programming, the hardware of the day, and the software tools of the day.

GAMESS has experienced two main periods of parallel development, the first being 1992-1993 when Theresa Windus was interested in this type of programming, and simultaneously, our DoD research projects gave us access to the Touchstone Delta project. The clear choice for parallel message passing at the time was TCGMSG, a library which arose from the chemical community and which possessed some features for dynamic load balancing still absent from MPI. Theresa's work led to a robust parallel SCF code program, running to 64 or 128 nodes, some exploration of parallel Hessian computation, and a MCSCF program capable of running on a limited number of nodes, perhaps 8.

The current DoD CHSSI program has provided funds for a renewed round of code development since 1996, primarily by Graham Fletcher. The emphasis is now on migrating from a replicated data model to the use of the total memory of the parallel machine, by development of a Fortran-to-MPI interface known as Distributed Data Interface. DDI's concepts owe much to the TCGMSG library, which is the famous Global Array tool. DDI relies on vendor extensions for memory locking and access to remote data in distributed memory architectures, so in 1998 we still are not able to write portable message passing applications, in spite of the welcome spread of MPI as a portable library. Some examples of a parallel MP2 gradient program running on the T3E using SHMEM extensions to MPI will be given.

Finally, prospects for the future improvement in scalability of SCF level analytic Hessian computation and MCSCF wavefunction computation will be described. We anticipate that the CHSSI support for our current coding efforts, the DDI tool, and sufficient hard work should permit all important portions of GAMESS to scale well on parallel machines. This will permit the chemist to choose the wavefunction model and type of run based on the complexity of the system, not on some artificial consideration about which parts of the code might scale and which won't.

References

Fletcher, G., M.W.Schmidt, and M.S.Gordon (1998). Developments in parallel electronic structure theory. *Adv. Chem. Phys.* To appear.

24 APPLICATIONS/TECHNOLOGY PUSH/PULL: BUILDING COMPUTATIONAL INFRASTRUCTURE

Peter Taylor

San Diego Supercomputer Center, U.S.A.

taylor@sdsc.edu

Abstract: A key function of NSF's new Partnerships for Advanced Computational Infrastructure is the development and deployment of software. We will outline how activities in applications science and computing technologies are coupled together.to their mutual benefit as part of the National Partnership for Advanced Computational Infrastructure.

VII NUMA Architectures

25 A VIRTUAL CACHE SCHEME FOR IMPROVING CACHE-AFFINITY ON MULTIPROGRAMMED SHARED MEMORY MULTIPROCESSORS

Dongwook Kim[1], Eunjin Kim[2] and Joonwon Lee[3]

[1] Samsung Electronics Co., Ltd., Korea
[2] IBM Korea, Inc., Korea
[3] Computer Science Department,
Korea Advanced Institute of Science and Technology, Korea

dwkim@tongky.sec.samsung.co.kr

Abstract: A number of cache-affinity scheduling policies for multiprocessors have been proposed. However, most of them do not fully exploit cache-affinity, and several schemes are too complex to implement. To overcome these limitations, we propose a new virtual cache scheme for exploiting cache-affinity effectively. The key idea of our scheme is adopting a flag to indicate the amount of cache-affinity by referencing scheduling information presented by the OS. Every process has a flag that indicates cache-affinity in its PCB. The OS sets this flag whenever the time quantum of the corresponding process expires. This means that a large amount of affinity of the process remains in the cache. When the OS finds a process in the ready queue to assign to a processor, the process whose flag is set will be preferred. Moreover, a victim cache line to be replaced is selected by referencing the PIDs of migrated processes or terminated processes, and thus more optimal cache line replacement can be achieved. The cache-affinity can be improved by the new cache line replacement policy. Trace-driven simulation results show that the new scheme is superior to other compared schemes.

26 STABLE PERFORMANCE FOR CC-NUMA USING FIRST-TOUCH PAGE PLACEMENT AND REACTIVE PROXIES

Sarah A. M. Talbot and Paul H. J. Kelly

Department of Computing,
Imperial College of Science, Technology and Medicine, Great Britain

samt@doc.ic.ac.uk
phjk@doc.ic.ac.uk

Abstract: A key problem for shared-memory systems is unpredictable performance. A critical influence on performance is page placement: a poor choice of home node can severely degrade application performance because of the increased latency of accessing remote rather than local data. Two approaches to page placement are the simple policies "first-touch" and "round-robin", but neither of these policies suits all applications. We examine the advantages of each strategy, the problems that can result from a poor choice of placement policy, and how these problems can be alleviated by using proxies. Proxies route remote read requests via intermediate nodes, where combining is used to reduce contention at the home node. Our simulation results indicate that by using reactive proxies with first-touch page placement, performance is always better than using either page placement policy without proxies. These results suggest that the application programmer can obtain stable performance without knowing the underlying implementation of cc-NUMA, and can avoid time-consuming performance tuning.

Keywords: cache coherence protocols, shared-memory, combining, page placement, CC-NUMA.

26.1 INTRODUCTION

Unpredictable performance anomalies have hampered the acceptance of cache-coherent non-uniform memory access (cc-NUMA) shared-memory architectures. One source of performance problems is the location of shared data: each page of shared data is allocated in distributed memory at a home node by the operating system when it is first accessed. The choice of home node for a page is commonly decided using a first-touch or round-robin algorithm, but neither of these policies is suited to all applications. A poor choice of page placement policy can have a marked effect on the performance of an application.

In this paper we examine the effects of simple page placement, and describe how a technique for reducing read contention in cc-NUMA machines can alleviate problems with inappropriate page placement. In the proxy protocol, we associate a small set of nodes with each location in shared memory, which act as intermediaries for remote read requests. Using execution-driven simulations, we show that using the *reactive* variant of proxies, in conjunction with first-touch page placement, yields performance which is always better than using either of the simple page placement strategies without proxies. This suggests that by using first-touch page placement with reactive proxies, application programmers can be confident that they will obtain stable performance.

The rest of the paper is structured as follows. Page placement is discussed in Section 26.2, and the proxy protocol is explained in Section 26.3. We describe our simulated architecture and experimental design in Section 26.4, and present the results in Section 26.5. The relationship to previous work is discussed in Section 26.6. In Section 26.7 we summarize our conclusions and give pointers to further work.

26.2 PAGE PLACEMENT POLICIES

In distributed shared memory multiprocessors, shared data is partitioned into virtual memory pages. Each page of shared data is then physically allocated to a home node by the operating system, as a result of the page fault that occurs on the first access to the page. The choice of home node is commonly decided using a first-touch or round-robin algorithm. First-touch allocates the page to the node which first accesses it, and this strategy aims to achieve data locality. It is important to distinguish between naïve first-touch and first-touch-after-initialization policies. A naïve policy will allocate pages on a first-touch basis from the start of program execution. This is a problem for applications where one process initializes everything before parallel processing commences, because all the pages end up on the same node (with overflow to its neighbors). It is better to use a first-touch-after-initialization policy, where shared memory pages are only permanently allocated to nodes once parallel processing has commenced, and this is the policy we use.

In the round-robin approach, allocation cycles around the nodes, placing a page in turn at each node. This approach distributes the data more evenly around the system. Unfortunately, for applications which have been written with locality in mind, it is likely that few, if any, of the pages accessed by a node will be allocated to it. As a result, first-touch is generally the default page placement policy, with round-robin being available as an option for improving the performance of some applications (*e.g.* on SGI's Origin 2000 (Laudon and Lenoski, 1997)).

The operating system may also provide facilities for pages to migrate to a new home node, or have copies of the page at other nodes. In recent years, there has been much discussion of dynamic page migration and replication schemes, mainly in the context of distributed virtual shared memory (DVSM), where the software implementation of shared memory on a message-passing distributed memory architecture mandates the movement and/or copying of pages between processing nodes (*e.g.* Munin (Carter *et al.*, 1995)). In contrast, where dynamic paging features are available on cc-NUMA systems, they are usually implemented as options; they are not the default because of the overheads of capturing and acting upon access patterns (Verghese *et al.*, 1996). There is the problem that if two or more processors update data on the same page, then the page may "ping pong" between the new homes, or alternatively the coherence traffic will increase greatly. Even when a page is migrated to the node which uses it most, the average memory access times of other nodes may increase, and reacting too late to the need to migrate or replicate a page may be completely useless, and even costly (LaRowe Jr and Ellis, 1991). In addition, even when all the migrations and replications are chosen correctly, but they occur in bursts, performance may suffer due to page fault handler, switch, and memory contention.

Given the pitfalls of dynamic page placement, can we get reasonable performance on cc-NUMA machines using simple page placement policies? The best performance for shared memory machines can be obtained by tuning programs based on page placement, but an important principle of shared memory is that it provides application programmers with a simple programming model, where they do not have to worry about the underlying machine, and which leads to more portable programs. We want to keep programmer involvement in page placement to a minimum, but still get reasonable performance.

26.3 PROXIES

Proxying is an extension to standard distributed directory cache coherence protocols, and is designed to reduce node controller contention (Bennett *et al.*, 1996). Normally, a read request message would be sent to the location's home node, based on its physical page address. With proxying, a read request is directed instead to another node controller, which acts as an intermediary. Figure 26.1 illustrates how more than one proxy could serve read requests for a given location, each handling requests from a different processor subset. In this paper, the nodes are split into three proxy clusters (*i.e.* subsets): this split was chosen after experimentation and represents the best balance for our simulated system between contention for the proxy nodes and the degree

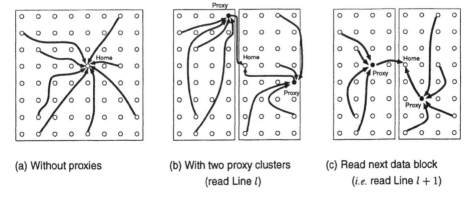

(a) Without proxies   (b) With two proxy clusters   (c) Read next data block
                           (read Line *l*)   (*i.e.* read Line *l* + 1)

Figure 26.1 Contention is reduced by routing reads via a proxy.

(a) First request to proxy has to be forwarded to the home node:

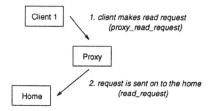

(b) Second client request, before data is returned, forms pending chain:

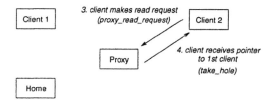

(c) Data is passed to each client on the pending chain:

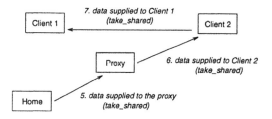

Figure 26.2   Combining of proxy requests.

of combining. The mapping of each data block to its proxy node ensures that requests for successive data blocks are served by different proxies. This balances the queuing of read request messages across the input buffers of all the node controllers.

If the proxy node has a copy of the requested data block in its cache, it replies directly. If not, it requests the value from the home node, allocates the copy in its own cache and replies to the client. Any requests for a particular block which arrive at a proxy before it has obtained a copy from the home node are added to a distributed chain of pending requests for that block, and the reply is forwarded down the pending chain, as illustrated in Figure 26.2. The current implementation is slightly unfair in that the first client will be the last to receive the data. We are investigating the tradeoff between increasing the hardware overhead to hold an additional pointer to the tail of each pending proxy chain, and any performance benefits.

Proxying requires a small amount of extra store to be added to each node controller. We need to identify the data blocks for which a node is currently obtaining data as a proxy, and hold the head of each pending proxy chain. The node controller also has to handle the new proxy messages and state changes. We envisage implementing this in software on a programmable node controller.

In the basic form of proxies, the application programmer uses program directives to mark data structures for handling by the proxy protocol – all other shared data will be exempt from proxying. If the application programmer makes a poor choice of data structures, then the overheads incurred by proxies may outweigh any benefits and degrade performance. These overheads include the extra work done by the proxy nodes handling the messages, proxy node cache pollution, and longer sharing lists. In addition, the programmer may not mark data structures that would benefit from proxying.

Reactive proxies, which take advantage of the finite buffering of real machines, overcome these problems and do not need application program directives. When a remote read request reaches a full buffer, it will immediately be sent back across the network. When the originator receives the bounced message and the reactive proxies protocol is in effect, the arrival of the buffer-bounced read request will trigger a proxy read (see Figure 26.3). A proxy read is only done in direct response to the arrival of a buffer-bounced read request, so as soon as the queue length at the destination node has reduced to below the limit, read requests will no longer be bounced and no proxying will be employed.

26.4 SIMULATED ARCHITECTURE AND EXPERIMENTAL DESIGN

This paper uses results obtained from our execution-driven simulations of a cc-NUMA system. Each node contains a processor with an integral first-level cache (FLC), a large second-level cache (SLC), memory (DRAM), and a node controller (see Figure 26.4). The node controller receives messages from, and sends messages to, both the network and the processor. The SLC, DRAM, and the node controller are connected using two decoupled buses. This decoupled bus arrangement allows the processor to access the SLC at the same time as the node controller accesses the DRAM. We simulate in detail the contention between the local CPU and the node controller for the buses, between

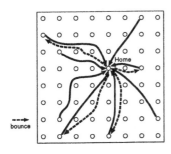

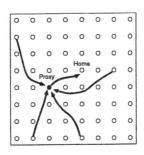

(a) Input buffer full, some read requests bounce   (b) Reactive proxy reads

Figure 26.3   Bounced read requests are retried via proxies.

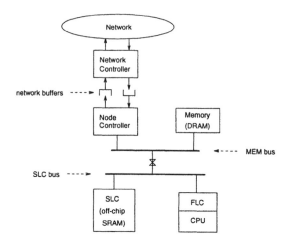

Figure 26.4   The architecture of a node.

the CPU and incoming messages for the node controller, and the use of the SLC to hold proxy data. Table 26.1 summarizes the specifications of the architecture. We simulate a direct-mapped cache, but note that its large size (4 Mb) will have a miss rate roughly equivalent to a 2-way associative cache of half that size (Hennessy and Patterson, 1996).

We simulate a simplified interconnection network, which follows the *LogP* model (Culler *et al.*, 1993). We have parameterized the network and node controller as follows:

- L: the latency experienced in each communication event: 10 cycles for long messages (which include 64 bytes of data, *i.e.* one cache line), and 5 cycles for all other messages.

- o: the occupancy of the node controller. We have adapted the LogP model to recognize the importance of the *occupancy* of a node controller, rather than just the overhead of sending and receiving messages (Holt *et al.*, 1995). The parameters are given in Table 26.2.

- g: the gap between successive sends or receives by a processor: 5 cycles.

- P: the number of nodes: set to 64.

We limit our finite length input message buffers to eight for read requests. There can be more messages in an input buffer, but once the queue length has risen above eight, all read requests will be bounced back to the sender until the queue length has fallen below the limit. This is done because we are interested in the effect of finite buffering on read requests rather than all messages, and we wished to be certain that all transactions would complete in our protocol. The queue length of $\sqrt{P}$, where P is the number of processing nodes, is an arbitrary but reasonable limit, and was chosen to reflect the limitations in queue length that one would expect in large cc-NUMA configurations.

Each cache line has a home node (at page level) which: (1) either holds a valid copy of the line (in SLC and/or DRAM), or knows the identity of a node which does have a valid copy (the owner); (2) has guaranteed space in DRAM for the line; and (3) holds directory information for the line (head and state of the sharing list).

The benchmarks and their parameters are summarized in Table 26.3. GE is a simple Gaussian elimination program, similar to that used to study eager combining (Bianchini and LeBlanc, 1994). We chose this benchmark because it is an example of widely-shared data and should benefit from using proxies, but we also wanted to observe how its performance would be affected by the two simple page placement policies. GE is interesting because it is a relatively long-running iterative code, where first-touch page placement becomes increasingly inappropriate over the execution time.

We selected six applications from the SPLASH-2 suite (Woo *et al.*, 1995), to give a cross-section of scientific shared memory applications. We used both Ocean benchmark applications in order to study the effects of page placement and proxies on the "tuned for data locality" and "easy to understand" variants. The Ocean-Contig implementation allows the grid partitions to be allocated contiguously and entirely in the local memory of the processors that "own" them, improving data locality but increasing algorithm complexity. In contrast, Ocean-Non-Contig implements the grids as 2-D arrays which prevents the partitions being allocated contiguously, but it is easier to understand and program. Other work which only refers to Ocean can be assumed to be using Ocean-Contig.

Table 26.1 Details of the simulated architecture.

Hardware	Details	
CPU	CPI	1.0
	Instruction set	based on DEC Alpha
Instruction cache	All instruction accesses assumed primary cache hits	
First level data cache	Capacity	8 Kbytes
	Line size	64 bytes
	Direct mapped, write-through	
Second-level cache	Capacity	4 Mbytes
	Line size	64 bytes
	Direct mapped, write-back	
DRAM	Capacity	Infinite
	Page size	8 Kbytes
Node controller	Non-pipelined	
	Service time and occupancy	See Table 26.2
	Cycle time	10ns
Interconnection network	Topology	full crossbar
	Incoming message queues	8 read requests
Cache coherence protocol	Invalidation-based, sequentially-consistent cc-NUMA. Home nodes allocated on a "first-touch-after-initialization" or "round-robin" basis. Distributed directory, based on the Stanford Distributed-Directory Protocol (Thapar and Delagi, 1990), using singly-linked sharing list.	

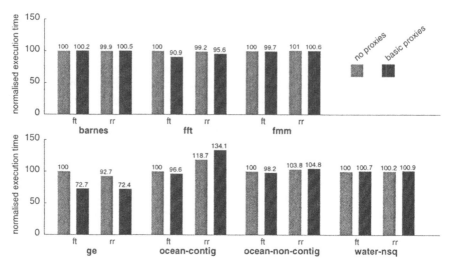

Figure 26.5   Relative performance, 64 nodes, infinite buffers.

26.5   EXPERIMENTAL RESULTS

In this section, we present the results obtained from our simulations, and discuss the benefits and potential drawbacks of using proxies in conjunction with a default page placement policy. Because first-touch is often the default page placement policy, we have normalized the results with respect to first-touch. The results for each benchmark are expressed as percentages, *e.g.* it will be 100% for the relative execution time of each benchmark running with first-touch page placement and without proxies. The "number of messages" reflects the total number of messages sent. The "remote read response" measures the delay from sending a read request message to the receipt of the data.

Table 26.2   Latencies of the most important node actions.

Operation	Time (cycles)
Acquire SLC bus	2
Release SLC bus	1
SLC lookup	6
SLC line access	18
Acquire MEM bus	3
Release MEM bus	2
DRAM lookup	20
DRAM line access	24
Initiate message send	5

Table 26.3 Benchmark applications.

Application	Problem Size	Shared Data Marked for Basic Proxying
Barnes	16K particles	all
FFT	64K points	all
FMM	8K particles	f_array (part of G_Memory)
GE	512 x 512 matrix	entire matrix
Ocean-Contig	258 x 258 ocean	q_multi and rhs_multi
Ocean-Non-Contig	258 x 258 ocean	fields, fields2, wrk, and frcng
Water-Nsq	512 molecules	VAR and PFORCES

26.5.1 Infinite Buffers

Without proxies, GE performs better with round-robin page placement, showing a
7.3% speedup over first-touch (see Figure 26.5). This was expected because the re-
sponsibility for updating rows of the matrix shifts during execution, so the first-touch
placement slowly becomes inappropriate. Also, the remote access bottlenecks get
worse towards the end of execution, as the accesses concentrate on fewer and fewer
home nodes. In contrast, the round-robin strategy will have worse locality at the start
of parallel execution, but it does not have the later problem of access concentration.
Figure 26.6 shows that for GE the overall number of messages increases by nearly
10% when round-robin is used, which reflects the increase in remote access requests
owing to the loss of locality. However, because it avoids first-touch's problem of an
increasing concentration of requests as the algorithm progresses, the service time for
remote reads improves by 19% (as shown in Figure 26.7).

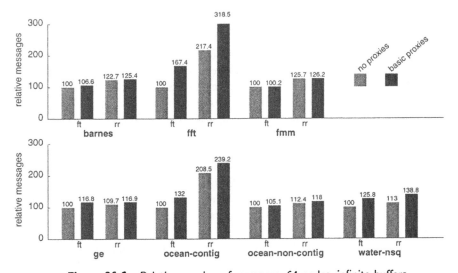

Figure 26.6 Relative number of messages, 64 nodes, infinite buffers.

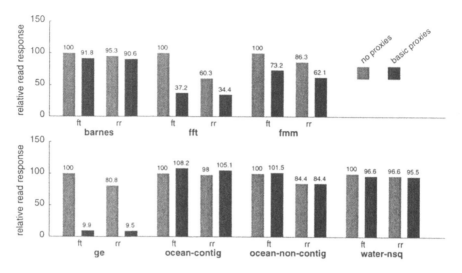

Figure 26.7   Relative remote read response times, 64 nodes, infinite buffers.

The performance of GE improves even more by using proxies. Both placement policies achieve more than 27% speedup over first-touch without proxies. Using proxies increases the overall number of messages by 17% in comparison to just first-touch due to the additional proxy read requests and acknowledgments. However, there is a dramatic reduction in the remote read response time of around 90%, because the combining of requests at proxies reduces the read messages queuing at the home node(s).

In contrast, the performance results for Ocean-Contig show first-touch as the best page placement policy, because this application has been written to exploit data locality. Round-robin leads to more remote reads, degrading performance to be nearly 19% worse than with first-touch. Using proxies makes the performance even worse for round-robin, increasing both mean remote read response time and the total number of messages, and these extra messages cause the network to overload. The best performance for the application is obtained using proxies with first-touch page placement.

FMM, Ocean-Non-Contig, and Water-Nsq perform marginally better with first-touch page placement, whereas Barnes and FFT perform marginally better with round-robin. In addition, FFT, FMM, and Ocean-Non-Contig get their best performance using first-touch page placement with proxies, with speedups of 9%, 0.3%, and 1.8% respectively. However, the performance of both Barnes and Water-Nsq suffers when basic proxies are used, showing slowdowns of 0.2% and 0.7%. These slight drops in performance illustrate the main pitfall of basic proxies, *i.e.* a poor choice of data marked for proxying.

The introduction of finite buffers favors the first-touch page placement policy (see Figure 26.8). The round-robin policy suffers because its lack of locality results in more remote access requests, which increases the chance that an input buffer already has eight or more messages. Hence, it is more likely that a read request will be bounced.

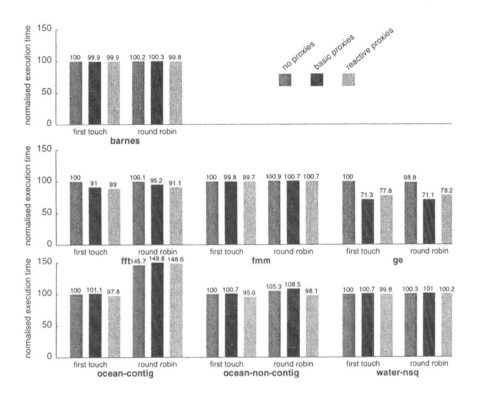

Figure 26.8   Relative performance, 64 nodes, finite buffers.

For six of the seven benchmark applications, round-robin page placement performs worse than first-touch, and for GE, the performance benefit has been cut to 1.2%.

26.5.2   Finite Incoming Message Buffers

Looking at the results with proxies, reactive proxies have the advantage that they handle all cases where contention occurs, but this is at the cost of a delay while the original read request is bounced by the home node. GE illustrates this change in behavior (see Figure 26.9), where the use of proxies results in fewer messages because they stop the repeated sending and bouncing of read requests. However, the relative read response time does not improve as much for reactive as it does for basic proxies, as is shown in Figure 26.10. This is because of the initial delay, of read request and bounce, before a proxy read request is made. In addition, for GE, the mean input buffer queuing cycles increase with the introduction of proxies. This is because there are now more messages being accepted into the buffers rather than bouncing (see Table 26.4).

Using reactive proxies in conjunction with first-touch page placement results in the best performance for six of the seven benchmarks. The exception is GE; as we have already noted it is particularly well-suited to the targeted approach of basic proxies. However, it still shows a marked performance improvement of 22.2% with reactive

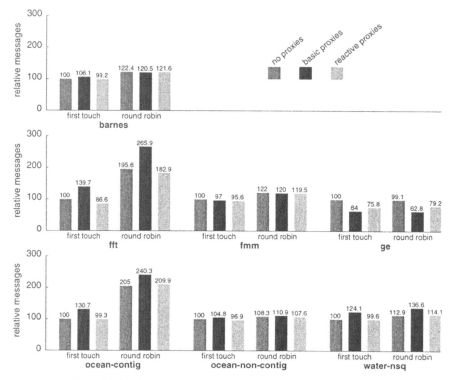

Figure 26.9   Relative number of messages, 64 nodes, finite buffers.

proxies. This suggests that a default policy of first-touch page placement with reactive proxies will give stable performance, and it avoids the more spectacular performance pitfalls that can occur using round-robin, such as occur for Ocean-Contig (Figure 26.8).

26.5.3 Summary

The choice of the best simple page placement policy, in the absence of any additional mechanism such as proxies or dynamic page migration/replication, depends on the individual applications. Some applications, such as our GE benchmark, suit the even distribution of shared data given by round-robin. Other applications, such as Ocean-Contig, have been specifically written to exploit data locality, and so suit the first-touch policy. This leaves the problem that the best performance may only be obtained by experienced programmers who know which page placement policy to choose, or by engaging in time-consuming performance tuning.

The use of proxies alters this situation. Proxies introduce a finer-grained sharing, at the level of data blocks rather than pages, and reduce queuing (as shown in Table 26.4). For basic proxies this can be detrimental to performance where inappropriate data structures are marked as "hot", because every load (for addresses subject to proxying) goes via a proxy, whereas without proxies no indirection would be involved.

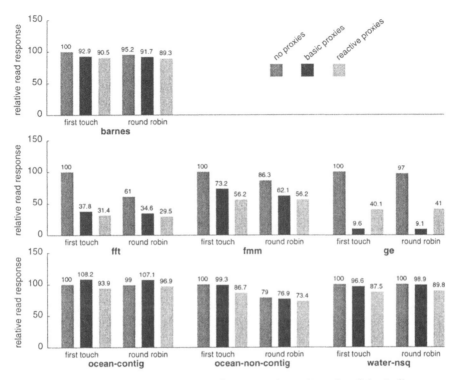

Figure 26.10   Relative remote read response times, 64 nodes, finite buffers.

Using reactive proxies has the benefit that proxies are only used when contention occurs at run-time: for six of our benchmarks this, in conjunction with first-touch page placement, resulted in their best performance. Water-Nsq illustrated this, where basic proxies degraded performance, whereas reactive proxies improved performance. Most importantly, first-touch page placement in conjunction with reactive proxies always resulted in performance that was better than either page placement policy without proxies. There are some applications, such as GE, where using reactive rather than basic proxies results in a smaller performance improvement because of the delay in invoking proxies. However, the application still showed a noticeable speedup over not using proxies.

26.6 RELATED WORK

The effects of page placement policy have been investigated for shared memory architectures where the coherence protocol is implemented in hardware (for CC-NUMA) or in software (for DVSM) (Marchetti *et al.*, 1995). Using a base policy of round-robin, they also considered a first-touch-after-initialization scheme, a dynamic-migration scheme, and three replication schemes. Their first-touch scheme improved the performance of all their applications compared to naïve first-touch, regardless of whether coherence was maintained in hardware or software. For their five applications, there was no per-

Table 26.4   Mean input buffer queuing cycles.

	Infinite Buffers				Finite Input Buffers					
	no proxies		basic		no proxies		basic		reactive	
	ft	rr	ft	rr	ft	rr	ft	rr	ft	rr
Barnes	3.49	2.55	2.57	2.45	1.70	1.55	1.88	1.71	1.55	1.48
FFT	70.98	28.83	7.74	7.01	11.83	7.82	6.96	6.13	7.00	5.92
FMM	19.87	11.82	11.47	5.58	5.15	4.52	4.30	3.98	3.93	3.78
GE	301.60	213.62	15.95	13.13	38.59	37.23	10.01	8.11	60.77	59.06
OceanC	5.80	5.21	6.01	5.28	4.61	4.80	5.17	5.19	4.40	4.89
OceanN	15.21	11.03	15.93	11.14	9.83	9.45	10.34	9.09	8.72	8.49
Water	6.37	4.82	4.89	4.07	4.20	3.71	3.67	3.48	3.95	3.55

formance benefit from the dynamic migration or replication schemes. Their choice of benchmarks was limited, in that they did not include an example of algorithms such as Gaussian Elimination which are better suited to round-robin page placement.

A study of dynamic page migration and replication was performed on the Stanford FLASH and distributed FLASH. The study considered three dynamic page placement policies, migration and/or replication, and three simple policies of round-robin, first-touch, and *post facto* (Verghese *et al.*, 1996). They found that first-touch always gave better performance than round-robin for their workloads, and that *post facto* was the best simple policy. Their dynamic policies generally obtained better performance than first-touch, and were never worse, even given the overheads associated with implementing the dynamic policies. However, three of their five workloads were multiprogrammed, and this put their first-touch policy at a distinct disadvantage. For example, in their SPLASH workload the jobs were redistributed across the processors as applications entered and left the system, but they did not re-invoke first-touch page placement when a job migrated.

Proxies allow read requests for data to be combined in controllers away from the home node. This is a restricted instance of the combining of atomic read-modify-write operations (*e.g.* as proposed for the NYU Ultracomputer (Gottlieb *et al.*, 1983)), although proxies retain the data in cache, which allows for more combining. Caching extra copies of data to speedup retrieval time for remote reads has been explored for hierarchical architectures (*e.g.* in the Swedish Institute of Computer Science DDM (Haridi and Hagersten, 1989)). The proxies approach is different because it does not use a fixed hierarchy. Instead it allows requests for copies of successive data lines to be serviced by different proxies.

Eager combining uses intermediate nodes which act like proxies for "hot" pages, *i.e.* the programmer is expected to mark data structures (Bianchini and LeBlanc, 1994). Unlike proxies, their choice of server node is based on the page address rather than data block address. In addition, their scheme eagerly updates all proxies whenever a newly-updated value is read, unlike our protocol where data is allocated in proxies on demand, which reduces cache pollution.

The GLOW extensions for widely-shared data are, like proxies, designed to be added to existing cache coherence protocols (Kaxiras and Goodman, 1996). GLOW uses agents to intercept requests for widely-shared data at selected network switch nodes.

At present, GLOW requires application program directives to identify widely-shared data.

26.7 CONCLUSIONS

We have used execution driven simulations to study the benefits of using proxies and simple page placement. Our results confirm that there is no ideal default policy for page placement. However, by using reactive proxies with first-touch page placement, we obtained better performance than using either page placement policy without proxies. This suggests that, with a default of first-touch page placement with reactive proxies, application programmers can be confident that they will obtain stable performance. The programmer will not have to worry about the cc-NUMA implementation, and will rarely have to do time-consuming performance tuning.

There are some overheads associated with proxies. As we noted in Section 26.3, there are the costs of implementing proxies: in hardware to hold the head of each pending proxy chain, and in software to handle the additional message types and state changes. In addition, there will be cache pollution (because allocating a proxy copy in the cache may displace another line), with invalidation overhead for the displaced line, and possibly a later cache miss. However, these costs may be balanced by the considerable benefits of performance stability, the promise of architecture-independent application programs, and the saving of performance tuning effort.

Study of further benchmarks will provide deeper insight into the trade-offs, and in particular we are looking for applications which have not been carefully optimized for existing architectures. We are currently investigating the cache pollution effect, by examining both a "no-allocate" proxy scheme (where nodes do not cache the proxy lines), and the use of a separate proxy cache. We plan to continue our simulation work to evaluate how changing the architectural balance (*e.g.* slower interconnection networks) affects our conclusions.

Acknowledgments

This work was funded by the U.K. Engineering and Physical Sciences Research Council: through the CRAMP project (GR/J 99117), and a Research Studentship. We would also like to thank Ashley Saulsbury for the ALITE simulator, and Andrew Bennett and the anonymous referees for their comments on this work.

References

Bennett, A., Kelly, P., Refstrup, J., and Talbot, S. (1996). Using proxies to reduce cache controller contention in large shared-memory multiprocessors. In Bougé, L. *et al.*, editors, *Euro-Par 96 European Conference On Parallel Architectures*, volume 1124 of *Lecture Notes in Computer Science*, pages 445–452, Lyons. Springer–Verlag.

Bianchini, R. and LeBlanc, T. (1994). Eager combining: a coherency protocol for increasing effective network and memory bandwidth in shared-memory multiprocessors. In *6th IEEE Symposium on Parallel and Distributed Processing (SPDP)*, pages 204–213, Dallas.

Carter, J., Bennett, J., and Zwaenepoel, W. (1995). Techniques for reducing consistency-related communication in distributed shared memory systems. *ACM Trans. on Computer Systems*, 13(3):205–243.

Culler, D., Karp, R., Patterson, D., Sahay, A., Schauser, K., Santos, E., Subramonian, R., and von Eicken, T. (1993). LogP: Towards a realistic model of parallel computation. *ACM Sigplan Notices (4th Symposium on Principles and Practice of Parallel Programming)*, 28(7):1–12.

Gottlieb, A., Grishman, R., Kruskal, C., McAuliffe, K., Rudolph, L., and Snir, M. (1983). The NYU Ultracomputer – designing an MIMD shared memory parallel computer. *IEEE Trans. on Computers*, C-32(2):175–189.

Haridi, S. and Hagersten, E. (1989). The cache coherence protocol of the Data Diffusion Machine. In *PARLE 89 Parallel Architectures and Languages Europe*, volume 365 of *Lecture Notes in Computer Science*, pages 1–18, Eindhoven. Springer–Verlag.

Hennessy, J. and Patterson, D. (1996). *Computer Architecture: A Quantitative Approach*. Morgan Kaufman, 2nd edition.

Holt, C., Heinrich, M., Singh, J., Rothberg, E., and Hennessy, J. (1995). The effects of latency, occupancy and bandwidth in distributed shared memory multiprocessors. Technical Report CSL-TR-660, Computer Systems Laboratory, Stanford University.

Kaxiras, S. and Goodman, J. (1996). The GLOW cache coherence protocol extensions for widely shared data. In *10th ACM International Conference on Supercomputing*, pages 35–43, Philadelphia.

LaRowe Jr, R. and Ellis, C. S. (1991). Experimental comparison of memory management policies for NUMA multiprocessors. *ACM Trans. on Computer Systems*, 9(4):319–363.

Laudon, J. and Lenoski, D. (1997). The SGI Origin 2000: A CC-NUMA highly scalable server. *Computer Architecture News (Proceedings of the 24th Annual International Symposium on Computer Architecture)*, 25(2):241–251.

Marchetti, M., Kontothanassis, L., Bianchini, R., and Scott, M. (1995). Using simple page placement policies to reduce the cost of cache fills in coherent shared-memory systems. In *9th International Parallel Processing Symposium (IPPS)*, pages 480–485, Santa Barbara.

Thapar, M. and Delagi, B. (1990). Stanford distributed-directory protocol. *IEEE Computer*, 23(6):78–80.

Verghese, B., Devine, S., Gupta, A., and Rosenblum, M. (1996). Operating system support for improving data locality on CC-NUMA compute servers. *ACM Sigplan Notices (ASPLOS-VII)*, 31(9):279–289.

Woo, S., Ohara, M., Torrie, E., Singh, J., and Gupta, A. (1995). The SPLASH-2 programs: Characterization and methodological considerations. *Computer Architecture News (22nd Annual International Symposium on Computer Architecture)*, 23(2):24–36.

27 A COMPARATIVE STUDY OF CACHE-COHERENT NONUNIFORM MEMORY ACCESS SYSTEMS

Gheith A. Abandah and Edward S. Davidson

Electrical Engineering and Computer Science Department,
University of Michigan, U.S.A.

gabandah@eecs.umich.edu
davidson@eecs.umich.edu

Abstract: We present a comparative study of three important CC-NUMA implementations—Stanford DASH, Convex SPP1000, and SGI Origin 2000—to find strengths and weaknesses of current implementations. Although the three systems share many similarities, they have significant architectural differences that translate into large performance differences. These include the number of processors per node, cache configuration, memory consistency model, location of memory in the node, and cache-coherence protocol. In this study, we evaluate the effects of these differences on cache misses, miss time, and local and internode traffic.

We first model the three systems according to their original parameters, and show that they have large performance differences due to using different component speeds and sizes. We then put the three systems on the same technological level by assigning them components of similar size and speed, preserving their organization and coherence protocol differences. Although the normalized Origin 2000 has the least average remote time, it spends the longest time satisfying its misses because most of them are remote. DASH's Illinois protocol and SPP1000's interconnect cache reduce their remote misses. The SPP1000 has the highest average remote time because its coherence protocol requires more signals to satisfy a miss than either of the other two protocols. DASH achieves lower miss time and its relaxed memory consistency model hides some of its miss time.

Keywords: CC-NUMA multiprocessors, performance evaluation, system architecture, coherence protocols, Stanford DASH, Convex SPP1000, SGI Origin 2000.

27.1 INTRODUCTION

An important approach in building scalable shared-memory multiprocessors is the *cache-coherent non-uniform memory access* (CC-NUMA) architecture. CC-NUMA systems use high-bandwidth, low-latency interconnection networks to connect powerful processing nodes that contain processors and memory (Lenoski and Weber, 1995). Each node has a *cache coherence controller* (CCC) that enables coherent data replication. When a cache line is updated, the CCC invalidates other copies, and ensures that a processor request always gets the most recent data. CC-NUMA multiprocessors provide the convenience of memory sharing with one global address space, some portion of which is found in each node. CC-NUMA has a good balance between programming complexity and hardware complexity. Several vendors are adopting it for their new high-end servers, *e.g.* HP/Convex SPP2000, Sequent NUMA-Q, and SGI Origin 2000.

In an effort to identify the strengths and weaknesses of current approaches, we evaluate three important CC-NUMA systems: (i) Stanford DASH, a prototype of the first project to demonstrate a scalable shared-memory system, (ii) Convex SPP1000, the first commercially available CC-NUMA system, and (iii) SGI Origin 2000 which represents today's state-of-the-art technology. Although the systems share many similarities, they have significant architectural differences that translate into large performance differences. They all cluster processors and memory into nodes that are interconnected by a low-latency interconnection network. Each node has a coherence controller, memory, and multiple processors. The systems differ in many respects; for example, in the number of processors per node, processor cache configuration, memory consistency model, location of memory in the node, and cache-coherence protocol. Sections 27.2, 27.3, and 27.4 present overviews of the three systems.

This study used a set of tools that were developed for analyzing shared-memory applications and supporting the design of new scalable shared-memory systems (Abandah, 1998). Running on a Convex SPP1600, a multiprocessor tracing tool, SMAIT, instruments shared-memory applications. At run-time, the instrumented applications pipe detailed traces to two analysis tools (CIAT and CDAT). CIAT's analysis, which is configuration independent, is used first to understand the inherent application properties (Abandah and Davidson, 1998). Then CDAT's configuration dependent analysis is used to simulate applications on models similar to the three case-study systems.

CDAT is a system-level simulator which assumes that each instruction takes one cycle as long as it does not generate a miss in the secondary cache; the processor stalls on these misses. CDAT pays special attention to secondary cache misses since they cause system traffic. CDAT simulates the signals among the system components needed to satisfy cache misses, and accurately models the cycles required by the critical path to satisfy each miss. CDAT assumes that other processors do not change the global state of a cache line while serving a processor miss for this line. This simplification has negligible performance effects on the applications used here as they do not suffer from false sharing. Although CDAT does not model contention, it characterizes system traffic to enable assessing potential contention. CDAT is a flexible tool that is easily retargeted to a new configuration by a configuration file. The configuration file allows the system designer to select among a wide range of design options.

In this study, we use two representative applications from Convex's implementation of the NAS Parallel Benchmarks. For a fair system comparison, we undid the SPP1600-specific optimizations in these benchmarks. The two applications, CG and SP, have interesting differences and are further characterized in (Abandah and Davidson, 1998). CG performs simple reduction operations on long vectors; its high degree of shared communication benefits the updating and reading of the partial products. SP performs computations on three-dimensional matrices, with a producer-consumer communication on the boundaries of matrix partitions. The data presented here is from the main parallel phase when solving the "A" problem size on 16 processors.

Section 27.5 presents a raw comparison where CDAT analyzes the performance of the two benchmarks on models that are like the three systems with their original parameters. It shows that the performance of the three systems spans a wide range because their components vary widely in size and speed, as these systems were introduced at different times with the then-prevailing technology. Section 27.6 presents a normalized comparison where the systems are put on the same technological level with similar components. We believe that the normalized comparison better exposes the performance differences due to system organization and cache-coherence protocol, rather than the underlying technology and component sizes. Some conclusions are drawn in Section 27.7.

27.2 STANFORD DASH

The Stanford DASH project started in 1988, and the first prototype was completed in 1991 (Lenoski *et al.*, 1992). The DASH is based on the SGI POWER Station 4D/340 which has 4 processor boards connected by a cache-coherent bus. As shown in Figure 27.1(a), the memory and the I/O interface are also connected to the shared bus. Two boards were added to handle remote memory access and global cache coherence.

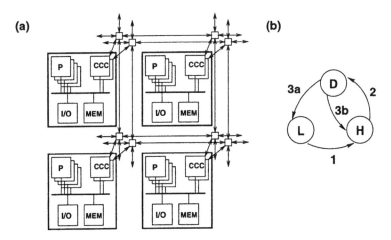

Figure 27.1 (a) Four-node DASH system. (b) Remote signals in the DASH to satisfy a cache miss to a line with home node H that is dirty in node D.

Each processor board includes a 33 MHz MIPS R3000 processor with 64 KB instruction cache, 64 KB first-level data cache, and 256 KB secondary data cache, which is a write-back snooping cache that interfaces with the shared node bus. The cache-line size is 16 bytes. The shared bus operates at 16 MHz and supports the Illinois MESI cache-coherence protocol (Papamarcos and Patel, 1984), where the highest priority processor with a valid copy supplies data on bus requests. This is different from the MESI protocols supported by modern processors where a processor only supplies data when it has a modified copy (Sweazy and Smith, 1986). The system interconnection is through two 2-D meshes: one for requests and another for replies. This dual arrangement reduces the possibility of deadlock. The network links are 16 bits wide with a maximum transfer rate of 60 MB/s per link.

The CCC functionality is implemented using two boards: the directory controller (DC) and the reply controller (RC). The DC generates node requests, and contains the directory of the global sharing state of the local portion of the memory and the x dimension network interface. The RC handles incoming requests and replies, and contains the y dimension network interface and a 128 KB *interconnect cache*. The ICC tracks outstanding requests made by the local processors and caches remote data.

Cache coherence is maintained by a combination of snoopy and directory protocols. The way in which a cache miss is satisfied depends on the home node of the missed line. When a processor generates a bus request for a local line, the other processors snoop this request and check their caches; the highest priority processor that has a valid copy supplies the line. At the same time, the DC accesses the directory to find the global sharing status of the line. If there is a remote node that has a modified copy, the DC blocks the request and generates a recall to the remote node; the processor is unblocked when the line is retrieved. In case the line is not cached locally and not modified in a remote node, the memory supplies the data.

When a processor generates a bus request for a remote line, the line is supplied by a local cache if there is a valid copy locally. Otherwise, the DC generates a request to the home node. When the RC of the home node receives the request, it echoes it on its local bus. A cache or the memory supplies the line unless it is modified in a third node. In the latter case, as shown in Figure 27.1(b), the home node (H) generates a recall signal (2) to the dirty node (D). When the dirty node receives the recall signal, its RC echos the recall on the local bus, the dirty cache supplies the line, and its DC forwards the line to the local node (3a), where it is inserted in the ICC. The ICC supplies the line when the processor retries its request. Additionally, the dirty node updates the home node (3b).

The *critical miss time* is from the start of the first processor request to the end of the reply that satisfies the processor miss. Thus, for this example, only the time of signal (3b) is not critical.

27.3 CONVEX SPP1000

The SPP1000 was introduced in 1994 and consists of 1 to 16 nodes that communicate via four rings (Brewer, 1995). As shown in Figure 27.2, each node contains four functional blocks interconnected by a crossbar (XBR). Each functional block interfaces with one ring and contains two 100 MHz HP PA 7100 processors and two memory

Figure 27.2   (a) Four SPP1000 nodes interconnected by 4 rings; each node has 4 functional blocks. (b) The functional block has 2 processors and 2 DRAM memory banks, and interfaces with the local crossbar and one ring.

banks. Each processor has 1 MB instruction and data caches. The processor pair share one agent to communicate with the rest of the machine. The memory has a configurable logical section that serves as an *interconnect cache*. The ICC is used for holding copies of shared data that are referenced by the local processors, but have a remote home.

Successive 64-byte lines are interleaved across the memory banks. A processor cache line is 32 bytes, and thus holds half a memory line. Each CCC is responsible for 1/4 of the memory space that a processor can address. When a processor has a cache miss, its agent generates a request through the crossbar to one of the four local CCCs. The CCC first accesses its memory (the ICC section for remote lines). If the attached memory does not have a valid copy, the CCC either contacts a local agent if any of the agent's processors has the valid copy in its cache, or contacts a remote CCC for service through its ring.

Each ring is a pair of unidirectional links with a peak bandwidth of 600 MB/s for each link. The rings support the Scalable Coherent Interface (SCI) standard (SCI, 1993), which coherently transfer one 64-byte line in response to a global shared memory access.

The coherence data is maintained in tags associated with the 64-byte memory lines. Each tag holds the local and global sharing state of the respective line. The local state part includes the local caching status of each of the two 32-byte halves of the line. The intra-node coherence protocol is a three-state MSI protocol (Astfalk *et al.*, 1994). The global sharing state is arranged in a doubly-linked list distributed directory rooted at the home node. The home node tag contains the global status and a pointer to the head of the sharing list. Each ICC tag in other sharing nodes contains the caching status of the line and pointers to the previous and next nodes in the list.

27.4   SGI ORIGIN 2000

The SGI Origin 2000 (Laudon and Lenoski, 1997) was introduced in 1996 and connects dual-processor nodes with an interconnect that is based on the SPIDER router chip (Galles, 1996). The CCC has multiple paths to interconnect the processor pair,

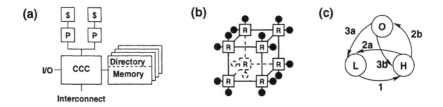

Figure 27.3 (a) Origin 2000 node. (b) Sixteen nodes connected in a cube configuration using 8 routers. (c) A speculative memory operation in the Origin 2000.

four memory banks with attached directory, the I/O interface, and the interconnect network (Figure 27.3(a)). Each processor runs at 195 MHz and has 32 KB on-chip instruction and data caches and a 4 MB, 2-way associative, combined secondary cache. The secondary cache line size is 128 bytes.

The bandwidths supported by the processor bus, the CCC, and the network are all matched and equal 780 MB/s. This perfect matching reduces latency in no-contention situations by allowing signals to be streamed directly between components; the bandwidth match eliminates the need for buffering.

The processor supports the MESI cache-coherence protocol, but does not snoop bus requests of other processors. Therefore, the Origin 2000 relies purely on a directory-based protocol and point-to-point signals for cache coherence. The directory is implemented in the memory and contains a sharing vector and a status field for each 128-byte line. The directory is accessed in parallel with memory accesses.

As shown in Figure 27.3(b), two nodes share one router. Each router has 6 bidirectional ports; two ports interface with the two nodes, and the other 4 ports can be used to interface with other routers. The figure shows eight routers—each using five of its ports—interconnecting 16 nodes in a cube configuration.

The Origin 2000 supports the sequential consistency model (Lamport, 1979) which supports a wide range of applications, but with minimal opportunities for hiding the latency of cache misses. For this reason, the Origin 2000 uses many protocol optimizations to reduce the local and remote latency. Figure 27.3(c) shows an example where the local node (L) has a load miss for a line which has a remote home in node (H) and is owned by node (O). When the home node receives the miss request, it accesses its directory and memory. Although the directory indicates that node (O) owns the line, the home node speculatively sends its line to the local node (2a) and generates a recall signal (2b). The local node saves the line until it receives a response from node (O). If node (O) has a modified copy, it is sent to the local node (3a) and the home is updated (3b). In this case, the saved speculative line is discarded. But if node (O) has a clean copy, it sends short negative replies to the local and home nodes, and the speculative line is used by the requesting processor.

27.5 RAW COMPARISON

This section presents the results of the raw comparison where the CDAT simulations use configuration files that select components of the same size and speed as those

Table 27.1 Signal latencies in nanoseconds. Multiple values are given when the latency depends on the signal length (see Subsection 27.6.5), latter values are for longer signals. "-" indicates that the signal is not relevant to the corresponding system.

Signal	DASH	SPP	Origin
From PRO to BUS	375/313[a]	70/120	100
From BUS to CCC	250	-	20
From BUS to MEM	0/188	-	-
From BUS to XBR	-	32/128	-
From XBR to CCC	-	32/128	-
From CCC to MEM	-	0/140	20
From CCC to NET	375	59/85/165/192	150
From MEM to CCC	-	172	80
From MEM to BUS	313	-	-
From NET to CCC	125/250	1432	10
From CCC to XBR	-	32/128	-
From CCC to BUS	125/376	-	20
From CCC to ICC	188	-	-
From XBR to BUS	-	30/90	-
From ICC to BUS	250	-	-
From BUS to PRO	0/188	40	20/50

[a] In the DASH, a data supply from the processor to the bus is a long signal and is faster than the short bus request signals generated by the processor on cache misses (see Subsection 27.5.1). Bus request signals are slower because they include bus arbitration.

used in the three case-study systems. The first-touch memory allocation policy is used throughout; a memory page is allocated in the node where the first reference to this page is made. Parameterizing the signal latencies part of the configuration files was relatively difficult because this information is not directly available from the literature that describes these systems. Nevertheless, based on analyzing the available information and some system calibration experiments via micro-benchmarking, we believe that the latencies used represent good approximations of the respective systems.

Parameterizing the DASH configuration file was the easiest because there are ample detailed DASH publications, e.g. (Lenoski and Weber, 1995). Parameterizing the Origin 2000 is based on the information available from the Silicon Graphics home page (O2KTune, 1996; R10000, 1997). As there is little publicly-available information about the SPP1000 latencies from Convex, we relied on the results of our calibration experiments (Abandah and Davidson, 1997). Table 27.1 summarizes the latencies used. The CDAT configuration files used in this study are listed in (Abandah, 1998).

We refer to the three systems evaluated in this section as DASH, SPP, and Origin. With these signal latencies, the misses satisfied from local and remote memories in the three systems are: 876 and 3,064; 564 and 3,964; and 290 and 610 nanoseconds, [1] respectively. The following subsections analyze the cache misses.

---

[1] A recent study reports higher Origin 2000 latencies (Hristea et al., 1997).

27.5.1   Miss Ratio

There are four main reasons for a processor to generate a bus request (other than prefetch operations which are not found in both the CG and SP applications):

- **Fetch** miss when there is an instruction code miss.

- **Load** miss when a load instruction misses in the secondary cache. Usually, the processor gets a shared copy (S) of the line. However, the MESI protocol gives the processor an exclusive copy (E) when the line is not cached in other processors.

- **Store** miss when a store instruction misses in the secondary cache.

- **Store/S** hit on a line in the shared state. The processor requests exclusive ownership of the line and, as for the store miss, the final state is modified (M).

Figure 27.4(a) shows the data miss ratio in the three systems due to the three reasons for generating bus data requests. The miss ratio is the number of data misses to the number of memory instructions.

The three systems have widely different miss ratios; especially with CG. DASH has the worst miss ratio followed by SPP. The large miss ratios of DASH and SPP are mainly due to capacity misses as the data working set is larger than the processor cache (Abandah and Davidson, 1998). Although CG's working set does not fit in the processor cache of SPP or DASH, SPP's miss ratio is smaller due to its wider cache. The large, wide, set-associative cache used in Origin succeeds in eliminating most of the capacity and conflict misses; thus it has a miss ratio that is less than 0.5%. The misses in the Origin are mainly coherence misses due to communication.

With CG, most of the misses are load misses due to CG's high percentage of load instructions used to perform the reduction operations. With SP, a large percentage of the misses is due to the producer-consumer communication; store/S miss produces and load miss consumes.

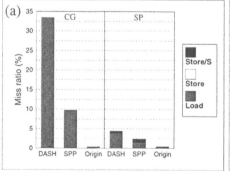
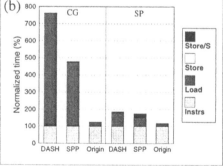

Figure 27.4   (a) The ratio of data misses to memory instructions. (b) Critical miss time relative to the instruction execution time.

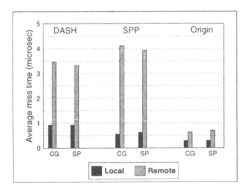

Figure 27.5   The average time of local and remote misses.

To illustrate the effect of the miss ratio on performance, Figure 27.4(b) shows the critical time spent satisfying misses relative to the instruction execution time, which is normalized to 100%. Notice that DASH with SP spends a relatively small time in satisfying its store/S misses. This is because DASH uses the relaxed memory consistency model (Gharachorloo *et al.*, 1991) that enables it to hide most of the remote time spent to satisfy this type of miss.

27.5.2   *Local and Remote Communication*

This subsection presents an analysis of the misses according to the place where they are satisfied. A local miss is one that is satisfied from a local cache or memory. A remote miss is satisfied from a remote cache or memory. Figure 27.5 shows the average time for the local and remote misses. In all cases, the average remote time is larger than twice the average local time. The NUMA factor is calculated as the ratio of the average remote time to the average local time. SPP has the highest NUMA factor (6.8), followed by DASH (3.7) and Origin (2.3).

In SPP, the local miss time with SP is larger than with CG because SP has a higher percentage of misses that are satisfied through processor supplies, which is more expensive than satisfying them from the memory. For DASH and SPP, the remote miss time with CG is higher than with SP. The first-touch memory allocation policy with SP is successful in reducing some of the more expensive remote accesses like the three-hop communication. Origin shows the opposite trend because the remote read misses with SP are expensive as they are generally satisfied from processor caches, in which case the speculative lines are useless.

27.6   NORMALIZED COMPARISON

In this section, we use CDAT configuration files that preserve the architectural and protocol differences, but put the three systems on the same technological level, *i.e.* same network speed, component sizes, and component speeds. CDAT is configured to simulate three derived systems: nDASH, nSPP, and nOrigin. They select modern component sizes and speeds similar to those used in the Origin 2000 (Table 27.1). The three derived systems, unlike the Origin 2000, have 4 processors per node and use cache lines that are 64 bytes.

As in the DASH node, the nDASH node has the memory on the local bus and its CCC interfaces with the bus, the network, and the ICC that participates in the local bus coherence protocol. nDASH uses an ICC that has a size and speed identical to the processor secondary cache. nSPP uses a 16 MB ICC that is a section of the node memory as in the SPP1000. However, nSPP node has four processors on one snoopy bus that is directly connected with the CCC, which interconnects the node bus, memory, and network interface. As in the SPP1600, nSPP allows caching local data in the exclusive clean state (MESI protocol). nOrigin has the same architecture as the Origin 2000, but has 4 processors per node and uses 64-byte cache lines.

In the following five subsections, we present analyses of the miss ratio, processor requests and returns, local and remote misses, where miss time is spent, and traffic exchanged between the system components.

27.6.1 Miss Ratio

Figure 27.6(a) shows the data miss ratio of the three systems. With CG, the three systems do not have noticeable miss differences and most of the misses are load misses. On the other hand, with SP, nSPP has a slightly higher load miss ratio due to conflicts in the ICC. When an ICC line is replaced, all the local copies of this line are invalidated; consequently, future processor accesses of this line generate misses. This behavior is not present in nDASH because its ICC does not maintain the inclusion property with respect to the locally cached remote lines. nOrigin has slightly fewer store/S misses than the other two systems because its global coherence protocol allows caching remote lines in the exclusive clean state.

Figure 27.6(b) shows the critical time spent satisfying misses relative to the instruction time. With CG, although the three systems have similar miss ratios, they spend different times satisfying these misses. nDASH reduces load miss time by satisfying more misses through snooping from neighbor processors. nSPP satisfies some misses

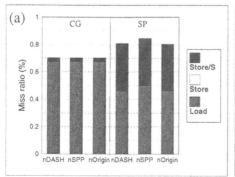
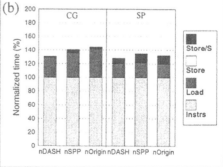

Figure 27.6 (a) The ratio of data misses to memory instructions. (b) Critical miss time relative to the instruction execution time.

locally from its ICCs. However, its load miss time is longer than that of nDASH due to its higher remote latency (see Subsection 27.6.3). Moreover, nDASH spends relatively less time in satisfying store/S misses due to its use of the relaxed memory consistency model. These misses are particularly expensive in nSPP with CG because they invalidate copies in multiple nodes, which the SCI protocol does serially.

27.6.2   Processor Requests and Returns

Figure 27.7 provides a closer look at the processor/bus interaction by showing the percentage of the four processor request signals described in Subsection 27.5.1, processor write-back (w/b) signals, and supply signals. As indicated by the small number of fetch and w/b signals, the three systems have negligible code and capacity misses; the secondary cache is large enough to contain the data and code working sets. However, as SP has more complex data structures than CG, it has more w/b signals due to conflict misses.

nDASH satisfies many misses from processor caches through supply signals. With SP, which has producer-consumer communication, almost all nDASH's misses are satisfied by supply signals. In nOrigin and nSPP, the data produced by a processor (visible as store/S) is communicated to other processors through an equivalent number of supply signals. With CG, which has larger sharing degree than SP, there is a larger ratio of load to store/S misses.

27.6.3   Local and Remote Communication

Figure 27.8 shows the average time of the local and remote misses. The average local miss latency in nDASH is the smallest because its memory is on the local bus, while in the other two systems the latency of traversing the CCC is added to the memory latency. The average local miss latency with SP is higher than with CG because SP has a higher percentage of misses with invalidation. This invalidation cost is more obvious in nSPP and nOrigin because of their use of stricter memory consistency models.

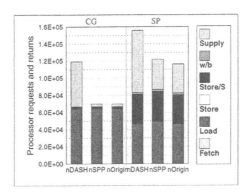

Figure 27.7   Processor/bus interactions; processor requests and returns.

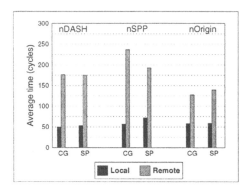

Figure 27.8   The average time of local and remote misses.

nOrigin's aggressive use of protocol optimizations and streaming of remote requests and responses pay off in achieving the lowest remote miss time. nDASH does not stream remote data to the processor; remote data is first inserted in the ICC, then the requesting processor gets it through snooping from the ICC. nSPP has the highest remote access latency because it checks the local ICC before generating a remote request, and its complicated global coherence protocol requires more network signals, often in a serial pattern, to satisfy remote misses.

The partial products generated by each processor are read by all processors. Consequently CG has mainly producer-consumers communication with a large sharing degree. In nOrigin, the remote miss time with SP is larger than for CG because SP has a higher percentage of remote load misses that are satisfied from remote caches which is more expensive in nOrigin than satisfying them from remote memory.

Figure 27.9(a) shows the percentage of local vs. remote misses for each system. nDASH and nSPP have fewer remote misses than nOrigin because they satisfy more misses locally through cache-to-cache transfers or from the ICCs. Figure 27.9(b) shows the percentage of time spent in satisfying local vs. remote misses. The high average remote miss time vs. local time makes the impact of remote misses exceed their ratio.

27.6.4   Where Time Is Spent

Figure 27.10 shows a breakdown of the time spent in satisfying processor requests and returns in the various system components for CG and SP. The time spent in each component type is split into three parts: critical time spent in local components, critical time spent in remote components, and non-critical time spent in both local and remote components. The figure shows seven component types: processor, bus, memory, cache-coherence controller, directory, interconnect cache, and network.

There are large differences between the three systems. For example, in nDASH the memory is less frequently used than the other systems. The CCC in nDASH is also less involved, and when it is involved, most of the time spent in it is not critical (either

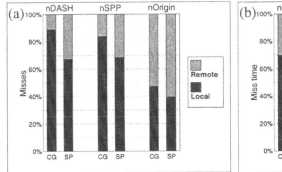
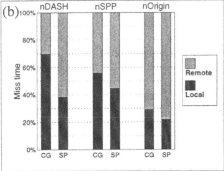

Figure 27.9 (a) Percentage of local vs. remote misses. (b) Breakdown of miss time.

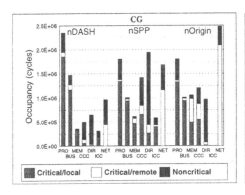
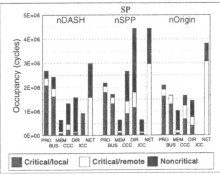

Figure 27.10 A breakdown of the time spent in satisfying processor requests and returns in the various system components.

because of the relaxed consistency model or because its time is overlapped with other bus activities).

Although most of the time spent accessing the directory in nSPP is not critical, it is the most involved component type in the system. This behavior is due to the SCI protocol which requires several directory accesses to maintain the distributed sharing linked-list. When we take into consideration the fact that nSPP implements both the directory and the ICC in the memory, it becomes clear that it has the most stressed memory subsystem.

nOrigin has a high network utilization because of its high percentage of remote misses. The network utilization in nSPP is also high because the SCI protocol uses more network signals to satisfy one miss.

The differences between CG and SP are due to their different communication patterns. The producer-consumer communication in SP exposes a higher percentage of the directory time as critical time due to the frequent need to find the node that owns the modified copy. SP also has relatively higher CCC and directory times than CG.

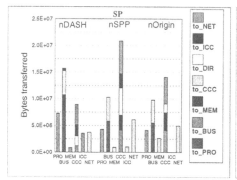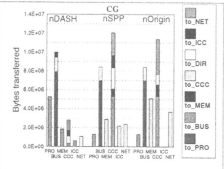

Figure 27.11   The number of bytes transferred from one component type to another.

27.6.5   Data Transfer

Figure 27.11 shows the number of bytes transferred from one component type to another for CG and SP. When reading this figure, remember that a component only communicates with those components that are directly connected with it. CDAT differentiates between two signal types: short signals that do not carry data, *e.g.* processor request; and long signals that carry one cache line, *e.g.* data response. In the three systems, the short signal is 16 bytes and the long signal is 80 bytes. However, nSPP's global coherence protocol has four signal types (24, 40, 88, and 104 bytes; CDAT does also model these four remote signal types). The shortest nSPP remote signal is 24 bytes. A signal becomes 16 bytes longer when it carries a node pointer, and 64 bytes longer when it carries data.

Relative to the other systems, nDASH's processors transfer more bytes to the bus as a result of the frequent local cache-to-cache transfers, and its busses carry higher traffic as they are the media of communication with the memory and the ICC. nDASH also transfers the least amount of data through the network due to its low percentage of remote misses. nSPP transfers more data over the network than might be expected from its remote miss percentage due to its protocol that uses more network signals per miss and larger remote signal sizes than the other protocols.

With SP, the percentage of bytes transferred to memory is larger than CG due to memory updates when processors supply dirty data.

27.7   CONCLUSIONS

The two benchmarks used in this study have different communication patterns. CG has repetitive producer-consumers communication with a large degree of sharing, thus most of its cache misses are load misses. In contrast, communication in SP has generally a single consumer, and SP's load and store misses are more balanced.

The raw comparison shows that there are large performance differences among the three systems due to their different component sizes and speeds. Their cache miss ratios are significantly different, and they have different NUMA factors (the average remote miss time to the average local time), with the Origin 2000 at 2.3, the DASH at 3.7, and the SPP1000 at 6.8.

In the normalized study, where we use same processor cache for the three systems, there are only small differences in the cache misses. Among the three normalized systems, the normalized SPP1000, nSPP, has slightly more cache misses due to conflicts in its ICCs. The normalized nOrigin has slightly fewer store/S misses due to allowing the exclusive state for caching remote lines.

nOrigin spends the longest time satisfying its misses because the majority of its misses are satisfied remotely. nDASH's Illinois protocol and nSPP's ICC reduce the percentage of remote misses. However, nDASH has less miss time than nSPP because its average remote time is smaller and its relaxed memory consistency model enables it to hide more of the miss time. nSPP has the highest average remote access time because it uses the SCI coherence protocol which requires more signals to satisfy a miss than the protocols used in the other two systems. The aggressive protocol optimizations used in nOrigin and its ability to stream remote requests and responses give it the least average remote access time.

Furthermore, nDASH has the smallest average local time because the memory, unlike other systems, is closer to the processor. nDASH memory is less involved in satisfying misses because nDASH satisfies more misses with cache-to-cache transfers.

The use of the complex SCI protocol in nSPP results in a large CCC involvement in satisfying misses and in its frequent directory accesses. nSPP also has the most stressed memory. Convex addresses this problem by having 4 CCCs in each SPP1000 node where each CCC has direct access to two memory banks. Although the nOrigin's large remote miss percentage results in heavy network utilization, the complex protocol in nSPP sometimes generates more network traffic for fewer remote misses.

In summary, nDASH outperforms the other normalized systems because of its coherence and consistency protocols, but it uses a bus coherence protocol that is not supported by modern processors. nSPP avoids some of its high remote access time by satisfying some misses from its ICCs. nOrigin, using the sequential consistency model, trades exposing more miss time for supporting a broader range of applications, and trades large remote miss percentage for small average remote time. It might be rewarding to look for new approaches that reduce the remote miss percentage and maintain a small average remote time. One such approach that uses a snoopy interconnect cache is investigated in (Abandah, 1998).

Acknowledgments

The development of the tools used in this study started in 1996 while Gheith Abandah was a research intern at HP Labs in Palo Alto, California, and the continuing research was supported by the National Science Foundation under Grant No. ACI–9619020. We would also like to thank the HP/Convex Technology Center for providing their implementation of the NAS Parallel Benchmarks. The simulation experiments of this study were carried out on the SPP1600 of the University of Michigan's Center for Parallel Computing.

References

Abandah, G. (1998). *Reducing Communication Cost in Scalable Shared Memory Systems*. PhD thesis, University of Michigan.

Abandah, G. and Davidson, E. (1997). Characterizing distributed shared memory performance: A case study of the Convex SPP1000. *IEEE Trans. Parallel and Distributed Systems*, 9(2):206–216.

Abandah, G. and Davidson, E. (1998). Configuration independent analysis for characterizing shared-memory applications. In *12th IPPS*, pages 485–491.

Astfalk, G., Brewer, T., and Palmer, G. (1994). Cache coherence in the Convex MPP. Technical report, Hewlett-Packard Co. http://www.hp.com/wsg/tech/technical.html.

Brewer, T. (1995). A highly scalable system utilizing up to 128 PA-RISC processors. In *COMPCON'95*, pages 133–140.

Galles, M. (1996). Scalable pipelined interconnect for distributed endpoint routing: The SGI SPIDER chip. In *HOT Interconnects IV*, pages 141–146.

Gharachorloo, K., Gupta, A., and Hennessy, J. (1991). Performance evaluation of memory consistency models for shared-memory multiprocessors. In *ASPLOS-IV*, pages 245–257.

Hristea, C., Lenoski, D., and Keen, J. (1997). Measuring memory hierarchy performance of cache-coherent multiprocessors using micro benchmarks. In *Supercomputing'97*. On CD-ROM.

Lamport, L. (1979). How to make a multiprocessor computer that correctly executes multiprocess programs. *IEEE Trans. on Computers*, C-29(9):241–248.

Laudon, J. and Lenoski, D. (1997). The SGI Origin: A ccNUMA highly scalable server. In *24th International Symposium on Computer Architecture*, pages 241–251.

Lenoski, D., Laudon, J., Gharachorloo, K., Weber, W.-D., Gupta, A., Hennessy, J., Horowitz, M., and Lam, M. (1992). The Stanford DASH multiprocessor. *Computer*, 25:63–79.

Lenoski, D. and Weber, W.-D. (1995). *Scalable Shared-Memory Multiprocessing*. Morgan Kaufmann.

O2KTune (1996). *Performance Tuning for the Origin2000 and Onyx2*. Silicon Graphics. http://techpubs.sgi.com/library/.

Papamarcos, M. and Patel, J. (1984). A low overhead coherence solution for multiprocessors with private cache memories. In *11th International Symposium on Computer Architecture*, pages 348–354.

R10000 (1997). *MIPS R10000 Microprocessor User's Manual*. MIPS Technologies Inc. Version 2.0.

SCI (1993). *IEEE Standard for Scalable Coherent Interface (SCI)*. IEEE Computer Society. IEEE Std 1596-1992.

Sweazy, P. and Smith, A. (1986). A class of compatible cache consistency protocols and their support by the IEEE Futurebus. In *13th International Symposium on Computer Architecture*, pages 414–423.

28 PARALLEL COMPUTATION OF THE DIAMETER OF A GRAPH

Jean-Albert Ferrez[1], Komei Fukuda[1,2], Thomas M. Liebling[1]

[1] Department of Mathematics,
Swiss Federal Institute of Technology, Lausanne, Switzerland
[2] Institute for Operations Research,
Swiss Federal Institute of Technology, Zurich, Switzerland

ferrez@dma.epfl.ch
fukuda@dma.epfl.ch
liebling@dma.epfl.ch

Abstract: The diameter of a graph is the maximum length of shortest paths between two vertices in the graph. It has some interesting theoretical properties, as well as a practical use as the lower or upper bound for various graph-based algorithms. Many methods to compute the diameter exist, and they can be classified in two categories: manipulation of the distance matrix and repeated use of a single source shortest path algorithm like Dijkstra's. We present several parallel implementations of some of these methods on parallel machines, and discuss their performance, focusing on large, sparse, polytopal graphs.

Keywords: graph diameter, shortest path, combinatorial optimization.

28.1 INTRODUCTION

Let $G = (V, E)$ be a graph with N vertices and M edges. The diameter of G is defined as the maximum length of the shortest path between two vertices of G. This value, together with other characteristics, gives a hint on the general shape or structure of the graph. Several algorithms use it as a lower or upper bound. In the context of parallel computing, the diameter of the network interconnecting the nodes of a distributed system is a measure of the worst case communication time. The bigger the value, the more you will wait if the nodes are far away. For many regular graphs (such as a ring, hypercube, and torus), the diameter is a function of the number of nodes. In general, however, it can take any value and is not known in advance.

Since the diameter is the greatest distance between any two vertices, we could simply compute those N^2 (in fact $\frac{N(N-1)}{2}$) distances and take the maximum. This can be done by manipulating a N by N matrix. For every pair of vertices (i, j), the idea is to keep either a value or a special mark that says "I do not know any path of that kind between i and j". Then these values or marks evolve towards the length of the shortest path between i and j. Ultimately, the greatest element in this matrix is the diameter. Section 28.3 describes two such algorithms.

However, we need to know only the longest distance in the graph, not the distance between every pair of vertices. Furthermore, as N grows it becomes impossible to store the N by N distance matrix. The idea here is to consider only the $N - 1$ shortest paths from one given source vertex to all other vertices, keep the longest as a candidate for the diameter, and iterate this process with another source, until all N possible sources have been considered. Section 28.4 discusses the underlying shortest path algorithm and its implementation, and presents some new ideas about reducing the number of sources to evaluate.

The computation of the diameter—also known as the all pairs shortest path—often serves as an example for online tutorials about parallelism in general, or a specific parallel programming model or library (see our web page for some links (Ferrez, 1998)). Furthermore, some theoretical results exist that give a very good expected number of operations for the computation assuming an unrealistic number of processors (Takaoka, 1995). With the opportunity to access two top-of-the-line parallel machines at the Scientific Computing Services of the Ecole Polytechnique Federale in Lausanne (EPFL), we have focused on the actual implementation of the algorithms, sacrificing reusability and portability for performance. MPI-based versions of the codes were also developed. Section 28.5 describes the machines, their environment, and the implementation of the algorithms.

Polytopal graphs are the wire-frame representation of a polytope in R^d. The set of vertices is the same, the set of edges is the same. In 1957 Hirsch conjectured that the maximal diameter of the graphs of polyhedra in R^d with n facets $\Delta(d, n)$ obeys the relation $\Delta(d, n) \leq n - d$ (Dantzig, 1963). (Klee and Walkup, 1967) proved it is false for unbounded polyhedra, but the conjecture is still open for polytopes. In this context, the actual length of an edge is not important, and the length of a path is given by the number of edges traversed, not by the sum of their length. This is why we focus in this work on undirected, unweighted graphs. If some non-negligible changes are required to support directed graphs, all the codes are designed to treat both weighted and unweighted graphs in the most efficient way. Section 28.6 gives some numerical results and compares the running time and scalability of the various algorithms on the different machines.

28.2 NOTATION

$G = (V, E)$ is a graph with vertex set V and edge set E. $N = |V|$ is the number of vertices and $M = |E|$ the number of edges of G.

The vertices are numbered from 1 to N and noted v_i. The edges are given two indices, those of their end vertices: e_{ij} connects v_i and v_j. The length of edge e_{ij} will be noted d_{ij}; for our purpose, $d_{ij} = 1 \ \forall i, j$. The length of a path is the sum of the lengths of the edges in the path. In our case it is equivalent to the number of edges in the path. L_i is the set of neighbors of vertex i, that is $v_j \in L_i \iff e_{ij} \in E$.

We use the term **distance** for the length of a *shortest* path between two vertices.

Uppercase letters are used for matrices, while the corresponding lower case letter is used for the elements of that matrix. For example, a_{ij} is an element of matrix A.

$\lceil x \rceil$ is the smallest integer greater or equal to x, $\lfloor x \rfloor$ is the largest integer smaller or equal to x. We use the symbol % for the modulo operator.

A processing element (PE) is a processor and its private memory. P is the number of PEs used, while p is the number of the current PE: $0 \leq p < P$. P will usually be a power of 2 between 1 and 64 on the T3D, and anything between 1 and 32 on the Origin 2000.

28.3 MATRIX APPROACH

Consider the $N \times N$ matrix $A^0 = \left[a_{ij}^{(0)} \right]$ where

$$a_{ij}^{(0)} = \begin{cases} 1 & \text{if } e_{ij} \in E \\ \infty & \text{otherwise} \end{cases} \tag{28.1}$$

This matrix is called the adjacency matrix of G^1. The matrix-oriented algorithms take this matrix as input and return a similar matrix where element a_{ij} is the distance between v_i and v_j. We explain first how the matrix is stored on a distributed memory system, and then what operations the various algorithms perform.

28.3.1 Repartition of the Matrix

The matrix approach needs large amounts of memory. One of the goals of the parallel implementation is to distribute the matrix over several PEs to allow larger graph sizes. This distribution is done row by row. That is, each PE has a subset of the rows, and one row is on only one PE. Since we consider only undirected graphs, all the matrices involved are symmetric so we could divide the memory utilization by two. However, we decided to sacrifice that memory for the following benefits: storing the whole symmetric matrix allows us to replace accessing a column (whose elements are to be retrieved over several PEs) with accessing a row (whose elements are stored in a contiguous portion of memory on one PE).

Only the elements in the upper half of the matrix are computed, then the corresponding elements in the lower half are updated. For any element a_{ij} the following rules apply:

- The row index i determines the PE p on which the element is stored.

- If $j > i$, the element is in the upper half of the matrix and is updated by the PE on which it is stored. To compute this, the PE is granted a read-only access to any element of the matrix.

---

[1]For the more general case of weighted edges, the elements $a_{ij}^{(0)}$ are initialized with the weight of e_{ij}.

- If $j < i$, the element is in the lower half of the matrix. Its column index j determines the PE that owns the symmetric element a_{ji}. That PE, after having updated a_{ji} in its own memory, is also responsible for updating a_{ij} wherever the latter is stored.

- The diagonal elements a_{ii} are equal to zero and are neither updated nor referenced.

These rules enforce data integrity:

1. Any element a_{ij} is stored at only one place, determined by i and j.

2. Any element a_{ij} can be read by any PE at any time.

3. Only one PE, determined by i and j, can update element a_{ij} and can do so at any time[2].

For the upper triangle of the matrix, the same PE is involved in 1) and 3). For the lower triangle, they may not be the same.

These rules imply that the number of elements to be computed in each iteration on row i is $N - i$. Allocating the rows to the PEs in blocks would yield a very bad repartition of the workload. The rows are therefore distributed to the PEs in an interleaved way (Figure 28.1), meaning that all PEs have some "large" rows and some "small" rows. The result is still not perfect, but doing better would cause additional hidden costs in computing the indices. In our case, the following values were used:

- $i\%P$ is the PE on which row i is stored.

- i/P locates row i on PE $i\%P$.

- if $i > j$, $j\%P$ is the PE that computes a_{ij}.

- if $i < j$, $j\%P$ is the PE on which to update a_{ji} with the same value as a_{ij}.

If P is a power of 2, these values are efficiently computed by bitwise operations.

28.3.2 Repeated Squaring

The definition of $a_{ij}^{(0)}$ given in Equation 28.1 is equivalent to $a_{ij}^{(0)} = 1$ if and only if there is a path using one edge between vertices i and j. Let

$$a'^{(1)}_{ij} = \min_k \left(a_{ik}^{(0)} + a_{kj}^{(0)} \right).$$

Clearly $a'^{(1)}_{ij} = 2$ if and only if there is a path using exactly two edges between vertices i and j. Now let

$$a_{ij}^{(1)} = \min \left(a'^{(1)}_{ij}, a_{ij}^{(0)} \right) = \min \left(\min_k \left(a_{ik}^{(0)} + a_{kj}^{(0)} \right), a_{ij}^{(0)} \right).$$

---

[2]This addresses concurrent access within the algorithm; concurrent access at a lower level is supposed to be addressed by the operating system, the communication library or the hardware.

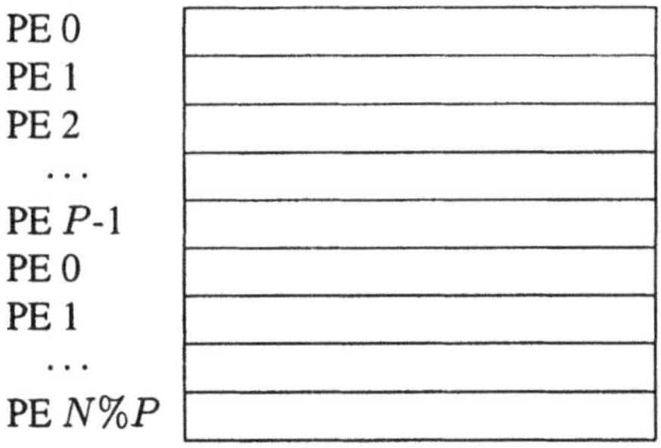

Figure 28.1   The interleaved repartition of the rows on the PEs.

In other words, $a_{ij}^{(1)}$ is the length of a shortest path using *at most* two edges between vertices i and j, and ∞ if no such path exists.

At the next iteration,

$$a_{ij}^{(2)} = \min \left(\min_k \left(a_{ik}^{(1)} + a_{kj}^{(1)} \right), a_{ij}^{(1)} \right).$$

$a_{ij}^{(2)}$ will be the length of a shortest path using *at most* four edges between vertices i and j, and ∞ if no such path exists.

The general iteration step,

$$a_{ij}^{(n)} = \min \left(\min_k \left(a_{ik}^{(n-1)} + a_{kj}^{(n-1)} \right), a_{ij}^{(n-1)} \right)$$

gives the length of a shortest path using *at most* 2^n edges between vertices i and j, and ∞ if no such path exists.

This process can be iterated until $a_{ij}^{(\lceil \log_2 N \rceil)}$ is the distance between i and j. If the graph is connected, there is (at least) one path using at most $N - 1$ edges between any pair of vertices. Furthermore, there can be no (simple) path using more than $N - 1$ edges. At this stage, the diameter of G is simply

$$\text{diameter} = \max_{ij} a_{ij}^{(\lceil \log_2 N \rceil)}.$$

In fact after at most $\lceil \log_2 \text{diameter} \rceil$ iterations the values in the matrix will not change, since all distances are computed.

Performing one iteration "in place", that is replacing the values in the current matrix instead of storing $A^{(n)}$ and $A^{(n-1)}$ separately, not only saves memory but also has an interesting *short-cutting* property. Suppose we are computing $a_{ij}^{(n)}$ and require the value of $a_{ik}^{(n-1)}$ (which has the length of the shortest path between v_i and v_k with at most 2^{n-1} edges) but we get $a_{ik}^{(n)}$ (which has the length of the shortest path between v_i and v_k with at most 2^n edges) instead. The computation remains correct since $a_{ik}^{(n)}$ "contains" a superset of the information in $a_{ik}^{(n-1)}$, and we get closer to the final result.

This short-cutting aspect grows with the number of processors, since it is more likely that a value in the matrix has already been updated by another PE.

Consider G, a ring of 1000 vertices with a diameter of 500. Typically, only 3 or 4 iterations are needed, depending on the number of PEs. In contrast, the theoretical value is 9 ($\lceil \log_2 500 \rceil = 9$).

The theoretical complexity of this first algorithm is $O(N^3 \log N)$. One iteration updates N^2 elements with $O(N)$ additions and comparisons each, and there are at most $\log N$ such iterations. A more efficient method exists, computing the same result in $O(N^3)$: Floyd's algorithm (Floyd, 1962).

28.3.3 Floyd's Algorithm

This method starts from the same matrix A^0 but goes on with a different iteration rule. It considers shortest paths between vertices i and j with *intermediate* vertices taken from a given set. $a_{ij}^{(0)}$ is the length of the shortest path between vertices i and j with no intermediate vertices. Let

$$a_{ij}^{(1)} = \min\left(a_{i1}^{(0)} + a_{1j}^{(0)}, a_{ij}^{(0)}\right)$$

where $a_{ij}^{(1)}$ is the length of the shortest path between vertices i and j with intermediate vertices in the set $\{v_1\}$. Then

$$a_{ij}^{(k)} = \min\left(a_{ik}^{(k-1)} + a_{kj}^{(k-1)}, a_{ij}^{(k-1)}\right) \tag{28.2}$$

is the length of the shortest path between vertices i and j with intermediate vertices in $\{v_1, v_2, ...v_k\}$. Ultimately, $a_{ij}^{(N)}$ will hold the length of the shortest path between vertices i and j using any vertex of V. This algorithm requires exactly N steps, but in each step the N^2 elements of the matrix are updated in constant time, yielding an $O(N^3)$ running time.

Computing $a_{ij}^{(k)}$ for any i,j requires the corresponding $a_{ij}^{(k-1)}$ and the kth row and column of $A^{(k-1)}$. As mentioned above, since $A^{(k-1)}$ is symmetric, column k is equal to row k. The latter is easy to fetch because it is stored in a contiguous portion of memory on PE $k\%P$. Therefore for an iteration of the algorithm, the PEs perform these two steps:

1. Fetch row k from PE $k\%P$.

2. Update a part of the matrix.

The second step follows the computation in Equation 28.2 and the repartition rules mentioned in Section 28.3.1.

The first step needs a closer look. The trivial approach where every PE will issue a remote read command to get the first row from PE 0 at the first iteration, then the second row from PE 1 at the second iteration, *etc.*, is guaranteed to cause a communication bottleneck. A cleverer approach is for the kth iteration to have the PE that owns row k broadcast its value to all other PEs. This relies on an efficient broadcast implementation in the communication library or in the hardware, and needs tighter synchronization between the processors. Experimental results showed a non-negligible improvement, especially when using many PEs.

28.4 NON-MATRIX APPROACH

For large sparse graphs, $O(N^3)$ cost of computing all the N^2 distances is too high. Even storing the adjacency matrix in its conventional form is not feasible due to memory limitations. Therefore, we keep for every vertex a list of its direct neighbors with the length of the corresponding edges. This results in an optimal $O(N + M)$ memory usage.

On top of this data structure, we implement Dijkstra's algorithm (Evans and Minieka, 1992) to solve the single source shortest path problem. The greatest distance returned by this algorithm gives a good candidate for the diameter. Once all the vertices have played the role of the source, the greatest known distance is the diameter.

28.4.1 Improvement: Bounding Techniques

The actual diameter will be reached for (at least) one specific source vertex. If this vertex happens to be the first to be chosen as the source, one could stop the computation at this point. Furthermore, for some classes of graphs, such as rings or hypercubes, this property holds for any source vertex. So the need for a stopping condition arises.

Any shortest path in G is a lower bound on the diameter. The diameter of any subgraph of G is an upper bound on the diameter. Whenever these two bounds meet, we are guaranteed to have the diameter and can stop evaluating the remaining sources. We are left with the problem of finding suitable subgraphs of G, in the sense of having a diameter that is easy to compute and as close as possible to the actual diameter of G.

The diameter of a tree is easy to compute by "hanging" the graph twice, as shown in Figure 28.2. The shortest path algorithm returns the shortest paths spanning tree, with the nice property that it is already hanging by the source vertex. By just choosing one of the bottom vertices and re-hanging, one gets the diameter of the tree.

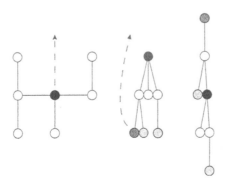

Figure 28.2 Computing the diameter of a tree by hanging it twice.

Unfortunately, the diameter of one such spanning tree is often much larger than the diameter of G. Furthermore, the resulting bound does not "converge" since those trees are independent of each other. There have been some examples, though, where this technique allowed us to stop the calculations after having evaluated only about 1% of the sources.

To further improve this bounding technique, we have tried to consider the union of the shortest paths spanning trees known so far. This subgraph converges to G^3, so its diameter will sooner or later reach that of G. Since the union is not a tree anymore, its diameter seems to be almost as hard to compute as the original problem. More theoretical work is needed in this area.

28.4.2   Implementation

Detailed descriptions of Dijkstra's algorithm can be found in any textbook on algorithms or graph theory (Evans and Minieka, 1992). Parallel implementations exist (Brodal *et al.*, 1997), but they involve a complex, distributed priority queue. In our case, it is easier to distribute the sources to the PEs and run independent sequential versions of Dijkstra's algorithm. This results in coarse-grain parallelism, low communication volume, loose synchronization and easy workload balancing.

28.4.2.1   Local Implementation. The underlying structure used to implement the priority queue has a great influence on the efficiency of a sequential Dijkstra implementation for general, weighted graphs. While a simple flat list gives an overall $O(N^2)$ complexity, the best theoretical value of $O(M + N \log N)$ achieved by a state-of-the-art Fibonacci heap has to be balanced by the overhead caused by elaborate data structure management (Goldberg and Tarjan, 1996). Since our graphs are sparse (small M), we have chosen a simple straightforward binary heap with *insert*, *remove_min* and *decrease_key* in $O(\log N)$, resulting in an overall $O(M \log N)$ complexity and a low overhead.

For the unweighted case we used a tweaked version of Dijkstra's algorithm based on a simple *first in first out* (FIFO) list. The Dijkstra code has been expanded to keep a record of the shortest paths spanning tree. This tree is then fed into the hanging procedure to compute the upper bound.

28.4.2.2   Global Implementation. The description of the graph (the $O(N + M)$ list of vertices and neighbors) is duplicated on every PE, substantially reducing communication. Only a few values have to be maintained globally: the current value of the diameter, the best current upper bound, and the list of sources that have not yet been evaluated. These values are stored on one processor, and shared memory and locking mechanisms are used to allow any PE to access and update them.

---

[3]In fact, in the weighted case, some edges of G may not appear in the union of all spanning trees, if those edges never appear in any shortest path. But in this case, they do not influence the diameter of G.

28.5   THE MACHINES AND LIBRARIES

This section evaluations our algorithms on a SGI Origin 2000, Cray T3D and a cluster of workstations.

28.5.1   The Cray T3D

The Cray T3D Massively Parallel subsystem has been installed at EPFL since early 1994. It has 256 Dec Alpha 21064 processors running at 150 MHz, each with 64 MB of private memory. The Alpha chip has two 8 KB primary level caches, but no secondary level cache is provided, which has a strong impact on performance. The system is setup in a way that, once you require (and obtain) a certain number of processors for your application, you get a dedicated partition where no time sharing occurs.

Although being a purely Distributed Memory machine, the T3D provides an alternative to message passing (PVM) programming: specialized hardware and the shmem library offer a simple and efficient pseudo shared memory programming model. It allows "one end communications" that is the remote_read and remote_write are performed by the PE issuing the request and the various hardware on the way, but the remote PE is not affected in any way[4]. This is to be compared with message passing libraries like PVM or MPI where every send (resp. receive) in the code must be balanced by a corresponding receive (resp. send). It is different from SMP-NUMA[5] in the sense that the exchanges always take place between the memory of the PEs: appropriate buffers must be explicitly provided, and the cost of local memory access—a critical issue on the T3D—is to be added to the total communication cost.

28.5.2   The SGI Origin 2000

The Cray-SGI Origin 2000 Distributed Shared Memory Parallel Server has been installed at EPFL since early 1997. It has 32 MIPS R10000 processors running at 195 MHz, each with 128 MB memory, giving a total of 4 GB of main memory. Each processor has two 32 KB primary level caches and a secondary level cache of 4 MB with a hardware machine-wide coherency mechanism. The standard message passing (PVM, MPI) are supported, but the machine can also be used as a shared memory system with non-uniform memory access (SMP-NUMA).

The system is sensitive to workload. All processes share the same pool of processors and if there are more than 32 active processes or threads, time-sharing occurs. This environment improves the overall throughput of the server, but can have an bad impact on the expected computation time for a given application.

When you read "distributed shared memory" in the description of a machine, you would like to stress the "shared" and forget about "distributed", if possible. However, although the hardware has been cleverly designed, access to remote memory remains 2-3 times slower than access to local memory (Chandra et al., 1997). Although the

---

[4]Well, not quite, since memory contention can happen there and slow things down.

[5]Symmetric Multi Processor, Non Uniform Memory Access

machine is able to dynamically move data around to achieve locality, to get the best performance the programmer should design and optimize his code knowing that the machine has distributed memory.

We first implemented the non-matrix method with shared memory: one common copy of the adjacency lists and several threads that would read this description to perform their part of the computation. Then we duplicated the lists so that every thread had its own copy in local memory. This resulted in a reduction of about 50% in computation time (on 4 PEs). The cost of this is, of course, more memory. However, unlike the matrix approach, memory is not an issue since the graph description and other data structures fit easily in the 128 MB available per PE.

28.5.3 *MPI*

The non-matrix method involves little communication, so an implementation on top of a message passing library would gain in portability without losing much in efficiency. We chose MPI (Message Passing Interface (MPI, 1998)) and ran this version both on the Origin 2000 (SGI's MPI 3.0) and on a cluster of 20 SGI O2 workstations (MPICH 1.1). The O2s have a MIPS R10000 processor running at 150 MHz, 128 MB memory, 1 MB secondary level cache, and are interconnected with 10 Mb/s Ethernet. The differences in the clock period and cache memory size between the Origin and the O2 yield, with exactly the same code and compiler options, a ratio of about 3/2 in single PE performance. That is, the times on the Origin should be multiplied by 1.5 before comparing them with the times on the O2.

The code was rewritten as a master-slave application on top of MPI. The master is responsible for distributing the sources to the slaves and maintaining the bounds. The slaves perform local, single-source shortest path computations and tree hanging. Clearly, the master has much less work to do than the slaves, but this issue can be addressed with a clever mapping of the tasks on the available PEs.

28.6 NUMERICAL RESULTS

In this section the results of various executions of the codes[6] on the different machines are shown. The diameters were always correct. The running times are shown in seconds and represent the wall clock time one has to wait to get the result, I/O (*i.e.* reading the graph from a file) *excluded*. On the Cray T3D, the timings are precise and repeated runs always take the same time. On the Origin using multi-threading, timings are fairly sensitive to the current workload, but out of several identical runs one can usually extract the "most significant value" that is given here. Furthermore, short runs on several PEs suffer from the time spent in setting up the sub-threads. Timings for the MPI versions are even less accurate, because of the lack of control on the load on the workstations and the network.

---

[6]Some of these codes are available from our web page (Ferrez, 1998).

Table 28.1   The matrix methods on the T3D (in seconds).

PEs	ring512		reg600	
	RepSq	Floyd	RepSq	Floyd
1	243.21	38.49	439.73	40.14
2	128.42	19.51	248.99	20.95
4	65.81	10.04	135.26	12.87
8	33.60	5.22	69.46	6.47
16	17.19	2.87	35.60	3.88
32	8.90	1.93	18.62	2.81
64	4.72	1.92	9.69	2.60

28.6.1   The Matrix Methods on the T3D

Table 28.1 shows the times for the two matrix methods on two different graphs on the T3D. The first graph is a simple ring with 512 nodes, the second is a regular graph with 600 nodes of degree 4 with a diameter of 15. Figure 28.3 shows the speedup and efficiency. The first, obvious observation is that Floyd's algorithm outperforms repeated squaring. However, Floyd scales worse, and the almost linear speedup observed until 16 PEs decreases afterwards. This shows that the most efficient sequential algorithm will not necessarily be the most scalable one, and that appealing speedup curves can hide poor actual performances.

28.6.2   The Matrix Methods on the Origin 2000

The same computations as above were then performed on the Origin using multithreading (see Table 28.2). A run on one Origin PE is approximately equivalent to a

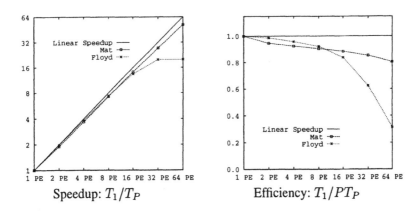

Speedup: T_1/T_P Efficiency: T_1/PT_P

Figure 28.3   The matrix methods on the T3D: speedup and efficiency.

Table 28.2   The matrix methods on the Origin 2000 (in seconds).

PEs	ring512		reg600	
	RepSq	Floyd	RepSq	Floyd
1	38	5	66	5
2	27	4	49	4
4	16	≈ 3	31	≈ 3

Table 28.3   The non-matrix method for the multi-threaded code (in seconds).

N	M	Origin 2000, multi-thread			
		1 PE	2 PEs	3 PEs	4 PEs
4004	20020	83	75	57	45
8736	48048	475	410	327	249
24752	148512	6656	4435	3088	2753

run using eight PEs on the T3D, but four Origin PEs are needed to match the performance of 16 T3D PEs. We have seen above that the scalability of the algorithm is not to blame. The distributed-shared-memory model shows its limits.

28.6.3   The Non-matrix Method on the Origin 2000 for Large Sparse Graphs

Table 28.3 gives the times to compute the diameter of 3 large sparse graphs[7] with the multi-threaded non-matrix code on the Origin 2000. For a given number of processors, they confirm the expected $O(N^2 \log N)$ complexity of the method. For a given graph, the scalability of the code is not very good, especially on small graphs.

Table 28.4 gives the times for the same computations with the MPI-based code on the Origin 2000. The numbers of PEs shown do *not* include the master. With this in mind, the MPI code turns out to be more scalable than the multi-threading code.

28.6.4   The Non-matrix Method on a Cluster of Workstations

Table 28.5 gives the times for the same graphs but on the cluster of O2 workstations. The master task was run on another workstation on the same network. Clearly, this approach is very sensitive to the setup overhead (≈ 2s. per task), but efficient for longer computations.

---

[7]These examples are the graphs of dual-cyclic polytopes of dimensions d with $2d$ facets, for $d = 10, 11, 12$. The dual cyclic polytopes attain the maximum number of vertices for a fixed dimension and a fixed number of facets. (See the famous upper bound theorem (McMullen and Shephard, 1971).) The diameters are not known in general but our code verified that the Hirsh upper bound $d = 2d - d$ is tight for these cases.

Table 28.4   The non-matrix method for the MPI-based code (in seconds).

N	M	Origin 2000, MPI		
		1 PE	2 PEs	3 PEs
4004	20020	91	43	30
8736	48048	530	238	160
24752	148512	7267	4026	2679

Table 28.5   The non-matrix method on a cluster of workstations (in seconds).

N	M	Cluster of O2		
		1 PE	6 PEs	20 PEs
4004	20020	65	55	80
8736	48048	760	171	145
24752	148512	10152	1753	671

28.7   CONCLUSION

From the algorithmic point of view, the various matrix approaches are clearly not optimal, both in terms of storage size and running time. This is true for relatively small values of M. But M can be as high as N^2 and in this case, the non-matrix method also needs $O(N^2)$ memory and at least $O(N^2 \log N)$ operations with much more overhead than the straightforward Floyd algorithm. But such very dense big graphs are not common, neither in theoretical research nor in industrial applications.

The matrix algorithms are very simple, easy to implement in pseudo-code, and not difficult to parallelize. However, they turn out to be tedious to optimize. To achieve high performance, the user needs to address issues such as reducing the communication, avoiding bottlenecks, maximizing cache utilization, reducing the overhead of indices computations, bailing out of loops as early as possible, and load balancing. It is very difficult to improve in one direction without losing in another.

The non-matrix method implemented in this work uses solid, well-known elements (Dijkstra's algorithm) as well as a new idea: the bounding of the diameter by the diameter of a tree (or union of trees). Clearly, this is a promising way to reduce, if not the worst case complexity of the problem, at least the actual computation time. The results we obtained in this work call for more theoretical research in order to obtain a rigorous demonstration of our expectations. Another area to explore is how to use part of the previously computed shortest paths spanning trees to speedup the subsequent iterations of Dijkstra's algorithm. This would require more communication between the threads and a finer grain of parallelism, but could also bring up interesting bounds.

Comparing the Cray T3D and the Origin 2000 is difficult and maybe not interesting. There is a difference of three years in the technologies of the two machines, a very long time in the world of high performance computing. Single-processor performance is, of course, much better on the Origin, but it is balanced by the fact that the T3D comes with more processors and—if properly set up—suffers less from time-sharing. The T3D seems to allow for a better scalability, while the Origin offers a much smoother transition for a newcomer to the world of high-performance parallel servers. Maybe the Cray T3E offers the best of both worlds.

The benefits of vector processors have not been considered at all in this study. It could be interesting to implement Floyd's algorithm on such a machine (Cray J90, NEC SX4), either on one vector processor or on several such processors using true, uniform shared memory.

References

Barriuso, R. and Knies, A. (1994). *SHMEM User's Guide for C (Rev. 2.2)*. Cray Research Inc.

Brodal, G., Träff, J., and Zaroliagis, C. (1997). A parallel priority data structure with applications. In *11th International Parallel Processing Symposium*, pages 689–693.

Chandra, R., Chen, D., Cox, R., Maydan, D., Nedeljkovic, N., and Anderson, J. (1997). Data distribution support on distributed shared memory multiprocessors. *SIGPLAN Notices (Proceedings of SIGPLAN Conference on Programming Language Design and Implementation (PLDI))*, 32(5):334–345.

Dantzig, G. (1963). *Linear Programming and Extensions*. Princeton University Press.

Evans, J. and Minieka, E. (1992). *Optimisation Algorithms for Networks and Graphs*. Marcel Dekker Inc., 2nd edition.

Ferrez, J.-A. (1998). Web page with links and codes. http://dmawww.epfl.ch/-roso.mosaic/jaf/pages/paradiam.html.

Fier, J. (1996). Performance tuning optimization for Origin2000 and Onyx2. Technical Report 007-3430-001, Silicon Graphics, Inc.

Floyd, R. (1962). Algorithm 97: Shortest path. *Communications of the ACM*, 5(6):345.

Goldberg, A. and Tarjan, R. (1996). Expected performance of Dijkstra's shortest path algorithm. Technical Report 96-070, NEC Research Institute, Inc.

Klee, V. and Walkup, D. (1967). The d-step conjecture for polyhedra of dimension $d < 6$. *Acta Math*, 133:53–78.

McMullen, P. and Shephard, G. (1971). *Convex Polytopes and the Upper Bound Conjecture*. Cambridge University Press.

MPI (1998). *The Message Passing Interface (MPI) standard*. http://www.mcs.anl.gov/Projects/mpi/.

Sedgewick, R. (1990). *Algorithms in C*. Addison-Wesley.

Seidel, R. (1992). On the all-pairs-shortest-path problem. In *24th Annual ACM Symposium on the Theory of Computing*, pages 745–749.

Takaoka, T. (1995). Sub-cubic cost algorithms for the all pairs shortest path problem. http://kaka.cosc.canterbury.ac.nz/~tad/.

VIII Computational Biology

29 CALCULATION OF SOLVATION FREE ENERGIES

Enrico O. Purisima

Biotechnology Research Institute,
National Research Council of Canada, Canada

rico@bri.nrc.ca

Abstract: Solvation effects play an important role in determining the properties and activities of biological molecules that function in a predominantly aqueous environment. To a good degree of approximation, a solvated molecule can be modeled as a set of point charges in a low-dielectric cavity that is embedded in a high-dielectric medium. The electrostatic free energy of the solvated molecule can then be obtained by solving the Poisson equation. One method for doing this is through a boundary element calculation. However, for large molecules such as proteins, this can involve solving a system of tens of thousands of linear equations. Using fast summation methods, we are able to reduce the calculation time to a matter of seconds.

Keywords: solvation, continuum dielectric, electrostatics, fast summation, multipole method, boundary element.

29.1 INTRODUCTION

Solvation effects can be a major determinant in the properties and behavior of most biological molecules. Detailed simulations of biomolecular systems often include explicit water molecules in order to properly incorporate solvation effects. Such simulations has yielded successful estimation of quantities such as free energies of binding, solvation free energies, relative pK_a values (Dang *et al.*, 1989; Jorgensen and Briggs, 1989; Mizushima *et al.*, 1991; Kollman, 1994). However, simulations with explicit water molecules have their drawbacks not the least of which is the added computational expense from the addition of hundreds to thousands of explicit water molecules. Explicit water molecules also make global conformational search calculations impractical due to the inordinate amount of time required for solvent re-equilibration after large scale movements in the solute.

Continuum dielectric models address some of these issues by replacing the solvent with a featureless bulk dielectric medium (Harvey, 1989; Davis and McCammon,

1990; Honig *et al.*, 1993; Rashin, 1993; Honig and Nicholls, 1995). In this approximation, the solute is treated as a low dielectric cavity embedded in a high dielectric medium representing the solvent and the Poisson or Poisson-Boltzmann equation is then solved numerically. The numerical solution is most commonly carried out using a finite-difference method in which the electrostatic potential is solved on a three-dimensional cubic grid enclosing the solute charges (Warwicker and Watson, 1982; Gilson *et al.*, 1987; Nicholls and Honig, 1991). The required electrostatic potentials are then calculated via a finite-difference solution of the Poisson-Boltzmann equation. An alternate method is the boundary element method (BEM) which replaces the effect of the dielectric by an appropriate charge density on the molecular surface of the solute (Miertus *et al.*, 1991; Zauhar and Morgan, 1985; Rashin and Namboodiri, 1987; Hoshi *et al.*, 1987; Zauhar and Morgan, 1988; Klamt and Schuurmann, 1993; Purisima and Nilar, 1995). The surface is represented by a mesh of elements or patches with corresponding charge densities. The electrostatic potential is then a composite of the potential due to the solute and surface charge densities. Boundary element methods are attractive in that they provide a more faithful representation of the shape of the molecule than is possible from finite-difference methods. They also allow a more direct calculation of reaction field and solvation free energies than with finite difference methods in which one requires special treatment for the self-energies of the solute charges.

The technical difficulty in this method lies in calculating the surface charge distribution. The process involves the solution of an n by n system of linear equations where n is the number of surface boundary elements. For a macromolecule n can be in the tens of thousands. This poses severe storage and computation time costs. Approaches to surmounting this problem by include multipole expansions and fast summation approaches (Bharadwaj *et al.*, 1995; Zauhar and Varnek, 1996; Purisima, 1998) as well as multigrid methods (Vorobjev and Scheraga, 1997). In this work, we present results for a fast summation method that reduces the computational cost of BEM calculations to the order of seconds on a regular workstation even for macromolecular systems such as proteins.

29.2   SURFACE GENERATION

In order to carry out a BEM calculation, we first have to generate a molecular surface (Richards, 1977; Connolly, 1983a; Connolly, 1983b; Connolly, 1985) or solvent-excluded surface as it is sometimes called (Greer and Bush, 1978). A detailed description of the history about work on molecular surfaces can be found in a recent review by (Connolly, 1996). Several methods exist for the generation of a tessellated molecular surface (Connolly, 1985; Pascual-Ahuir *et al.*, 1994; Zauhar, 1995; Chan and Purisima, 1998; Vorobjev and Hermans, 1997). For our BEM calculation, the surface generation program must provide a surface that it be smooth and free of singularities, as well as supply the surface normal and area of each surface element. For the results presented in this work, we used the Smooth Invariant Molecular Surface (SIMS) program of (Vorobjev and Hermans, 1997). Their algorithm constructs a fairly uniform distribution of points on the molecular surface with the surface cusps removed.

29.3 BEM EQUATIONS

In the BEM calculation, the solute molecule is treated as a collection of atom-centered point charges inside a cavity with dielectric constant, D_{in}, surrounded by an external continuum with a dielectric constant, D_{out}. In this model, the molecular nature of the solvent and accompanying detailed interactions, such as hydrogen-bonding, are lost. Both the internal and external regions are treated as *macroscopic* dielectric media. The entire electrostatic effect of solvation is contained in the modification of the electrostatic potential of the solute charges induced by the dielectric interface. The equations describing the interactions of the solute charges with the dielectric have been discussed in detail by various workers (Miertus *et al.*, 1991; Zauhar and Morgan, 1985; Rashin and Namboodiri, 1987; Zauhar and Morgan, 1988). Summarized below are the main concepts and equations relevant to this work.

In this continuum dielectric model, the presence of a dielectric boundary results in an accumulation of an induced surface charge density at the dielectric interface in response to the solute charge distribution. This surface charge density produces a reaction field that is superimposed on the solute-solute coulombic interactions. The reaction field energy of a solute molecule immersed in a medium of different dielectric can be calculated as:

$$G_{rf} = \frac{1}{2} \sum_k q_k \int_S \frac{\sigma(r)}{|r - r_k|} dS \qquad (29.1)$$

where σ is the induced surface charge density, and q_k and r_k are the partial charge and position of atom k, respectively. In the boundary element method, we represent the surface by a discrete mesh with an effective uniform charge density within each surface element. The reaction field energy can then be approximated as:

$$G_{rf} = \frac{1}{2} \sum_k \sum_j q_k \frac{\sigma_j A_j}{|r_j - r_k|} \qquad (29.2)$$

where σ_j and A_j are the charge density and area of patch j, respectively.

The electrostatic free energy of solvation is obtained by calculating the difference in the free energy of the solute for the two states: $D_{out} = D_{solv}$ and $D_{out} = 1$. This gives:

$$\Delta G_{solv} = \frac{1}{2} \sum_i q_i [\Phi_i^\sigma - (\Phi_i^\sigma)_0] \qquad (29.3)$$

where the subscript, 0, refers to the state where the external medium is a vacuum. In Equation 29.3 we assume that there is no change in the position and magnitude of the solute charges between the two states.

The values for σ_j are obtained by solving the system of linear equations (Zauhar and Morgan, 1985; Rashin and Namboodiri, 1987; Zauhar and Morgan, 1988):

$$(\mathbf{I} - f \, \mathbf{K})[\sigma] = f \, [E] \qquad (29.4)$$

where

$$f = \frac{1}{2\pi} \left(\frac{D_{in} - D_{out}}{D_{in} + D_{out}} \right) \tag{29.5}$$

and D_{in} and D_{out} are the dielectric constants in the solute and solvent, respectively. I is the n by n identity matrix, $[\sigma]$ the column vector of patch charge densities, and $[E]$ the column vector of the normal component of the electric field at each of the patch centers due to the solute charge distribution. The dimension of the matrix is the number of surface patch elements.

$$E_j = \frac{1}{D_{in}} \sum_k \frac{(r_j - r_k) \cdot n_j}{r_{jk}^3} q_k . \tag{29.6}$$

The elements of the matrix $\mathbf{K}$ depends only on the geometry of the dielectric interface and the mesh used to define it. The off-diagonal elements of the matrix $\mathbf{K}$ are computed as (Zauhar and Morgan, 1985; Rashin and Namboodiri, 1987; Zauhar and Morgan, 1988; Purisima and Nilar, 1995):

$$K_{ij} = \frac{(r_i - r_j) \cdot n_j}{r_{ij}^3} A_j . \tag{29.7}$$

Equation 29.7 is obviously not valid for the diagonal elements where $i = j$. We have shown previously that we can express the diagonal elements as a linear combination of the off-diagonal ones (Purisima and Nilar, 1995):

$$K_{ii} = 2\pi - \sum_{j \neq i} K_{ji} \frac{A_j}{A_i} . \tag{29.8}$$

Equation 29.8 is solved using Gauss-Seidel iterations.

29.4 VALIDATION AND CALIBRATION

For a BEM calculation, we have to decide on a choice of partial atomic charges and radii. The use of electrostatic potential fit charges is appropriate for the boundary element approach. These charges reproduce well the electrostatic potential of a molecule in the vicinity of its van der Waals surface. They are therefore well-suited for calculating the solute potential at the cavity interface from which the induced surface charge distribution may then be obtained. In our work, we use partial charges obtained from 6-31G*ESP fit charges.

The choice of cavity radii is not straightforward since the location of the dielectric boundary is not well-defined. For our purposes, we have opted to define cavity radii based on AMBER 4.0 (Pearlman et al., 1991) atom types. We use AMBER van der Waals radii scaled by a factor of 0.93.

The experimentally measured free energy of solvation is, of course, not due entirely to electrostatic effects. In fact, for aliphatic alkanes our calculated electrostatic component is quite small (less than 0.1 kcal/mol). For the non-electrostatic part of the free energy of hydration we use a term linearly related to solvent-excluded surface area.

Table 29.1   Comparison of experimental and calculated solvation free energies.

	Calculated	Experiment
acetate	-77.86	-79.9
methylammonium	-78.45	-71.3
trimethylammonium	-57.07	-56.6
acetamide	-9.50	-9.71
acetic	-6.90	-6.70
water	-6.81	-6.3
methanol	-4.69	-5.12
acetone	-3.20	-3.85
benzene	0.19	-0.87
toluene	0.24	-0.89
ethanol	-4.39	-5.01
4-methylphenol	-4.36	-6.14

Table 29.1 summarizes the calculated and experimental (Kang *et al.*, 1987a; Kang *et al.*, 1987b) solvation free energies for a number of molecules. The solute and solvent dielectric constants were taken to be 2 and 78.5, respectively. The good agreement with these small molecules gives us some confidence in applying the method to polypeptides and proteins for which no experimental solvation free energies are available.

For protein-sized molecules, the molecular surface description generated consists of thousands to tens of thousands of points. Application of BEM to proteins requires the efficient solution of a system of equations of high dimensionality. This is accomplished using fast summation multipole methods.

29.5   MULTIPOLE METHOD

(Zauhar and Varnek, 1996) have described a multipole method to the solution of Equation 29.4. In their formulation, they decompose $\mathbf{K}$ into two components $\mathbf{K}_F$ and $\mathbf{K}_N$:

$$(\mathbf{I} - f \, \mathbf{K}_N)[\sigma] = f([E] + \mathbf{K}_F[\sigma]) \qquad (29.9)$$

where $\mathbf{K}_N$ and $\mathbf{K}_F$ contain those K_{ij} corresponding to near and far interactions, respectively, based on the distance, r_{ij}, between points i and j. The right hand side of Equation 29.9 is an n-dimensional vector whose elements are the sum of the normal components of the electric field at each surface point due to the solute charges and due to the surface charge density from far away patches. $[E]$ is calculated once and $\mathbf{K}_F[\sigma]$ can be rapidly calculated using multipole approximations (Zauhar and Varnek, 1996). The idea is to make the left hand side of Equation 29.9 a sparse matrix affording rapid solution of the system of equations as well as significantly reducing memory costs.

From Equation 29.7, the ith element of the vector $\mathbf{K}_F[\sigma]$ is:

$$(\mathbf{K}_F[\sigma]) = \left(\sum_{far \ j} \frac{r_i - r_j}{r_{ij}^3} \sigma_j A_j \right) \cdot n_i \tag{29.10}$$

It is the summation in parenthesis on the right hand side of Equation 29.10 that is computed via a multipole approximation. In a multipole approximation, a cubic grid of cells is superimposed on the system and points are assigned to their respective cells. The electric field due to far cells can then be approximated by monopole, dipole and quadrupole contributions from the centers of the cells. Equations 17-19 of (Zauhar and Varnek, 1996) give the standard multipole expressions in cartesian coordinates. We have shown that a monopole approximation is sufficient for the summation in the right hand side of Equation 29.10 (Purisima, 1998). Thus, the approximation of the field of cell v felt at surface element i is given by

$$\mathbf{E}_v(r_i) \approx q_v \frac{r_i - r_j}{r_{iv}^3} \tag{29.11}$$

$$q_v = \sum_{k \in v} \sigma_k A_k \tag{29.12}$$

where r_v is the center of cell v.

A further approximation that can further reduce computational cost can be obtained by using a hierarchical multipole approximation and local Taylor expansion (Greengard and Rokhlin, 1987). In this approach, a multilevel cubic grid is defined. The lowest useful level is a $4 \times 4 \times 4$ grid with higher levels being $2^n \times 2^n \times 2^n$ grids. In going from one level to the next, each cell is subdivided into eight child cells by halving the grid spacing along each axis. The idea in this multilevel approach is that rather than carrying out a point-cell interaction calculation as in Equation 29.11, we calculate a cell-cell interaction at the lowest (coarsest) level and iteratively propagate the interaction to child cells at the higher levels via Taylor expansions. Only at the highest level is the field at individual surface points in the cell calculated. It is calculated via Taylor expansions about the center of the cell. The use of a hierarchical multipole approach has been described by (Bharadwaj et al., 1995). Detailed descriptions of the fast multipole method, local Taylor expansions and hierarchy of cells applied to the calculation of long-range nonbonded interactions in molecules can be found in the literature (Ding et al., 1992; Board Jr. et al., 1992; Shimada et al., 1994; Fenley et al., 1996) and are based on the work of (Greengard and Rokhlin, 1987). Using this fast summation method, we are able to attain dramatic speedups compared to a full-matrix calculation (Purisima, 1998).

As an example of the efficiency improvements we can attain, we calculated the reaction field energy of the protein crambin with a series of surfaces represented by an increasingly fine meshing. Table 29.2 summarizes the comparative CPU times for the fast summation method versus the full matrix solution. We obtain 42- and 85-fold speedups for the surfaces with 9,505 and 19,590 points, respectively. Even for the

Table 29.2   CPU time comparison of fast summation vs full matrix (in seconds).

Method	Number of Points		
	9505	19590	37703
Fast summation	9.9	22.2	54.8
Full matrix	419.9	1863.4	

surface represented by 37,703 points, the calculation was completed in under a minute on a regular workstation (an SGI with an R10000 CPU). The full-matrix calculation was not carried out on the highest density surface, but we extrapolate that the full-matrix calculation would take 170 times longer than the fast summation solution.

CPU times are for the calculation of the reaction field energy of the protein crambin with 653 atoms and a molecular surface represented by the shown number of points. The full matrix calculation with 37,703 points was not attempted due to the significant CPU requirements. All times are for an SGI workstation with an R10000 CPU.

29.6 CONCLUSIONS

The estimation of reaction field and solvation free energies is an important element in understanding the behavior of biological molecules. The fast summation boundary element method provides rapid yet high quality estimates of those quantities. For small molecules to polypeptides these energies can be calculated interactively. For proteins, it can be obtained in a few seconds to under a minute. This will facilitate the inclusion of realistic solvation terms in applications such as conformational search and docking.

References

Bharadwaj, R., Windemuth, A., Sridharan, S., Honig, B., and Nicholls, A. (1995). The fast multipole boundary element method for molecular electrostatics: An optimal approach for large systems. *Journal of Computational Chemistry*, 16:898–913.

Board Jr., J., Causey, J., Leathrum Jr., J., Windemuth, A., and Schulten, K. (1992). Accelerated molecular dynamics simulation with the parallel fast multipole algorithm. *Chemical Physics Letters*, 198.

Chan, S. and Purisima, E. (1998). Molecular surface generation using marching tetrahedra. *Journal of Computational Chemistry*. To appear.

Connolly, M. (1983a). Analytical molecular surface calculation. *Journal of Applied Crystallography*, 16:548–558.

Connolly, M. (1983b). Solvent-accessible surfaces of proteins and nucleic acids. *Science*, 221:709–713.

Connolly, M. (1985). Molecular surface triangulation. *Journal of Applied Crystallography*, 18:499–505.

Connolly, M. (1996). Molecular surfaces: A review. *Network Science*, 2. http://www.netsci.org/Science/Compchem/feature14.html.

Dang, L., Merz Jr., K., and Kollman, P. (1989). Free energy calculations on protein stability: Thr-157 to Val-157 mutation of T4 lysozyme. *Journal of the American Chemical Society*, 111:8505–8508.

Davis, M. and McCammon, J. (1990). Electrostatics in biomolecular structure and dynamics. *Chemical Reviews*, 90:509–521.

Ding, H.-Q., Karasawa, N., and Goddard, W. (1992). Atomic level simulations on a million particles: The cell multipole method for Coulomb and London nonbond interactions. *Journal of Chemical Physics*, 97:4309–4315.

Fenley, M., Olson, W., Chua, K., and Boschitsch, A. (1996). Fast adaptive multipole method for computation of electrostatic energy in simulations of polyelectrolyte DNA. *Journal of Computational Chemistry*, 17:976–991.

Gilson, M., Sharp, K., and Honig, B. (1987). Calculating the electrostatic potential of molecules in solution: Method and error assessment. *Journal of Computational Chemistry*, 9:327–335.

Greengard, L. and Rokhlin, V. (1987). A fast algorithm for particle simulations. *Journal of Computational Physics*, 73:325–348.

Greer, J. and Bush, B. (1978). Macromolecular shape and surface maps by solvent exclusion. *Proceedings of the National Academy of Sciences, USA*, 75:303–307.

Harvey, S. (1989). Treatment of electrostatic effects in macromolecular modeling. *Proteins: Structure, Function, and Genetics*, 5:78–92.

Honig, B. and Nicholls, A. (1995). Classical electrostatics in biology and chemistry. *Science*, 268:1144–1149.

Honig, B., Sharp, K., and Yang, A.-S. (1993). macroscopic models of aqueous solutions: Biological and chemical applications. *Journal of Physical Chemistry*, 97:1101–1109.

Hoshi, H., Sakurai, M., Inoue, Y., and Chûjô, R. (1987). Medium effects on the molecular electronic structure. i. The formulation of a theory for the estimation of a molecular electronic structure surrounded by an anisotropic medium. *Journal of Chemical Physics*, 87:1107–1115.

Jorgensen, W. and Briggs, J. (1989). A priori pKa calculations and the hydration of organic anions. *Journal of the American Chemical Society*, 111:4190–4197.

Kang, Y., Nemethy, G., and Scheraga, H. (1987a). Free energies of hydration of solute molecules. 2. Application of the hydration shell model to nonionic organic molecules. *Journal of Physical Chemistry*, 91:4109–4117.

Kang, Y., Némethy, G., and Scheraga, H. (1987b). Free energies of hydration of solute molecules. 3. Application of the hydration shell model to charged organic molecules. *Journal of Physical Chemistry*, 91:4118–4120.

Klamt, A. and Schuurmann, G. (1993). COSMO: A new approach to dielectric screening in solvents with explicit expressions for the screening energy and its gradient. *J. Chem. Soc. Perkin Trans.*, 2:799–805.

Kollman, P. (1994). Theory of macromolecule-ligand interactions. *Current Opinion in Structural Biology*, 4:240–245.

Miertus, S., Scrocco, E., and Tomasi, J. (1991). Electrostatic interaction of a solute with a continuum. direct utilization of ab initio molecular potentials for the prevision of solvent effects. *Chemical Physics*, 55:117–129.

Mizushima, N., Spellmeyer, D., Hirono, S., Pearlman, D., and Kollman, P. (1991). Free energy perturbation calculations on binding and catalysis after mutating threonine 220 in subtilisin. *Journal of Biological Chemistry*, 266:11801–11809.

Nicholls, A. and Honig, B. (1991). A rapid finite difference algorithm, utilizing successive over-relaxation to solve the poisson-boltzmann equation. *Journal of Computational Chemistry*, 12:435–445.

Pascual-Ahuir, J., Silla, E., and non, I. T. (1994). GEPOL: An improved description of molecular surfaces. iii. A new algorithm for the computation of a solvent-excluding surface. *Journal of Computational Chemistry*, 15:1127–1138.

Pearlman, D., Case, D., Caldwell, J., Seibel, G., Singh, U., Weiner, P., and Kollman, P. (1991). *AMBER Version 4.0*. San Francisco: University of California.

Purisima, E. (1998). A fast summation boundary element method for calculating solvation free energies of macromolecules. *Journal of Computational Chemistry*. To appear.

Purisima, E. and Nilar, S. (1995). A simple yet accurate boundary element method for continuum dielectric calculations. *Journal of Computational Chemistry*, 16:681–689.

Rashin, A. (1993). Aspects of protein energetics and dynamics. *Progress in Biophysics and Molecular Biology*, 60:73–200.

Rashin, A. and Namboodiri, K. (1987). A simple method for the calculation of hydration enthalpies of polar molecules with arbitrary shapes. *Journal of Physical Chemistry*, 91:6003–6012.

Richards, F. (1977). Areas, volumes, packing and protein structure. *Annual Review of Biophysics and Bioengineering*, 6:151–176.

Shimada, J., Kaneko, H., and Takada, T. (1994). Performance of fast multipole methods for calculating electrostatic interactions in biomacromolecular simulations. *Journal of Computational Chemistry*, 15:28–43.

Vorobjev, Y. and Hermans, J. (1997). SIMS: Computation of a smooth invariant molecular surface. *Biophysical Journal*, 73:722–731.

Vorobjev, Y. and Scheraga, H. (1997). A fast adaptive multigrid boundary element method for macromolecular electrostatic computations in a solvent. *Journal of Computational Chemistry*, 18:569–583.

Warwicker, J. and Watson, H. (1982). Calculation of the electric potential in the active site cleft due to a-Helix dipoles. *Journal of Molecular Biology*, 157:671–679.

Zauhar, R. (1995). SMART: A solvent-accessible triangulated surface generator for molecular graphics and boundary element applications. *Journal of Computer-Aided Molecular Design*, 9:149–159.

Zauhar, R. and Morgan, R. (1985). A new method for computing the macromolecular electric potential. *Journal of Molecular Biology*, 186:815–820.

Zauhar, R. and Morgan, R. (1988). The rigorous computation of the molecular electric potential. *Journal of Computational Chemistry*, 9:171–187.

Zauhar, R. and Varnek, A. (1996). A fast and space-efficient boundary element method for computing electrostatic and hydration effects in large molecules. *Journal of Computational Chemistry*, 17:864–877.

30 MOLECULAR DOCKING WITH A VIEW: THE INTEGRATION OF A MONTE CARLO DOCKING PROGRAM INTO A VIRTUAL REALITY ENVIRONMENT

Trevor N. Hart[1], Richard E. Gillilan[2], Ryan Lilien[2],*
Steven R. Ness[1]† and Randy J. Read[1]

[1] Department of Medical Microbiology and Immunology,
University of Alberta, Canada
[2] Cornell Theory Center,
Cornell University, U.S.A.

trevor.hart@ualberta.ca

Abstract: Molecular docking has become a fundamental tool in both the process of drug discovery and the understanding of protein structure and function. While much effort has gone into faster and more effective automated docking algorithms, the visualization of molecular structures remains an important tool for the correct interpretation of structural predictions. The integration of the Research docking algorithm into the Virtual Reality facility at the Cornell Theory Center offers a unique tool for understanding of protein-ligand interactions. The docking program Research is a fast Monte Carlo docking method employing a full force-field interaction model to represent molecular interactions. The program is integrated as a function call within the VR/3D workspace environment and connected with a sophisticated molecular viewer that displays docking results in real time. The workspace viewer allows the user to change perspective, style of molecular display, and parameters of the Research algorithm during the docking procedure. Multiple docking processes can be run simultaneously and interactively attached or detached from the viewing workspace. The integrated docking/viewing environment is not only a practical tool for drug discovery and design, but also offers an inside look at the molecular docking process.

Keywords: molecular docking, drug design, protein structure, molecular graphics, virtual reality.

---

*Current address: Department of Computer Science, Dartmouth College, U.S.A.
†Current address: Department of Biochemistry, University of British Columbia, Canada.

30.1 INTRODUCTION

The problem of molecular docking is that of determining, by computation, the preferred mode of interaction between two given molecular structures. When only a small number of atoms is involved, there are few possible configurations and the problem is relatively easy. For systems with large numbers of atoms, biological molecules in particular, the problem is much more difficult because of the enormous number of possible modes of interaction. Molecular docking has become an important problem in structural biochemistry, the solution of which will have impact not only on drug design but on our fundamental understanding of protein structure and function.

Proteins are large macromolecules, typically containing hundreds to thousands of atoms, and are fundamental to the function of life as we know it (Stryer, 1981). While DNA represents the blueprint of life, proteins are the workers: they build, maintain and recycle the chemical structures necessary for biological function. Perhaps the most important aspect of protein structure is the specificity of protein shape: proteins adopt specific shapes[1] in order to perform specific functions (Cantor and Schimmel, 1980). Enzymes are proteins that perform specific chemical reactions by recognizing complementary shapes. Drugs, such as antibiotics, often act by binding to specific enzymes in the pathogen, thereby disabling their function.

The basic purpose of docking methods is to predict and study the binding of a molecule to a protein. Protein structures lend themselves well to computational analysis, since the protein is represented as a set of three-dimensional coordinates, specifying a position for each atom in the protein. A docking algorithm will then model the interaction between the bound molecule (the *ligand*) and protein, searching for optimal binding modes. Current algorithms are limited both by the computational expense of computing interactions and by the ability to search for modes efficiently (Blaney and Dixon, 1993; Hart and Read, 1994).

While docking algorithms have been an active area of research, particularly in the past eight years, the use of graphical representations of structures has remained an important aspect of predicting protein/ligand interactions. As recent results from blind docking tests show (Dixon, 1997), docking methods are still not completely reliable in determining correct ligand binding, although they can effectively reduce the problem to a small number of possibilities (Hart *et al.*, 1997). Other methods must be used to discriminate between correct and incorrect results, the primary being visualization by the trained structural biochemist. Thus, *ability to examine solutions to a docking problem is almost as important as generation of the solutions themselves.*

Our research groups at the Cornell Theory Center and the University of Alberta have developed an interactive molecular docking and viewing environment for Virtual Reality (VR). VR is being used in an increasing number of scientific disciplines for visualization of data, design and remote manipulation (Ihlenfeldt, 1997). As more

---

[1]This is a simplification, since not only the shape but also the chemical nature of the molecule is important for specificity. The specific atoms, bonds and shape of a molecule give rise to a distribution of electric charge; regions of negative charge on one molecule prefer to associate with positive regions on another molecule, *etc*.

scientifically oriented groups explore this growing technology, the unique things that it has to offer are becoming more clearly defined.

The integration of VR with advanced docking algorithms explores a new approach to using visualization together with simulation methods for drug design. The VR facility at the Cornell Theory Center is an immersive graphical environment, using a two-wall projection system based on CAVE^{TM} technology (Cruz-Neira *et al.*, 1993). The docking program Research, developed at the University of Alberta, is a Monte Carlo-based docking method. It ranked well in predictive ability compared with other docking methods as evaluated at the CASP2 protein structure prediction challenge (Dixon, 1997) and is now incorporated into the commercially available docking package, DockVision.

We will begin by discussing the docking problem generally and aspects of the Research algorithm specifically. This will be followed by a brief introduction to virtual reality with emphasis on how we take advantage of its unique characteristics. Finally, we will present a detailed description of the interface and a docking application.

30.2 DOCKING SIMULATION

Docking seeks to understand and predict how inhibitors and substrates interact with proteins. There are two main applications where docking is used: first, to understand the binding of a known ligand and to predict its binding *mode* (the term *mode* meaning any configuration for the ligand specifying position, orientation and possibly internal conformation); and second, to determine which, if any, of a collection of molecules might bind or possess inhibitory activity against a particular protein. In either case, a docking algorithm must generate and evaluate alternative binding modes in some way and select one from the various possibilities. The computational challenge of docking is to perform efficient evaluation of binding modes and at the same time to limit the necessity of an intensive search, which multiplies the number of evaluations that are required.

There are various reasons why docking is a difficult computational and scientific problem. First, protein-ligand interactions are highly complex and require large amounts of computer time for the most sophisticated simulations. In docking calculations, where hundreds or thousands of alternate binding modes must be evaluated, this computational overhead is impractical (although even the most intensive calculations still have many limitations). Thus, simpler models that are computationally cheaper but more poorly represent the real situation must be used. Second, there is an inherently large number of local energy minima in protein-ligand interactions. Since proteins and typical ligands comprise irregular and unsymmetric shapes, the overall shape of the energy hypersurface (in the ligand's degrees of freedom) is also highly irregular and complex. Each local energy minimum will be surrounded by regions of very high energy due to "steric" interactions as atoms are forced to be too close to each other. The net result is an extremely complex energy hypersurface, where low energy regions are in close proximity to regions of very high energy. Searching for the global minimum is very much like looking for a needle in a haystack.

Depending on the size and complexity of the ligand, there are probably hundreds of low-energy binding modes for the ligand. Selecting the correct binding mode from among these is often a very difficult problem, even for the trained structural biochemist. Thus, it is very difficult to dock molecules "by hand", using graphical methods alone, without the aid of automated or partially-automated docking methods.

The Research docking method (Hart and Read, 1992; Hart *et al.*, 1997; Hart *et al.*, 1998), developed at the University of Alberta, is a simulation method using Monte Carlo simulated annealing. Monte Carlo is a natural technique to use for docking because it has been shown to be effective for many problems that, like docking, have many non-optimal local minima (Kirkpatrick *et al.*, 1983; Subbiah and Harrison, 1989; Wilson and Cui, 1990). Furthermore, because molecules constitute a real thermodynamic system, there is a natural interpretation of classical Monte Carlo parameters, such as the annealing temperature. Experience has shown that the most effective docking strategy is to generate a large number of random starting modes, and perform relatively short local searches, as opposed to running very long searches from a single starting mode. This does two things: it avoids the computationally expensive energy calculations in very unfavorable regions, and it removes bias towards the starting ligand mode. In a typical docking simulation, the results of a number of trials (normally between 100 and 1,000 for a typical ligand) are ranked, the top several being the most likely candidates to correspond to the true answer. The Research algorithm actually employs several different scoring functions at different stages within a given trial so that the computational effort is focussed on the energetically favorable regions of the protein surface.

As an example, we consider the problem of understanding a particular inhibitor to the important drug-design target, HIV-1 proteinase. The HIV-1 inhibitor, known by the undistinguished name CGP 53820, is a pseudo-symmetric, peptide-analogue inhibitor of the aspartyl proteinase from HIV-1. Its structure bound to that protein has been solved by experimental methods (x-ray crystallography, Protein Data Bank code: 1HIH) (Priestle *et al.*, 1995). The flexible ligand binds in the primary binding pocket of the proteinase, as would be expected, so it is reasonable to search within this pocket for the docking simulations. The ligand was docked from a set of 60 alternative low-energy starting conformations. The total runtime per conformer for the Research docking simulations was approximately 2 minutes per conformer on a 200 MHz Pentium computer running Linux. The correct structure was predicted by the simulation, scoring much better than the alternative configurations (Hart *et al.*, 1998).

Although docking has been the focus of extensive study by many research groups worldwide, complete success in docking remains elusive. The fact that no current docking method can give the correct answer all the time is illustrated by the result of the recent CASP2 (Critical Assessment of Structure Prediction, round 2) conference, where leading researchers were invited to submit predictions to "blind" tests in a number of protein structure categories, including molecular docking (Dixon, 1997). The primary problem seems to be the ability of the score function to correctly single out a particular low-energy docking as being correct: in several cases a good answer was among the better ranked dockings, but was not ranked best by any criterion (Hart *et al.*, 1997).

Thus, purely automated methods are not presently sufficient to solve every docking problem completely and effort needs to be directed towards analyzing docking results. The primary method for this task is graphical, interactive visualization. New graphical methods, such as VR, can play an important role in interpreting docking results. Furthermore, by allowing docking and visualization to work together within a common interface, the user cannot only understand automated docking results, but also directly control and affect the outcome.

This, briefly, summarizes some of the important aspects of docking and our approach to the problem. We next turn to the details of the virtual reality implementation, in hardware and software, and the application of these leading-edge techniques to the docking problem.

30.3 VIRTUAL REALITY AT CTC

The term "virtual reality" is not particularly well defined, and can refer to a great diversity of hardware and software technologies that are now emerging (Ihlenfeldt, 1997). Most commonly associated with the term are the so-called head-mounted displays that create a truly immersive experience. At the other extreme are languages and programming paradigms like the Virtual Reality Markup Language (VRML), which are not associated with any particular hardware. The key feature that characterizes true immersive virtual reality is that *real space is made to visually coincide with the computer-generated space.*

Modern computer graphics displays can accomplish the alignment of space in a variety of ways. Head-mounted displays (HMD's) are most widely known and used. Our application takes advantage of recent projection-based technology. The CAVE^{TM} is a room-sized rear-projection system originally developed at the University of Illinois at Chicago (Cruz-Neira, 1995; Cruz-Neira et al., 1993). With the commercialization of this technology, numerous facilities now exist throughout the world. Cruz-Neira has continued projection-based VR research at Iowa State with the recent construction of C2, a new generation CAVE-like device driven with an underlying object-oriented software design.[2]

The VR facility at the Cornell Theory Center supports a modest two-wall projection device compatible with CAVE and C2 software (see Figure 30.1). Projection-based virtual reality has several unique advantages that make it attractive for group-oriented work. Although perspective is calculated from the head position of a single operator, we have found that nearly a dozen people can have a reasonable view of a virtual world even within our 8ft×8ft space.

30.4 MOLECULAR VIEWING AND VR

The problem of macromolecular crystallography, because of its complex three-dimensional nature, has been a major driving force in the development of modern interactive graphics. Now that stereoscopic interactive graphics on the desktop has be-

[2]See http://www.icemt.iastate.edu/Labs/se.html.

Figure 30.1   Cornell Theory Center's virtual reality facility. The two 8ft×8ft walls at the left are translucent rear-projection screens. Because of limited space, the projection path is doubly folded by two glass mirrors per wall (right). Active LCD stereo eyewear (lower right) and electromagnetic tracking (not visible) combine to create the effect of immersive virtual reality.

come commonplace, what does immersive virtual reality have to add? Modern graphics has mostly replaced the physical models of earlier years, but interaction of humans with those models remains confined to lower-dimensional devices such as dials (1D) and mouses (2D). Additionally, the visual field of a conventional computer monitor is quite limited. The net result is that molecular modeling is most often performed with the model residing behind the narrow monitor window. Interaction with the model is necessarily indirect since the viewer cannot reach what is seen. As a corollary, transformations are usually applied to the model coordinate system rather than to the camera coordinate system. In other words, the object is being moved and the viewer is stationary.

Virtual reality expands the possibilities of interaction and perception in two ways. First, when the model space coincides with the viewer's physical space, direct three-dimensional interaction becomes possible. For example, most modeling packages emulate the physical rotation of an object by coupling the screen axes (x,y in the screen and z out of the screen) with either a mouse (with the x,y plane corresponding to the mouse's x and y motion, and z accessed by a different mouse button) or dials (with a different dial for each of x, y and z). These interactions can be defined as 2D×1D and 1D×1D×1D, respectively. In either case, there is a separation in the coordinate system, and it becomes virtually impossible to perform a simple rotation that is not within the separated coordinate space. In an immersive virtual world, however, all three degrees of freedom are accessible and there is no assumption of separability: the object can be rotated about any coordinate axis in a single operation, exactly as we naturally interact with objects in the real world.

The second way in which virtual reality expands interaction and perception is related to the field of view. When the field of view is unrestricted, it becomes practical for the first time to view nearby objects that are larger than the observer. Consider asking a person to perform a task, like sharpening a pencil in an unfamiliar office. If the person suffers from tunnel vision (as a hypothetical observer through a monitor would), the task would be difficult. Identifying the location of the pencil sharpener would require visually scanning the room, a task that grows quadratically larger as the field of view decreases. The observer must remember the arrangement of the room in order to navigate to the sharpener, but will frequently trip over objects not in the visual field. With a full field of view, movement of the viewer is brought onto an equal footing with movement of the object. Navigation can become an important means of finding an optimal view of an object.

Why view an object from the inside when you can just as easily hold it in your hand? The optimal view for a particular object or task is likely to be somewhere in between these two extremes. Immersive virtual reality gives the viewer a full range of perspectives from which to choose.

Another reason for placing objects around oneself is to improve spatial organization of tasks. Two-dimensional windowing systems were an innovation for desktop computing because they allowed users to have multiple applications simultaneously within the field of view. The ability to iconify windows is a further innovation that helps to reduce clutter caused by excessive stacking of windows, though at the expense of hiding information. The popularity of multi-headed displays (desktops spread across more than one monitor) should underscore the need users have for more screen real estate in which to organize their work. The amount of real estate available in a virtual environment is limited only by how far a user is able to reach within the confines of the tracking space. When navigation tools are provided, objects need not even be within reach.

30.5 DOCKING STRATEGY AND VR INTERFACE DESIGN

The WorkSpace windowing toolkit (Wood et al., 1996) gives us an application-development environment built around the concept of three-dimensional windowing. When an application is launched, it initially appears as a one-cubic-foot movable, resizable window in our virtual environment. In much the same way that a conventional desktop window can be resized to fill the screen, our virtual windows can be resized to fill the room and become immersive virtual worlds. Most windows in a typical application are not visible to the user, but serve as programming constructs for organizing how mouse events are processed. Mouse events, such as button-clicks and movements, are handled in a hierarchical parent-child fashion just like a conventional windowing environment. We have used the WorkSpace Windowing Toolkit to construct an interface to the Research docking program (Hart and Read, 1992; Hart et al., 1998). This section expands upon the initial work, presented elsewhere (Wood et al., 1996), where the emphasis was tele-collaboration. Here, we concentrate on aspects of the simulation interface and the nature of human interaction with real-time docking.

Virtual reality interfaces for molecular modeling and simulation have been investigated by several groups. Early work involved static models and precomputed results (Brooks *et al.*, 1988; Bergman *et al.*, 1993; Brooks *et al.*, 1990). More recently, molecular mechanics and dynamics simulations have been explored (Surles. *et al.*, 1994; Leech *et al.*, 1996), but the requirement of real-time performance has limited both the quality of the models and the type of problem that can be examined. The Virtual Biomolecular Environment (VIBE) utilizes parallel computing to overcome these limitations and has achieved 15 animation frames per second on HIV protease/drug docking calculations (Cruz-Neira *et al.*, 1996).

Research is an ideal simulation code for real-time control since the computational work required for Metropolis-Monte-Carlo updates to a small drug along a fixed protein surface is quite modest. The VR interface accesses the "inner loop" of the docking procedure, so the user is actually interacting with the ligand as the algorithm is at work. The basic simulation parameters, *i.e.* the annealing temperature, and rotation and translation step sizes, are controlled dynamically within the interface. Since the VR interface is dealing with one molecule at a time, the ligand is initially placed manually in the interface and is docked from that starting position. During the run, both the current conformation and the lowest energy conformation (for that run) are visually displayed. Once a low-energy conformation is fully refined, it may be added to a database, and a new conformation selected for annealing. The control panel (see Figure 30.2) is a movable, resizable 3D window with initial dimensions of one cubic foot. Temperature is visually displayed as a bar graph and manually adjusted by grabbing the sliding scale to the left of the bar. The range of temperature is between zero (cold) and 2.0 kcal/mole (hot) with the actual numerical value displayed below the slider. Monte-Carlo step sizes are adjusted similarly with the two remaining sliders. With some practice, novice users accustomed to 2D interfaces begin to learn that the 3D control panel need not be placed in front of the molecular models but can be placed out of the way and walked to when needed.

In any docking run, the user must select an appropriate annealing schedule with step sizes chosen at each temperature to maximize the amount of conformational space explored with a minimum number of steps. This process can be done visually with our interface. The acceptance ratio (accepted vs. total steps) is dynamically updated and displayed on the right side of the panel.

Navigation tools are provided for changing orientation and for traveling. Each tool takes the form of a 3D graphical icon floating above the viewer's hand. In the original version of our application, the user could scroll through these tools with a click of the 3D mouse button. More recently, we have found that a pull-down menu is more convenient for the task. The menu appears with the click of a mouse button and is always within reach. As mentioned previously, we have found that rotating large objects is most naturally accomplished by setting a pivot point and then rotating the world about that point by grabbing and pulling any other point in space. A triangle is drawn from the pivot point to the starting hand position and then to the current hand position as a guide. The actual rotation axis is perpendicular to the plane defined by the triangle. By standing at a greater distance from the pivot, finer control over the rotation can be achieved. We have used this type of control recently to manually

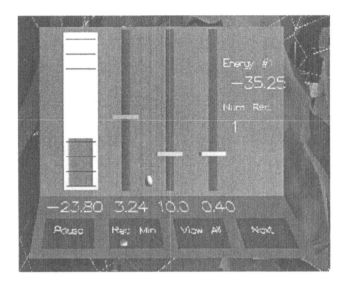

Figure 30.2   The simulation control window. The wide vertical bar gives a visual indicator of binding energy while sliders to the left control temperature, rotational and translational step sizes respectively. Buttons at bottom control database and display modes.

dock large protein domains (Kersten *et al.*, 1997). This type of rotation is fine for large objects such as protein domains, but has proven awkward for placing small drug molecules. Attaching the hand (position and orientation) directly to the pivot point is more appropriate in this case. Translation of the viewer also exhibits a duality related to scale. For traveling large distances (when the viewer is small), we have used an airplane-like tool which permits banking and turning. When the viewer is of the same size or larger than the data, it is easier to grab the world and drag it. In this way, a user can literally swim through a scene.

30.6   APPLICATION

Our earliest application was motivated by the work of Faerman and collaborators (Ponasik *et al.*, 1995) on inhibitor design for the treatment of Chagas' disease. Chagas' disease is a parasitic illness that ranks just below malaria and schistosomiasis as a major health concern in Central and South America. The disease-causing parasite *T. cruzi* is a single-celled organism that effectively evades the human immune response and replicates within muscle and nerve cells over a period of many years. Much work in this area concentrated on developing inhibitors to trypanothione reductase (TR), an enzyme that maintains an adequate supply of the protective molecule trypanothione within the parasite (Schirmer *et al.*, 1995). Obtaining good crystals of the enzyme in complex with drugs has proven difficult experimentally. The medium-resolution (2.8Å) crystal structure of the drug Mepacrine complexed with TR provided us with

Figure 30.3   Molecular model of trypanothione reductase (TR). Ribbons trace the polypeptide backbones of the protein. The large circular patch is a part of the molecular surface containing the active site. Drug conformations are represented by the stick figures within the active site. Some important amino acid side-chains are also shown.

an opportunity to test our interactive approach to docking (Jacoby *et al.*, 1996). Tele-collaborative aspects of the project have already been documented (Wood *et al.*, 1996).

The model of the protein was generated using Cornell Theory Center's Chemistry Modules, a suite of extensions to the IBM Data Explorer visualization program[3] (Gillilan and Ripoll, 1995). Molecular coordinates from the published structure of TR solved by S. Bailey, K. Smith, A.H. Fairlamb and W.N. Hunter were converted into a standard ribbon representation tracing out the polypeptide backbones of the two monomer units (Hunter *et al.*, 1992). To visualize the shape of the active site pocket where the natural substrate binds, we computed a molecular surface using an algorithm due to Varshney (Varshney *et al.*, 1994) in the immediate region of the pocket. Figure 30.3 shows a ribbon representation of the protein, the molecular surface, and models of mepacrine in the binding pocket. The entire protein model geometry was written out in VRML 1.0. Our virtual reality interface reads arbitrary VRML files and displays them in register with a simple ball-and-stick model of the ligand.

We chose a procedure where the ligand would start from a single random conformation and allow the operator to explore the conformational space using only temperature and step-size but not step-direction adjustments. This arrangement gives the operator the opportunity to bypass unfavorable or chemically uninteresting regions that would normally trap an automated procedure. By setting the temperature to a high value,

---

[3]See http://www-i.almaden.ibm.com/dx/.

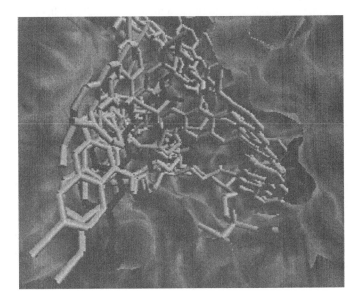

Figure 30.4 Distribution of low-energy drug molecule conformations (light rods) obtained from a single starting position by interactively controlling temperature and step size but not step direction. Experimental conformation is represented by the dark rods on the left.

the simulation simulation provides an on-going random exploration of conformational space. When an interesting region is entered, by setting the temperature to a low value, the simulation moves the ligand into the nearest low-energy minimum. Our initial calculation on mepacrine bound to TR required about 2 hours of interactive work to obtain 15 candidate conformations. Three of the 15 lowest energy conformations had ring orientations that were close to the crystal structure (see Figure 30.4).

This work serves to illustrate, briefly, how the VR/docking interface can be used to approach real problems. The brevity reflects our emphasis on the description of the overall philosophy and techniques involved in developing VR techniques for an applied, multi-disciplinary scientific problem. Specific applications using the interface are ongoing and will be reported elsewhere.

30.7 SUMMARY

We have created a novel virtual reality interface to the docking program Research and touched upon how virtual reality differs from conventional behind-the-screen computer graphics. It offers directness of interaction, greater range of perspective because of increased field of view, and better availability and use of space. The conventional process of docking involves a visual inspection of conformations in order to choose the final candidates. We have demonstrated one way in which human interaction can be integrated at a deeper level into an algorithm to overcome a technical limitation. Moving beyond this initial experiment, we have implemented a multi-process version of the simulation that allows users to run and manage a small number of simultaneous docking jobs. The program Research is inherently parallel, but the tools necessary to supervise the annealing of many random initial conformations remain to be developed.

Graphics, display and tracking hardware are rapidly becoming inexpensive and widely available. Though projection-based VR technologies like the CAVE have had the dubious distinction of being among the most expensive, this is unlikely to continue to be a problem given the current pace of technology. Such setups are particularly appropriate at the departmental and institutional level where collaboration is important.

The importance of this work lies in seeing what is possible with advanced molecular graphics technology. While we can easily assess the effectiveness of docking algorithms, or present the technical specifications of a VR interface, it is more difficult to truly measure the effectiveness of a viewing interface on a user's ability to understand and investigate a scientific problem. However, only by developing interfaces and having them used can we gain an understanding of their inherent potential and of the most effective interface design strategies. It took many years for molecular graphics on workstations to evolve to its present form; it will no doubt take an equal time and effort for VR-molecular graphics to similarly evolve to widespread use.

Acknowledgments

The Cornell Theory Center (CTC) is funded by New York State, the National Center for Research Resources at the National Institutes of Health (grant #P41 RR04293), Cornell University, and members of the Corporate Partnership Program.

This work is also funded by the Medical Research Council of Canada and the Alberta Heritage Foundation for Medical Research.

References

Bergman, L., Richardson, J., and Brooks, F. (1993). VIEW: An exploratory molecular visualization system with user-definable interaction sequences. In *SIGGRAPH'93*, pages 117–126.

Blaney, J. and Dixon, J. (1993). A good ligand is hard to find: Automated docking methods. *Persp. Drug Dis. Des.*, 1:301–319.

Brooks, C., Karplus, M., and Pettitt, B. (1988). *Proteins: A Theoretical Perspective of Dynamics, Structure and Thermodynamics*. Advances in Chemical Physics, Vol 71. Wiley & Sons, NY.

Brooks, F., Ouh-Young, M., NJ, J., and Kilpatrick, P. (1990). Project grope: Haptic displays for scientific visualization. In *SIGGRAPH'90*, pages 177–185.

Cantor, C. and Schimmel, P. (1980). *Biophysical Chemistry, Part I: The Conformation of Biological Macromolecules*. W.H. Freeman and Company, New York.

Cruz-Neira, C. (1995). *Virtual reality based on multiple projection screens: the CAVE and its applications to computational science and engineering*. PhD thesis, University of Illinois at Chicago. University Microfilms International #9532383.

Cruz-Neira, C., Langley, R., and Bash, P. (1996). VIBE: A virtual biomolecular environment for interactive molecular modeling. *Computers and Chemistry*, 20:469. http://www.cs.uiowa.edu/~cremer/sive95/.sive-sessions.html.

Cruz-Neira, C., Sandin, D., and DeFanti, T. (1993). Surround-screen projection-based virtual reality: The design and implementation of the CAVE. In *SIGGRAPH'93*.

Dixon, J. (1997). Evaluation of the CASP2 docking section. *Proteins, Suppl.*, 1:198–204.

Gillilan, R. and Ripoll, D. (1995). Vizualizing enzyme electrostatics with IBM visualization data explorer. In Bowie, J. E., editor, *Data Visualization in Molecular Science*. Addison-Wesley, New York.

Hart, T., Ness, S., and Read, R. (1997). Critical evaluation of the Research docking program for the CASP2 challenge. *Proteins, Suppl.*, 1:205–209.

Hart, T., Ness, S., and Read, R. (1998). Monte Carlo docking of flexible ligands using multiple conformations. Manuscript in preparation.

Hart, T. and Read, R. (1992). A multiple-start Monte Carlo docking method. *Proteins*, 13:206–222.

Hart, T. and Read, R. (1994). Multiple-start Monte Carlo docking of flexible ligands. In Merz, K. and LeGrand, S., editors, *The Protein Folding Problem and Tertiary Structure Prediction*, pages 71–108. Birkhäuser, Boston.

Hunter, W., Bailey, S., Habash, J., Harrop, S., Helliwell, J., Abogye-Kwarteng, T., Smith, K., and Fairlamb, A. (1992). Active site of trypanothione reductase: a target for rational drug design. *J. Mol. Biol.*, 227:322–333.

Ihlenfeldt, W. (1997). Virtual reality in chemistry. *J. Mol. Model.*, 3:386–402.

Jacoby, E., Schlichting, I., Lantwin, C., Kabsch, W., and Krauth-Siegel, R. (1996). Crystal structure of the Trypanosoma *cruzi* trypanothione reductase-mepacrine complex. *Proteins*, 24:73–80.

Kersten, S., Reczek, P., and Noy, N. (1997). The tetramerization region of the retinoid X receptor is important for transcriptional activation by the receptor. *J. Biol. Chem.*, 272:29759–29768.

Kirkpatrick, S., Gelatt, C., and Vecchi, M. (1983). Optimization by simulated annealing. *Science*, 220:671–680.

Leech, J., Prins, J., and Hermans, J. (1996). SMD: Visual steering of molecular dynamics for protein design. *IEEE Computational Science and Engineering*, pages 38–45.

Ponasik, J., Strickland, C., Faerman, C., Savvides, S., Karplus, P., and Ganem, B. (1995). Selective inhibitors of crithidia-faciculata trypanothione reductase. *Biochemical Journal Part 2*, 311:371–375.

Priestle, J., Fassler, A., Rosel, J., Tintelnot-Blomley, M., Strop, P., and Grutter, M. (1995). Comparative analysis of the x-ray structures of HIV-1 and HIV-2 proteases in complex with CGP 53820, a novel pseudosymmetric inhibitor. *Structure*, 3:381–389.

Schirmer, R., Muller, J., and Krauth-Siegel, R. (1995). Disulfide-reductase inhibitors as chemotherapeutic agents: The design of drugs for trypanosomiasis and malaria. *Angew. Chem. Int. Ed. Engl.*, 34:141–154.

Stryer, L. (1981). *Biochemistry*. W.H. Freeman, San Francisco.

Subbiah, S. and Harrison, S. (1989). A simulated annealing approach to the search problem of protein crystallography. *Acta Cryst.*, A45:337–342.

Surles., M., Richardson, J., Richardson, D., and Brooks, F. (1994). Sculpting proteins interactively: Continual energy minimization embedded in a graphical modeling system. *Protein Science*, 3:198–210.

Varshney, A., Brooks, F., and Wright, W. (1994). Linearly scalable computation of smooth molecular surfaces. *IEEE Computer Graphics Applications*, 14:19–25.

Wilson, S. and Cui, W. (1990). Applications of simulated annealing to peptides. *Biopolymers*, 29:225–235.

Wood, F., Brown, D., Amidon, R., Alferness, J., Joseph, B., Gillilan, R., and Faerman, C. (1996). Workspace and the study of Chagas' disease. *IEEE Computer Graphics and Applications*, 16:72–78.

31 STRUCTURAL RIBONUCLEIC ACID MOTIFS IDENTIFICATION AND CLASSIFICATION

Patrick Gendron[1], Daniel Gautheret[2] and Francois Major[1]

[1]Département d'Informatique et de Recherche Opérationnelle,
Université de Montréal, Canada
[2] Département de Biologie,
Université Aix-Marseille II and CNRS, France

gendrop@iro.umontreal.ca
gauthere@gauss.cnrs-mrs.fr
major@iro.umontreal.ca

Abstract: A structural ribonucleic acid (RNA) motif is a recurrent subset of nucleotide arrangements in secondary or tertiary structure. RNA motifs are represented by graphs of relations where the nodes represent the nucleotides and the edges represent structural relations. In order to identify all RNA motifs from all known secondary and tertiary structures, an incremental search algorithm was developed. Given a list of motifs of size n, the algorithm builds the motifs of size $n + 1$. The building of an RNA motif database, the development of RNA secondary structure prediction energy potentials, and the integration of RNA motifs in 3-D modeling are discussed.
Keywords: RNA motif, database, graph isomorphism, molecular modeling.

31.1 INTRODUCTION

Since the structure of a macromolecule dictates its function, researchers invest considerable efforts in the determination of their 3-D structures. Determining the secondary structure of a ribonucleic acid (RNA) from sequence data is the first step towards the 3-D structure. The secondary structure of an RNA encodes the majority of its nitrogen base interactions (see Figures 31.1 and 31.2). The second step consists of determining the tertiary structure which includes all inter-nucleotide interactions found in the 3-D structure. The last step builds the 3-D structure from the secondary and tertiary structures. All steps involve combinatorial enumerations.

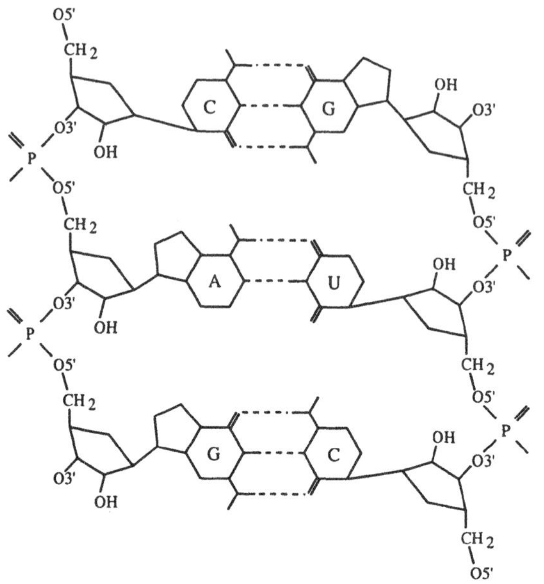

Figure 31.1   Molecular structure of the largest recurrent subgraph in yeast tRNA^{Phe}. The dashed lines indicate H-bonds.

Structural recurrence suggests functional conservation. Knowledge about RNA motifs help us to identify functional sites, to improve secondary structure prediction, and to simplify three-dimensional modeling. In this paper, we present a method for the detection of unknown RNA motifs. Section 31.2 describes the molecular structure of RNAs, and a relational graph representation that will be used in the construction of an RNA motif database. The following sections are respectively dedicated to the motif search algorithm and its application to the yeast tRNA^{Phe}, and 16S and 23S ribosomal RNA (rRNA) subunits.

31.2   REPRESENTATION

The primary structure of an RNA is a single chain polymer formed of nucleotide units linked together by phosphodiester bonds which connect the O3'-end of one nucleotide to the O5'-end of the next one. Each nucleotide consists of a nitrogen base, a ribose sugar, and a phosphate group (see Figure 31.1). There are four types of nitrogen bases, namely the purines adenine (A) and guanine (G), and the pyrimidines cytosine (C) and uracyl (U). The nitrogen bases contain hydrogen bond (H-bond) donors and acceptors. An RNA three-dimensional (3-D) structure is stabilized by the formation of H-bonds between pairs of nitrogen bases. Most of these interactions are Watson-Crick base pairs, such as in deoxyribonucleic acid (DNA), that is the $C - G$ and $A - U$ base pairs. The determination by x-ray crystallography and NMR spectroscopy of new RNA structures has revealed many non-canonical base pairs, such as $G - U$,

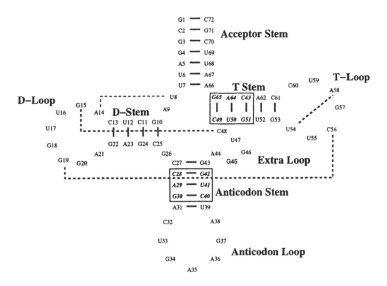

Figure 31.2 The clover leaf secondary structure of yeast tRNA^Phe. The base pairs of the secondary structure are indicated by bold lines linking both nitrogen bases. The tertiary interactions between two nitrogen bases are indicated by dashed lines. The longest subgraph that appears at least twice is shown in boxes.

$G - A$ and $U - U$. RNA 3-D structures are composed of double-helical regions (similar to the DNA double-helix), hairpin loops, multi-branched loops, and bulges that result, in part, from the nitrogen base interactions (Major and Gautheret, 1996) (see Figure 31.2).

Secondary structure information is encoded in a graph of relations, $G = (V, E)$, where V is the set of vertices labeled by one of {A,C,G,U} representing each type of nucleotides, and E is the set of edges representing the structural relations between two nucleotides. In the case of adjacent nucleotides in the sequence, the direction of the edges follows the phosphodiester linkage from 5'→3', and undirected edges are used for non-adjacent nucleotides, such as for H-bonding relations (see Figure 31.3). A linear representation was developed to simplify the computer storage of relational graphs. In these graphs, *helical* relations are indicated by vertical bars (|), adjacent *non-helical* relations by slashes (/), *Watson-Crick* base-pairing relations by dashes (−), *other than Watson-Crick* base pairing relations by dots (.), and all other *tertiary* relations by tildes (∼). Figure 31.3 shows the largest recurrent subgraph in yeast tRNA^Phe using the graph and linear representations.

31.3   SEARCHING MOTIFS

The problem in finding structural motifs in RNA secondary structures can be divided into two distinct tasks. First is an enumeration of all possible subgraphs and, second, is their isomorphic classification.

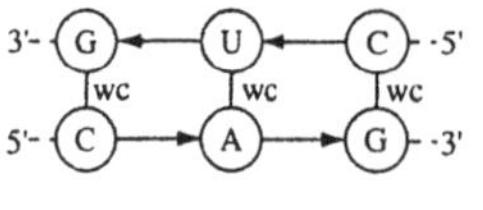

A29 C28 C40 G30 G42 U41 ;
(2-1) (1-4) (4I3) (2I5) (1I6) (3-6) (6-5)

A64 C49 C63 G51 G65 U50 ;
(3-1) (1-5) (2I5) (4I3) (6I1) (2-6) (6-4)

a.   b.

Figure 31.3   The largest recurrent subgraph in yeast tRNA^{Phe}. (a) Graph of relations. (b) A linear representation of the motif shown in (a) as used in the motif database. Note that the linear representation is not unique for each motif since the nucleotides can be ordered in many different ways.

31.3.1   Enumeration

The size of interesting RNA sequences range from several tens to more than three thousand nucleotides. The number of subgraphs grows rapidly as its size increases, and therefore a straightforward enumeration is often impractical. Different heuristics have been developed to cope with this overwhelming data, and must be used in the identification of RNA motifs. One of them is to limit the motif size to no more than 15 nucleotides. Another one consists of retaining only the most *significant* motifs. One possible definition of significance is the number of occurrences. For example, we can identify a motif if the corresponding subgraph occurs at least p times in all considered secondary structures. For this demonstration, we decided to eliminate all the subgraphs that appear only once by fixing the value of p to 2 (see Figure 31.4). However, in the building of the actual RNA motif database, a more precise evaluation of the significance will be made, based on the relevance of a motif's constituents.

Central to the motif identification process is the notion of incremental enumeration. In order to find the subgraphs of size n, we consider the subgraphs of size $n - 1$. The subgraphs of size $n - 1$ are extended by connecting the nodes that are connected to it from the secondary structure. This approach allows us to find all motifs since any subgraph of size n occurring q times contains at least one occurrence of a subgraph of size $n - 1$ that occurs at least q times.

31.3.2   Graph Isomorphism

The classification of subgraphs requires an efficient graph isomorphism algorithm. However, the graph isomorphism problem is one of the long-standing intractable problems in computer science. Given two graphs, $G_A = (V_A, E_A)$ and $G_B = (V_B, E_B)$, the problem is to establish the existence of a one-to-one mapping function, f, from V_A to V_B such that $(i, j) \in E_A$ if and only if $(f(i), f(j)) \in E_B$.

This problem belongs to NP, the class of problems that can only be decided by a *nondeterministic polynomial time algorithm* and hence, no subexponential running time algorithm is known to solve it. Nevertheless, specific RNA secondary structure information allows us to split the isomorphism determination in three stages of increasing complexity. First, a comparison is made between two subgraph vertices, based on their respective type and number of relations, or *degree*. Then, if the subgraphs contain the same nucleotides, their edges are compared, and if they are equal, a

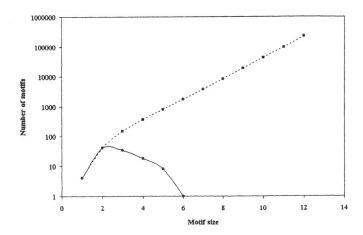

Figure 31.4   Changing the definition of significance. The graph shows the reduction in the number of motifs by using a different definition of significance in yeast tRNA^{Phe}. The dashed line shows the number of motifs occurring at least once, $p = 1$. The plain line shows the number of motifs occurring at least twice, $p = 2$.

depth-first search is finally applied to verify their isomorphism. The algorithm for this depth-first search was adapted from (Ullmann, 1976) and can be described as follows.

Let M_A and M_B be the adjacency matrices of each subgraphs, A and B, defined similarly as:

$$M_k[i][j] = \begin{cases} t_{ij} & if\ edge\ (i,j) \in E_k \\ 0 & otherwise \end{cases}$$

where $k \in \{A, B\}$, and $t_{ij} \in \{-, |, /, ., \sim\}$ is the type of relation between vertices i and j. Let M_0 be the matrix of possible equivalences between subgraphs, A and B, where $M_0[i][j] = 1$ if node i in A and node j in B are the same nucleotides and have the same degree. The algorithm generates all permutations of equivalent vertices in the subgraphs based on M_0, stores them in the form of a vector of equivalences H, and tests for isomorphism for each permutation. The two subgraphs are isomorphic if and only if $M_A[i][j] = M_B[[H[i]][H[j]]$ for all i and j. The details of the third stage of the algorithm are shown in Figure 31.5.

A simple analysis of the algorithm allows us to evaluate the time required during the first two stages. It is $O(max(|E|, |V|) \log(max(|E|, |V|)))$, due to the sorting of the vertices and edges. The third stage takes time in $O(|V|^2)$ for each permutation. Section 31.4 gives an estimate of the number of permutations based on the results of applying the algorithm on three different secondary structures.

Several alternative approaches to the depth-first search algorithm were considered. All required modifications of the graph structure in order to meet the characteristics of

```
GraphIsomorphism(depth): Boolean
  finished := false;
  i := 1;
  while ( !finished and i <= graphSize ) do
    if ( M0[depth][i] = 1 and G[i] = 0 )
      H[depth] := i;
      G[i] := depth;
      if ( depth < graphSize )
        finished := GraphIsomorphism( depth+1 );
      else
        return ( TestIsomorphism ( MA, MB, H ) );
      H[depth] := 0;
      G[i] := 0;
    i := i + 1;
  return finished;

TestIsomorphism ( MA, MB, H ): Boolean
  for j := 1 to graphSize do
    for k := 1 to graphSize do
      if ( !( MA[j][k] = MB[[H[j]][H[k]] ) )
        return false;
  return true;
```

Figure 31.5 Pseudo-code for the third stage in the isomorphism evaluation. The function is initially called with depth=1.

the concerned graphs. For instance, (Hopcroft and Wong, 1974) showed that isomorphism of planar graph can theoretically be tested in linear time. If tertiary interactions are not considered, a secondary structure becomes a planar graph which would allow the use of the approach proposed by Hopcroft and Wong.

31.4 APPLICATIONS

Prior to the construction of a complete RNA motif database, our algorithms were tested on a reduced number of secondary structures, namely, the yeast tRNA^{Phe} (Kim *et al.*, 1974), and *Escherichia coli* 16s and 23s rRNAs (Noller and Woese, 1981).

31.4.1 Secondary structure motifs

For the considered structures, a minimum occurrence of two ($p = 2$) was used to define the significance of a motif. In fact, without this distinction, the number of motifs grows exponentially and becomes intractable even for the small secondary structure of the yeast tRNA^{Phe} (see Figure 31.4).

The largest motif identified in the yeast tRNA^{Phe} is composed of three base pairs and appears twice, as shown in Figure 31.3. The rRNA secondary structures of *Escherichia*

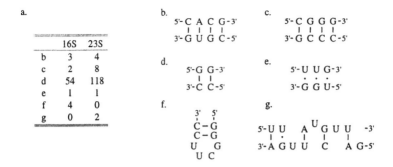

Figure 31.6 Interesting motifs found in E. Coli 16S and 23S rRNAs. (a) Their respective occurrences, (b) and (c) stems, (d) double G-C pairing, (e) triple G-U mismatches, (f) hairpin loop, and (g) one of the longest motifs found.

coli contains 1542 (16S) and 2904 (23S) nucleotides. These secondary structures were first treated separately, and then together. A large number of motifs (see Fig 31.6) was found, but no motifs of size larger than 14 nucleotides were found.

Several stems, such as those shown in Figures 31.6(b) and 31.6(c), as well as more complex motifs such as the hairpin loop of Figure 31.6(f) which appears four times in the 16S rRNA, were among the motifs found by our algorithm. The double guanine-cytosine Watson-Crick base-pairing (see Figure 31.6(d)) appears twice as many times as any other motif of the same size. This suggests that this particular tandem of base pairs might have been selected for a peculiar stability or function. Also of interest is the occurrence in both rRNAs of a triple G-U mismatches (see Figure 31.6(e)), which embed the double G-U mismatches that was identified by (Gautheret *et al.*, 1995).

Many more motifs were found in these preliminary studies, and their significance will be determined precisely when the RNA motif database will be constructed.

31.4.2 Algorithm Efficiency

In order to determine the efficiency of the algorithm, a simple benchmark was used. First, the time required by the construction of a given set of motifs was evaluated. Second, the mean number of evaluated permutations for each subgraph pair was determined. These evaluations were made using the *E.coli* rRNAs, as above, on a Silicon Graphics Origin 2000 equipped with R10000 CPUs.

The 34,350 identified motifs that range from size 1 to 14 were found in less than seven minutes. This is rather fast considering that more than one million subgraphs were tested for isomorphism. For all motif sizes, the last stage of the isomorphism algorithm, involving a depth-first search of all possible permutations of vertices, took $O(|V|)$ trials. It is worth noting that more computation is required on average to compare two non-isomorphic subgraphs than to compare two isomorphic subgraphs since in the former case all permutations need to be evaluated, whereas the evaluation of only a subset is required in the latter case. The complexity of the entire algo-

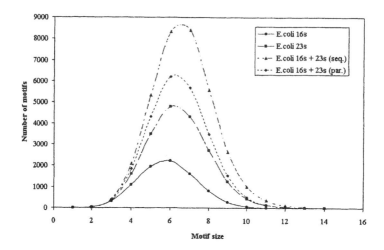

Figure 31.7   Variation of the number of motifs found in Escherichia Coli rRNAs.

rithm would then be $O(max(|E|, |V|) \log(max(|E|, |V|)))$ for the first two stages, and $O(|V|^3)$ otherwise. This satisfies Corneil and Gotlieb's criterion, that is, a graph isomorphism algorithm is efficient if the running time is polynomial (Corneil and Gotlieb, 1970). Over 5000 secondary structures will be considered for the construction of the RNA motif database. Since larger motifs are expected to be found, we predict a running to be bounded by a higher degree polynomial.

31.5   PERSPECTIVES

The described algorithm defines the basic component for the creation of an RNA motif database. This algorithm allowed us to identify and classify the structural motifs of three secondary structures.

Two approaches can be adopted for the building of a motif database. First is sequentially, that is by finding all significant motifs for each RNA one after the other. This approach has the downside that even when the most relevant motifs are present in a majority of secondary structures, one motif may be missed if it is not found at least p times in one secondary structure. Second is in parallel, that is by considering all secondary structures at once. In this case, all motifs would be identified but a very efficient external memory would be required since the number of subgraphs that must be considered for each motif size is so large that they cannot all be kept in main memory. Both approaches can be parallelized since the subgraph enumeration can be done independently for each structure and motif size. The curves in Figure 31.7 illustrate the number of motifs that would have not been identified by treating the structures independently. These results suggest that the first approach is not appropriate for the study of RNA motifs.

Several improvements will be necessary before we can apply our algorithm to a large number of secondary structures. The parallelization of the enumeration and isomorphism evaluation procedures will be considered. The only other option would be to neglect tertiary information and use a planar graph representation.

Knowledge of RNA motifs will allow us to develop secondary structure energy potentials, such as those used in Zuker's algorithm (Zuker and Sankoff, 1984). The basic assumption used is that a secondary structure containing motifs would be favored. We would even be in a position to evaluate how much motifs a secondary structure must contain to be valid. Finally, motif three-dimensional information would provide a library of building blocks for three-dimensional modeling.

Acknowledgments

We thank Elie Hanna for his early work on this project and useful discussions. This work is funded by the Medical Research Council (MRC) of Canada. Patrick Gendron holds a FCAR scholarship. Francois Major is an MRC fellow.

References

Corneil, D. and Gotlieb, C. (1970). An efficient algorithm for graph isomorphism. *J. ACM*, 17(1):51–64.

Gautheret, D., Konings, D., and Gutell, R. (1995). G.U base pairing motifs in ribosomal RNA. *RNA*, 1(8):807.

Hopcroft, J. and Wong, J. (1974). Linear time algorithm for isomorphism of planar graphs. In *6th Annual ACM Symposium on Theory of Comput.*, pages 172–184, Seattle, WA.

Kim, S. *et al.* (1974). Three-dimensional tertiary structure of yeast phenylalanine transfer RNA. *Science*, 185:435.

Major, F. and Gautheret, D. (1996). Computer modeling of RNA three-dimensional structure. In Meyers, R. A., editor, *Molecular Biology and Biotechnology*, pages 847–850. VCH Publishers Inc., New York.

Noller, H. and Woese, C. (1981). Secondary structure of 16S ribosomal RNA. *Science*, 212:403–411.

Ullmann, J. (1976). An algorithm for subgraph isomorphism. *J. ACM*, 23(1):31–42.

Zuker, M. and Sankoff, D. (1984). RNA secondary structures and their prediction. *Bulletin of Mathematical Biology*, 46(4):591.

32 COMPUTATIONAL PROTEIN FOLDING

John R. Gunn

Département de Chimie, Université de Montréal
Centre de Recherche en Calcul Appliqué, Canada

gunnj@cerca.umontreal.ca

Abstract: The protein-folding problem presents an enormous challenge for computer simulation due to the exponentially large number of possible conformations available to a protein molecule. Some methods will be described which we have recently developed to find minimum-energy structures for simple reduced models of proteins, with an emphasis on finding ways to reduce the scaling of the computational effort with increasing system size. The simulation is based on a combination of Monte Carlo Simulated Annealing and Genetic Algorithm techniques which are integrated into a single framework. The selection cycle of the genetic algorithm is carried out at the same temperature as the mutations or, alternatively, the crossover cycle can be considered as a type of Monte Carlo trial move, such that each temperature annealing step corresponds to a new generation. The scaling aspect is addressed by using a hierarchical approach in constructing and evaluating new structures. The sequence is divided up into segments, and the mutation step consists of replacing an entire segment with a choice from a pre-selected list. This list is in turn constructed from a list of smaller segments, and the number of overall conformations can thus be pruned at each level of selection. The scoring function is evaluated by first calculating an approximate value using only the positions of segment endpoints, and the complete distance matrix is then only calculated for structures which satisfy a series of cutoffs applied to the approximate functions. The implementation of this algorithm on a parallel computer is also discussed.

Keywords: protein folding, genetic algorithm, scaling, screening, parallelization.

32.1 INTRODUCTION

Proteins are linear polymers of amino acids which adopt unique compact structures with a specific biological activity. Each amino acid consists of a common repeat unit (the backbone) and one of twenty naturally-occurring functional groups (the side-chains). The classic problem of protein folding consists of trying to determine the three-dimensional native structure of a protein from its amino-acid sequence. This is a problem with tremendous implications because there is a rapidly-growing number of sequences being identified (from DNA sequencing, for example), but the experimental determination of structure remains a difficult and time-consuming process. Structure-based drug design, however, relies on knowledge of the structure of a target protein in order to design suitable activators or inhibitors.

From the point of view of computer simulation, it is generally assumed that the native structure minimizes a (more-or-less) known thermodynamic potential (the "free energy"), and that the problem therefore reduces to one of function minimization. The difficulty of the problem lies in the astronomical number of possible structures, with each amino-acid unit (residue) able to adopt any of several distinct conformations. For a medium-sized protein on the order of 150 residues, this leads to in excess of 10^{100} combinations. Furthermore, protein folding falls into the class of NP-complete problems, growing exponentially with the number of residues, so that current methods which work reasonably well for smaller molecules are helpless in the face of sequences which can reach several hundred residues.

The problem is further complicated by the nature of the potential function which consists typically of a large number ($\mathcal{O}(N^2)$) of interactions between chemically similar side-chains, which leads to a highly degenerate surface with many comparable local minima. In addition, there are steep barriers between minima, which gives rise to a highly frustrated system at many different scales (Bryngelson et al., 1995).

The goal of our current work is to develop strategies which will improve the efficiency of the conformational search and help to overcome the scaling problem for larger proteins. This includes a multi-level screening algorithm for evaluating the potential function in order to avoid unnecessary calculations for unlikely structures, as well as a multi-level build-up algorithm for constructing new trial structures using fragments which have been screened according to their geometries. The function minimization is carried out using a genetic algorithm which also permits large-scale changes in the generation of cross-over structures. These techniques will be described following a brief description of the protein representation and potential function.

32.2   THE PROTEIN MODEL

We use a reduced model representation of the protein molecule introduced previously (Gunn, 1996; Gunn, 1997). The description of each residue consists of the positions of six atoms, N, H, C_α, C_β, C', and O. All internal coordinates are held fixed at standard geometries with the exception of the ϕ and ψ dihedral angles around the N–C_α and C_α–C' bonds, respectively, which are thus the only degrees of freedom in the model. The possible values of ϕ and ψ are further restricted to a discrete set of points on the two-dimensional Ramachandran map, corresponding to representative conformations in the energetically allowed regions. For the present work we have chosen a uniform distribution within the allowed regions with a spacing of roughly seven degrees. Since there is no explicit potential energy term which depends on ϕ and ψ, this corresponds to a hard-wall-type effective torsional potential which is used to bias the choice of trial moves. The conformation of the molecule can thus be simply represented by a series of integers labeling the conformational state of each residue. The secondary structure is imposed simply by assigning specified residues to conformations corresponding to idealized helix or strand geometries ($\{-65, -40\}$ and $\{-147, 145\}$, respectively) and holding them fixed throughout the simulation. In this model all residues (including glycine) are equivalent from a geometric perspective, corresponding to a physical model of poly-alanine (thereby neglecting the ring closure and possible *cis* conformation of proline.) This is not however a necessary restriction, and is used here simply for convenience.

32.3 POTENTIAL TABLES AND ENERGY SCREENING

The potential function has the form of a generalized contact potential between residues based on selected interatomic distances. Five distances were used in the present work, namely: $|\mathbf{r}_{C_\beta,i} - \mathbf{r}_{C_\beta,j}|$, $|\mathbf{r}_{C_\beta,i} - \mathbf{r}_{C',j}|$, $|\mathbf{r}_{C_\beta,i} - \mathbf{r}_{N,j}|$, $|\mathbf{r}_{C',i} - \mathbf{r}_{C_\beta,j}|$, and $|\mathbf{r}_{N,i} - \mathbf{r}_{C_\beta,j}|$. The N, C', and C_β atoms are used as reference points, with the C_α atoms being necessarily located at the pyramidal positions in the middle. Besides the $C_\beta - C_\beta$ distance, the potential makes use of the distances between the C_β atoms and the other backbone atoms to provide an indication of the backbone orientation. The potential is thus a five-dimensional look-up table with, in principle, an independent value for each relative orientation of each amino-acid pair and with a resolution determined by the discretization of the distance indices.

The potential function consists of three components: an empirical contact potential, an excluded volume interaction, and a pressure term to generate compact structures. The contact potential is derived from a statistical analysis of the Protein Data Bank (PDB). Using the so-called quasi-chemical approximation, the distribution of amino acids in a protein is assumed to follow the canonical distribution, $\rho(r) = e^{-E(r)/kT}$. The potential of mean force is therefore defined as the logarithm of the observed histogram of distances for each pair of amino acids. In principle, this potential should take into account all of the forces which produce the observed distribution, so that there is no need to explicitly consider side-chain conformation or solvent interactions. The parameters used here were developed by (Bryant and Lawrence, 1993) for use in threading calculations.

Because the threading potential was parameterized for evaluating the compatibility of different sequences with known crystal structures, it does not take into account differences in density and packing, for which additional terms in the scoring function are required. The excluded volume is represented by a penalty term which is assigned to any combination of the five distances which cannot be satisfied without creating atomic overlaps. This penalty is a variable parameter which can be modified during the course of the minimization. The final term is simply proportional to the radius of gyration of the molecule, and is used to ensure that compact structures will be formed with the correct overall size. The use of variable weights for the excluded volume and pressure terms, a strategy called parameter annealing, allows the shape of the potential surface to be smoothed out in the initial stages of the simulation. By removing local barriers due to packing effects, the algorithm can more easily move across the potential surface and determine the topology of the structure using the long-range empirical potential before refining the structure to take into account local interactions.

During the course of a simulation, most trial structures will be very high in energy and will be rejected. The most time-consuming part of the calculation is the evaluation of these energies which requires computing the distances between all pairs of residues. Methods used in molecular dynamics calculations, such as neighbor lists, cannot be used in this case because the interaction distance is comparable to the size of the molecule. The idea of the hierarchical energy screening is to screen new structures using a cheaper scoring function which will permit most structures to be rejected without the need for the full potential evaluation. This is similar in principle to a related procedure (Johnson *et al.*, 1992) which can be implemented exactly for a model where there is a short-range cut-off and the screening potential can be used to rigorously rule out any possible interactions between distant segments. It was shown that the hierarchical algorithm in that case leads to a linear scaling of computational effort with chain length. In order to accommodate long-range forces, there is necessarily a trade-off between increased efficiency and the premature rejection of trial moves which would otherwise be accepted.

Rather than establish a rigorous bound on the energy, the approach taken here is simply to estimate the energy and to bias the set of trial moves evaluated with the full potential towards those more likely to have good scores. In order for the efficiency to be improved. it is only necessary that the trial moves be "enriched" at each step, so that the acceptance rate will be increased and a larger number of trial moves can be attempted in the same total time. Within this framework, the full potential is the five-index table as described above, and the screening functions are coarse-grained averages of the full potential where only a subset of the distances are used and the energy is averaged over the remaining coordinates.

The first level of coarse-graining is the reduction of the five-index table to a single-index residue-residue potential, based on the C_α interatomic distances. This is done by calculating all possible combinations of the five distances which are compatible with a given bin for the C_α distance. The value of the potential is then a weighted average of all such possible values of the full potential. The average is weighted in this case to better represent the minima of the full potential within bins which cover a wide range of energies, and also as a means of including the overlap penalty. Since this penalty varies during the simulation, the fraction of overlaps in each bin is stored along with the average of non-overlapping energies so that the total weighted average can be recalculated. This allows for the parameter annealing to be carried out without having to repeat the coarse-graining procedure.

Higher levels of approximation were calculated as potentials acting between multi-residue segments of the chain, and were constructed by using the first-level approximation as a residue-residue potential. There is therefore no explicit calculation of the overlap penalty since the first-level approximation already includes the relative contribution of the overlap energy as a function of distance. The segments are defined by including only every nth C_α in a distance matrix and interpolating the positions of the skipped residues. The simplest approximation would be a single-index potential acting between the remaining residues, where the interaction at each is a sum of all the amino acids closest to it. This is a plausible approximation to the hydrophobic energy, but fails badly at estimating the number of overlaps since it would count n^2

each time two C_α's were too close and zero otherwise. A compromise is used in the present work, where the potential is a two-index function acting between a point and a pair of consecutive points defining a segment. The total contribution of each residue pair is determined by assigning its energy proportionally to the interactions among its neighboring segment endpoints. This is clearly not an accurate model of the real spatial relationships of the residues, but is intended simply to smooth out the potential in such a way that the total error in the energy of a molecule is more evenly distributed. In particular, the error in counting overlaps should be reduced to $\mathcal{O}(n)$.

Three levels of screening function are used in this work. At each step, the approximate energy is compared to a fixed threshold value, and only those energies which are below that value are evaluated at the next level. The threshold values are determined at each iteration by calculating all approximate energies for the current ensemble of structures in order to establish the maximum acceptable error at each level. This error can be adjusted in order to optimize the overall efficiency. Cut-offs which are too generous do not reject a sufficient number of structures to allow a significantly larger number of trial moves, and cut-offs which are too strict will erroneously reject too many good structures for the minimization to converge properly.

Tests were carried out with various choices of the coarse-graining scheme and higher level approximations. Using every third residue and every ninth residue, respectively, to calculate distances were found to be the best. Larger scale coarse-graining was found to be too inaccurate to be of much use, and more closely-spaced approximations (e.g. doubling the gap at each step) did not offer enough speed difference between levels to justify using all of them.

The algorithm was tested on the following three examples of small globular proteins: sperm whale myoglobin (1MBO), 146 residues, 8 helices and 25% loop; L7/L12 ribosomal protein C-terminal domain (1CTF), 66 residues, 3 helices, 3 strands, and 26% loop; and phage 434 repressor N-terminal domain (1R69), 60 residues, 5 helices, and 33% loop.

The scaling of the performance with sequence length was evaluated by carrying out similar experiments for each protein. In all cases, the choice of the (9,3,1) potential approximations was still found to be superior, suggesting that the efficiency depends on the spacing of the approximations, independent of the overall size. The fact that the approximate potentials were evaluated with fewer segments in the smaller molecules was reflected in the need for more generous cut-off criteria to obtain good results. The results of this comparison are summarized in Table 32.1 where the times per move are shown with and without screening. In all cases, lower energies were obtained in less time with the screening, however the relative advantage was greater for 1MBO than for the smaller molecules. While the time per move in the unscreened case is very nearly proportional to N^2, it is reduced to close to linear with the screening.

32.4   SEGMENT LISTS

In the hierarchical algorithm, trial moves are generated and evaluated in three different steps. At the simplest level, segments of three residues (triplets) are generated by choosing three conformational states at random. Each triplet is immediately accepted or rejected according to whether or not the orientation of its endpoints falls into an al-

Table 32.1 Comparison of CPU times with and without screening for different sequence lengths.

Molecule		1MBO	1CTF	1R69
Number of Residues		146	66	60
CPU time [ms]/ trial move	no screening	23.5	5.04	4.36
	with screening	3.45	1.78	1.19
	ratio	6.8	2.8	3.7
CPU time [s]/ accepted move	no screening	1.00	0.365	0.439
	with screening	0.365	0.291	0.332
	ratio	2.7	1.3	1.3

lowed region of triplet conformational space. Three residues contain a sufficient number of degrees of freedom to make this non-trivial in the sense that different sequences of backbone dihedral angles can produce the same endpoints, however the segment is small enough that the screening of random conformations can be done fairly cheaply. The second level consists of complete loop segments as determined by the secondary structure. These loops are evolved from previously existing structures by using the set of triplets from the first level as trial moves and by evaluating new loops based on the difference in overall geometry from the starting loop. The final level then corresponds to the entire molecule, for which the trial moves consist of substituting entire loops with the new loops generated in the second level. It is only at this final level that the structure is evaluated by calculating the full potential using the table of pairwise contact energies.

For the purposes of evaluating triplets and loops in the first two levels, an endpoint geometry is introduced which is specified by five coordinates that determine the orientation of the final residue relative to the first one. The objective is to estimate the "lever-arm" effect, that is, the amount by which the ends of the molecule will move following a change in the conformation of a series of interior residues, since this will dominate the total energy of the molecule. The trial moves in the simulation are carried out by using the atomic coordinates of the rigid $H-N-C_\alpha$ unit to define an absolute coordinate system for each residue. This allows the net rotation between any two residues to be determined and therefore makes it trivial to splice together different structures and incorporate new segments into existing structures, including the necessary rigid-body rotations.

The advantage of using the hierarchical algorithm is that the Monte Carlo step size, in this case the maximum difference between loops, can be controlled by the choice of cutoff in the loop generation thereby eliminating trial moves which would be too large to have a reasonable chance of being accepted. Although the set of loops in the list is a limited subset of all possible conformations satisfying the same cutoff, it is still many orders of magnitude larger than the number of attempted Monte Carlo moves and therefore imposes a negligible restriction on the conformational sampling. The hierarchical screening of loops and triplets by geometry, however, is a much more efficient method of restricting the trial moves than simply applying a geometry cutoff to randomly generated loop conformations. The acceptance rate at each level is adjusted by tightening the selection criteria as the number of degrees of freedom increases.

32.5   THE GENETIC MONTE CARLO MINIMIZATION

The genetic algorithm consists of three main components: mutation, hybridization, and selection. The "genome" corresponding to each member of the ensemble of structures is the series of loop conformations defined by the loop lists as described above. The mutation step involves making a random change to the genome of an individual structure, and is therefore equivalent to a standard Monte Carlo move in which the trial moves consists of changing the conformation of a single loop. The new loop is accepted with a probability of one if the new structure is lower in energy, and with a probability of $\exp(-\beta\Delta E)$ if the new structure is higher in energy, with β being the inverse annealing temperature.

The hybridization and selection steps, on the other hand, couple different members of the ensemble and give the genetic algorithm its additional advantage. Hybridization consists of creating a new structure by combining the genomes of two existing structures. One of the secondary structural elements is randomly selected as the splice point, and a new structure is created by taking all loop conformations on the N-terminal side of the splice point from one parent structure and all loop conformations on the C-terminal side from the other, with at least two loops coming from each parent. The two parent structures are chosen randomly among the existing ensemble according to a Boltzmann distribution. A population of new structures is created by keeping those distinct hybrids whose energies are less than a maximum cut-off determined by the average energy of the two parent structures.

A selection algorithm is then used in which members of the newly created hybrid population are allowed to compete with and displace existing structures from the original ensemble of parents, thereby modifying the new ensemble which will be passed on to the next iteration. This is carried out during an equilibration cycle in which two structures, one from each of the old and new populations, are randomly chosen and their exchange is accepted or rejected according to the probability $\exp(-\beta\Delta E)$ exactly as in the mutation step. Trial exchanges are attempted until the two populations have been completely scrambled and the total energy of the new ensemble is stable. The new generation is thus a combination of old and new structures selected from a Boltzmann distribution, and will not generally correspond to the set of lowest-energy structures. However, the best structures will have increasingly higher probabilities of being selected, with the single lowest-energy structure almost certain to be one of those in the new ensemble.

One iteration is defined as a complete cycle of loop-list generation followed by one generation of the genetic algorithm. Each iteration is carried out at a specified temperature and within a limited region of the total conformational space. This region is defined by the cut-off used to select new loops for the loop lists, and the space within this region is further discretized by selecting a fixed number of loops to be used during the subsequent iteration. At the beginning of each iteration, the ensemble is used to construct new loop lists. The mutation and selection steps of the genetic algorithm then constitute a trajectory within this space which satisfies the Metropolis algorithm at a specified temperature. The end result is a new ensemble located somewhere within this space which is then used as a starting point for the following iteration. As the annealing temperature is reduced the population of structures converges to a set of low-energy local minima.

In order to provide a reference potential for which the native structure is known *a priori* to be a unique global minimum, a simplified version was developed in which the empirical contact potential was replaced with a calculation of the r.m.s. deviation from the native structure. The same five-index table was used as above to assign an energy penalty to sterically forbidden conformations and the term proportional to the radius of gyration was used as well to provide a model for the hydrophobic pressure of the solvent and to compensate for the repulsive nature of the excluded volume interaction. The overall potential is thus a combination of smooth long-range terms (the calculation of the r.m.s. deviation and the radius of gyration) and stiff short-range terms (the look-up table penalty), such that the r.m.s. deviation is minimized within realistic constraints.

The hierarchical algorithm was tested using the same three proteins used earlier (1R69, 1CTF, and 1MBO), and also a larger molecule, porcine cytosolic adenylate kinase (3ADK), 193 residues, 10 helices, 5 strands, and 32% loop. Results are shown in Table 32.2 for different variations of the algorithm. For each test protein, the potential was minimized using four different algorithms:

1. Simple Monte Carlo in which trial moves consist of changing the conformational state of a single residue,

2. Loop-list Monte Carlo in which trial moves consist of one- and two-loop substitutions from the loop lists,

3. Simple genetic algorithm in which simple Monte Carlo is combined with hybridization and selection cycles, and

4. The full hierarchical algorithm.

The number of iterations in each case was adjusted in order to give approximately equal total run times.

In each case, the hierarchical algorithm shows a net improvement. For the larger molecules, this is clearly more than the sum of the effects of using the loop lists and the genetic algorithm separately. The superior convergence of the algorithm is also shown by the much narrower distributions which are obtained. These results are close to the limit of what can be obtained with the simplified model being used which has fixed residue structures (excluding ϕ and ψ), idealized rigid helices and strands, and a finite set of possible conformations.

32.6   DATA STORAGE AND PARALLELIZATION

The size of the potential table depends on the distance resolution used in defining the indices, and grows as the fifth power of the number of distance bins if all distinct

Table 32.2   Results of minimizing the reference potential for different elements of the hierarchical algorithm. Average RMS values are shown in each case along with the standard deviation.

		1R69	1CTF	1MBO	3ADK
Simple MC	Ave. RMS	3.83	3.56	5.69	11.09
	Std. dev.	0.58	0.49	0.70	1.13
Looplist MC	Ave. RMS	2.78	2.57	4.35	9.46
	Std. dev.	0.32	0.35	0.68	1.19
Simple GA	Ave. RMS	2.95	2.83	4.70	10.27
	Std. dev.	0.61	0.58	0.94	1.52
Hierarchical	Ave. RMS	2.05	1.89	2.59	4.74
	Std. dev.	0.17	0.07	0.18	0.33

combinations are used. However, due to the geometric constraints of the bonding and the excluded volume of the side-chain atoms, most of the possible combinations of the five indices will in fact never be observed. In order to avoid having to store values for all possible combinations in the table, the lookup is carried out sequentially, and the number of possibilities is pruned with the addition of each new index. The initial lookup thus consists of a table of pointers with a default value for all combinations which have no non-zero entries for any choice of the remaining indices, and unique labels for the remainder. Adding the next index then only requires a table of length $N \times$(the number of distinct combinations), rather than N^3. The final step will thus be a pointer directly to the non-zero entries in the table. While this procedure would not be particularly useful for a random distribution of data, because of the geometrical relationship of the indices the potential table in this case has a banded structure making this a reasonable strategy. In addition, a further simplifying assumption is made that the structure of the potential table will be the same for each amino-acid pair. Since the side-chains are of different sizes, this requires the inclusion of a number of zero values in the potential table, but at the same time reduces the size of the pointer arrays by a substantial fraction. The pointer arrays in this case were generated so as to preserve a unique label for any combination of the distance indices which had a non-zero entry for at least one amino-acid pair. In the implementation used here, there are a total of 16 indices for each distance yielding a total number of addressable energies of 2.20×10^8. The procedure described here used pointer arrays with a combined length of 5.80×10^4 and a potential table of length 2.89×10^6 to represent 2.37×10^6 non-zero entries.

There are three important ways in which parallelism can be exploited in the hierarchical algorithm, depending on the type of platform being used. The first and most obvious is at the data-parallel level where the distinct structures in the ensemble are distributed across the nodes and the trial moves are carried out independently. Relative performance can be improved in this case simply by increasing the size of the ensemble, and virtually no communication is required. This structure can also be adapted to include the use of the genetic algorithm as has been done previously (Gunn *et al.*,

1994). There are, however, two instances where this simple approach is limited, and other types of parallelism need to be introduced. The first concerns the segment list, for which the advantage lies in the number of independent structures making use of the same list. What is in principle a globally accessible array must therefore be distributed for use in data-parallel trial moves. The second concerns the calculation of inter-residue distances where the data-parallel structure can never overcome the N^2 scaling of the distance matrix and thus reduce its cost relative to the other parts of the program. This must be addressed by invoking a multiple-instruction model in which different nodes are allowed to evaluate different parts of the distance matrix of the same structure. Finally, these issues are coupled in the case of the potential table. The large size of the table means that it should be distributed across the nodes rather than copied in its entirety to each one. That however requires that different nodes calculate different contributions to the energy of each structure. The major difficulties in parallelization are thus related to combining these different models into a single program and trying to avoid excessive inter-node communication.

The algorithm was originally implemented on a Thinking Machines CM-5, which consists of a partition manager (PM) that controls the execution, and a number of nodes that control communication and memory access. Each node in turn consists of four vector processors (VUs) each of which has its own dedicated memory. Execution can be carried out in either a data-parallel model in which each node carries out the same instructions and communication is invoked automatically for array operations which span more than one VU, or in a local-node model in which each node carries out different instructions while retaining a data-parallel structure at the level of the VUs. Communication in the latter case can be handled through the use of explicit message passing between nodes, or by using a global program to move data and call subroutines which then execute locally on the nodes. The optimum performance on the CM-5 is obtained in the VUs, with the major bottlenecks coming from inter-node communication. The VU performance is further dependent on the vector length of each array operation thus requiring further data parallelism within each VU.

The generation of the triplet lists was carried out in parallel across all of the VUs, with structures being generated independently and then screened using the same geometry mask. Trial moves were then carried out with each node only using the triplet list stored locally in order to eliminate all inter-node communication in this step. Periodically during the simulation, the entire triplet list was transferred from one node to the next in a ring pattern. In this way, each node eventually had access to all triplets in the list and the entire list was rebuilt when it had completed a full tour. The calculation of the residue distance matrix needed for the potential evaluation was somewhat more complicated. An earlier version of the program which did not use the energy screening functions was used, so that all distances were needed for all structures. For this step, the coordinates were copied between nodes, so that each had a copy of all structures. Each distance was thus calculated simultaneously for all structures in order to use the largest possible vector length on the VUs. Each node was assigned a list of distances to calculate, and these were carried out independently on each node. The gain in performance due to the parallelization was offset by the inter-node communication needed to ensure that all structures were up-to-date on each node, since the trial moves were carried out for each structure only on a specified node. The optimum performance was obtained using groups of two nodes, with each member of each pair working on half of the structures.

The serial implementation of the algorithm on a single workstation spends 80–85% of the total time in the potential evaluation routine, whereas for the parallel version on the CM-5 this is reduced to 41%. This suggests that the parallelization can be effective, but that the overall performance is then limited by other elements of the algorithm.

32.7 CONCLUSIONS

In order to overcome the exponential growth of the conformational sampling problem as a function of sequence length, techniques must be developed which attack the scaling of the problem. The hierarchical algorithm represents an important step in that direction, with a multi-level model for both the geometry of the molecule and the evaluation of the scoring function. The use of loop lists allows trial segments to be screened independently using local geometrical criteria in order to improve the acceptance rate of the overall mutations. Hierarchical energy evaluation allows unlikely structures to be cheaply rejected thereby reducing the average cost of screening trial moves. Each method divides the molecule up into segments so that the problem remains tractable at each step.

Further improvement will come from the implementation of an adaptive strategy, in which different levels of approximation are used only where necessary and the representation of the structure is calculated only to the level of resolution needed to evaluate it for a given level of detail in the scoring function. Work is currently being carried out in our lab along this direction.

Acknowledgments

This work has been supported financially by FCAR (Québec) and NSERC (Canada), whose contributions are gratefully acknowledged.

References

Bryant, S. and Lawrence, C. (1993). An empirical energy function for threading protein sequence through the folding motif. *Proteins Struct. Funct. Genet.*, 16:92.

Bryngelson, J., Onuchic, J., Socci, N., and Wolynes, P. (1995). Funnels, pathways, and the energy landscape of protein folding - a synthesis. *Proteins Struct. Funct. Genet.*, 21:167.

Gunn, J. (1996). Minimizing reduced-model proteins using a generalized hierarchical table-lookup potential function. *J. Phys. Chem.*, 100:3264.

Gunn, J. (1997). Sampling protein conformations using segment libraries and a genetic algorithm. *J. Chem. Phys.*, 106:4270.

Gunn, J., Monge, A., Friesner, R., and Marshall, C. (1994). Hierarchical algorithm for computer modeling of protein tertiary structure. *J. Phys. Chem.*, 98:702.

Johnson, L., Monge, A., and Friesner, R. (1992). A hierarchical algorithm for polymer simulations. *J. Chem. Phys.*, 97:9355.

IX Applications

33 DISTRIBUTED COMPUTING OVER HIGH-SPEED LOCAL AREA NETWORKS

Sherali Zeadally

Electrical Engineering Department,
University of Southern California, U.S.A.

zeadally@usc.edu

Abstract: Technological advances in networking, computer architecture, VLSI, and processor performance have led to the emergence of a wide range of multimedia applications, particularly those with high bandwidth, low latency, and real-time requirements. Typical examples include video conferencing, visualization, industrial applications, and scientific applications. In this paper, we first describe the design of a distributed visualization application. Then, we focus on the impact of different host architectures, high-speed networks, and operating systems on the performance of the distributed visualization application. Doing so highlights the performance that can be delivered by state-of-the-art technologies to end-user applications. Finally, we demonstrate the use of high-performance low-cost hardware (Pentium PCs)hardware and network technologies, such as Asynchronous Transfer Mode (ATM), to support high-bandwidth applications such as visualization in a distributed, networked environment.

Keywords: ATM, FDDI, latency, performance, throughput, multimedia, visualization.

33.1 INTRODUCTION

The recent development of powerful inexpensive desktop personal computers and high-speed networks have paved the way for the emergence of a wide range of network applications such as video distribution, computer imaging, distributed scientific computation, multimedia conferencing, and visualization. A key characteristic of these applications is that they often involve the transfer and manipulation of large amounts of data. In the past, high-performance applications such as visualization and scientific simulations used specialized expensive workstations and supercomputers. However, the decreasing costs and increasing performance of both computer hardware and networks now offer great potential for distributed network computing. Clusters of workstations can be connected to a high-speed Local Area Network (LAN) and be applied to solve different types of scientific problems (Sterling *et al.*, 1996; Lin *et al.*, 1995). With distributed network computing, a cluster of workstations can collaborate with each other to solve some common task (Tanenbaum and Renesse, 1985). One of the major benefits of distributed computing is its scalability in terms of the amount of computing power and resources available for large-scale applications. In this paper, we explore the performance of a distributed visualization application with different networks, host and processor architectures using experimental clusters of workstations and low-cost Personal Computers (PCs).

The rest of the paper is organized as follows. Section 33.2 describes the implementation of a distributed visualization application. The performance experiments and results obtained by experimenting with different networks and workstation architectures is discussed in Section 33.3. The impact of new technologies, such as ATM and low-cost Pentium architectures, on the performance of our visualization application are discussed in Section 33.4. Finally, Section 33.5 offers some concluding remarks and presents suggestions for future work.

33.2 A DISTRIBUTED VISUALIZATION APPLICATION

In order to experiment with high-bandwidth applications, we have developed multimedia applications such as video conferencing using high-quality uncompressed video and distributed visualization. In this paper, we report on performance results and experiences of a real-time interactive distributed visualization application we have developed running over high-speed LANs, 100 Megabits per second (Mbits/s) or higher.

The visualization application discussed in this work uses 8 bit/pixel, gray scale, volume-rendered CAT-scan medical images of a child's head. The head can be rotated and viewed at different angles. The image set consists of 20 image frames, each of size 512×512 pixels (*i.e.* almost 2.1 Mbits per frame) and stored in pixmap format. Using pixmap makes direct display by the X server possible without requiring any further manipulation. The graphical user interface of the visualization application supports simple operations such as play, stop, and rewind which allows a user to view the rendered image from different angles. Thus, a user such as a doctor can rotate the head and cut a slice window of the volume image. In addition, the user can zoom into the slice selected at a certain Z-depth chosen by the user and explore the internal structure of the head.

The algorithm used by the distributed visualization application is illustrated in Figure 33.1. It uses two servers. One server accepts requests (both the head image and slice window information) from the end user via the client and renders the volume image of a child's head (this server is henceforth referred to as the *head* server). The rendered images serve as orientation for the slices of the internal images. The head server also accepts window slice requests from the client and sends them to another server (henceforth referred to as the *compute* server) which calculates the slice window requested by the user. After receiving the slice window information, the head server combines it with the volume rendered image and sends the resultant image to a client process for display.

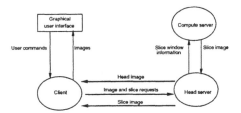

Figure 33.1   The distributed visualization algorithm.

There are many challenges that need to be addressed when supporting distributed high-performance applications. In the case of our distributed visualization application, the client and servers can either all reside on the same machine or be running on different physical machines in spatially separated locations. Professionals using these visualization applications demand high-resolution, lossless images because the quality of the images determines the quality of the diagnosis. In addition, the timely delivery of the images is also essential to the efficiency of the professional. For instance, long delays will impact the time spent with a patient. These characteristics place considerable demands on Quality of Service (QoS) parameters, such as high bandwidth and low-latency support for such professional applications. With the diverse requirements of desktop applications using a combination of media streams such as data, voice, and video, it is becoming increasingly important to provide QoS support at different levels namely, application, operating system, and network (Nahrstedt and Smith, 1994).

In previous work (Zeadally *et al.*, 1997), we have showed experimentally how various optimizations can be applied to multimedia applications to achieve optimal end-to-end application performance. These optimizations include gigabit networks, single-copy protocols, and X-shared memory support. In this paper, we have applied those optimizations to our distributed visualization application and studied the impact that host architectures and networks have in delivering high performance to end-user applications running in a *distributed heterogeneous* environment where computing, storage, and display platforms are physically distributed and connected via a high-speed LAN.

33.3 EXPERIMENTAL MEASUREMENTS AND RESULTS

This section describes the experimental arrangements of the different hardware and network configurations used in our tests. We used two different networks in our experiments, namely Fiber Distributed Data Interface (FDDI) and Jetstream. FDDI is a 100 Mbits/s LAN which uses a token ring protocol to schedule station access. Jetstream is a gigabit per second token ring LAN which uses copper coaxial or fiber optic cable for the physical link. The Jetstream network adapter for the HP 9000 Series 700 workstations is made up of two cards. One is called Afterburner which is equipped with 1 Mbyte of video random access memory used in a dual ported configuration. Afterburner is used to interface to the host system bus. The other card, Jetstream, is the link adapter used for network connectivity. The shared memory present on the Afterburner board enables the support of *single-copy* implementations of network protocols such as TCP/IP and UDP/IP. Briefly, the single-copy implementation of TCP with Afterburner allows network data to be copied *directly* from the network adapter to the user application. This is in contrast to the case of *two-copy* implementations where network data is first copied from the network adapter to a kernel buffer, followed by a second data copy from the kernel buffer to a user buffer. Standard UNIX kernels usually use two-copy TCP/IP or UDP/IP protocol stacks. Further details on Afterburner and Jetstream are given in (Dalton *et al.*, 1993) and (Watson *et al.*, 1994), respectively.

33.3.1   Latency

Many applications often involve the exchange of small packets of information, typically ranging from 64 to 512 bytes (Keeton *et al.*, 1995). For instance, client-server transaction-type applications and interactive video applications require fast response to small requests sent over the network. Thus, to ensure optimal response time especially for interactive applications, it is therefore important to ensure that the latency overhead introduced by both the operating system and communication layers be minimized. Some interesting works describing latency reduction schemes at the end system are described in (Damianakis *et al.*, 1996; Eicken *et al.*, 1995).

In this section, we first investigate the latency benefits obtained when using single-copy compared to two-copy protocol stacks. Then, using two-copy protocol stacks, we compare round-trip latency results of 100 Mbits/s FDDI with the gigabit/s Jetstream network. The round-trip latency measurements experiments were carried out between two HP 9000 Series 700 workstations (125 MHz PA-RISC). Both machines were running HP-UX 9.01 and were equipped with FDDI and Jetstream network adapters from Interphase Corporation (Interphase, 1996) and Hewlett Packard Laboratories, respectively. In the case of Jetstream, measurements were performed using single-copy TCP/IP and UDP/IP stacks, as well as standard two-copy TCP/IP and UDP/IP stacks. With FDDI, only two-copy TCP/IP and UDP/IP stacks were used.

We measured user-to-user round-trip latency between two machines using Netperf (Jones, 1993). Basically, a client machine sends an M-byte message to the server (timing starts) and waits to receive the M-byte message back. The interaction was repeated N times between the client and server, after which timing stops. From the N readings obtained, an average round-trip time for exchanging an M-byte message between the two workstations was calculated.

We first measured the user-user round-trip latency over the Jetstream gigabit network for TCP/IP and UDP/IP protocols. The variation of latency with message size using single-copy and two-copy TCP/IP and UDP/IP is shown in Figure 33.2. We note from Figure 33.2 (top) that the latency is almost independent of whether single-copy and two-copy protocols are used for small message sizes in the range 4 to 512 bytes. The inherent latency that all messages (regardless of size) incur is around 500 microseconds with both TCP and UDP, although UDP latencies are slightly better than TCP for larger messages. The effects of message size on latency start to show with message sizes above 2 kilobytes (KB). From Figure 33.2 (top), we observe that single-copy TCP/IP or UDP/IP stacks give around 1.3 to 1.4 times better latency than standard two-copy protocol stacks.

We also measured the round-trip latency for both TCP/IP and UDP/IP over a FDDI network between the same workstations using standard two-copy protocol stacks. To

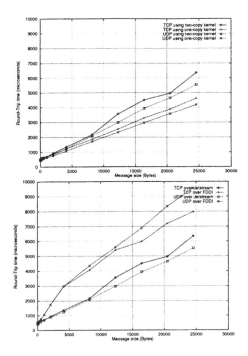

Figure 33.2   Variation of round-trip time with message size for single-copy and two-copy TCP/IP, UDP/IP protocol stacks (top graph). Variation of round-trip time with message size for two-copy TCP/IP, UDP/IP protocol stacks using FDDI and Jetstream networks (bottom graph).

allow fair comparison we also used the same TCP/IP and UDP/IP stacks (*i.e.* two-copy stacks) with Jetstream. As Figure 33.2 (bottom) shows, for both Jetstream and FDDI networks, the latencies for small messages (up to 512 bytes) are the same and are around 500 microseconds. Above 4 KB message sizes, latency over Jetstream is 1.2 to 1.5 times better than FDDI with TCP and about two times better with UDP. It was not possible to explain the better latency with TCP compared to UDP in the case of the FDDI network when the message sizes were above 4,096 bytes (Figure 33.2 (bottom)). We speculate that this behavior is partly due to the effect of the maximum transmission unit used by FDDI which is 4,500 bytes. We plan further experiments in the future to explain this result.

33.3.2   Impact of server distribution on application performance

In all performance experiments of our distributed visualization application, we used average playback rate as our quantitative metric to characterize application performance. Moreover, in all experiments the visualization application was run as a normal user process, along with the usual system processes and daemons in the background. In addition, all tests were performed using unloaded networks which carried no other traffic except that of the visualization application under investigation. CPU usage was

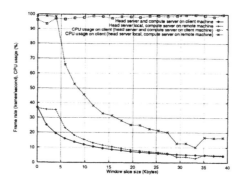

Figure 33.3   Effect of server distribution on frame rate performance.

measured using *Glance*, a performance monitor tool that comes with the standard HP-UX operating system.

Figure 33.3 shows the variation of frame rate and CPU usage at the client with window slice size for the following two cases. In the first case, both the head and compute servers (Figure 33.1) run on an HP-9000/755 (125 MHz) workstation which is also running the client for displaying the images. Without slicing, only the head images are served by the head server to the client and all communication is local. We achieved a frame rate of 37 frames/second with the CPU usage reaching its limit. As the window slice increases to expose more internal structure, the frame rate diminishes considerably. The reason for the gradual slow down in frame rate as the slice size increases is due to the increased time for calculating the slice image. As Figure 33.3 shows, the CPU is saturated.

In the second case, we used a two-machine configuration. The compute server is migrated to another workstation (HP-9000/735), connected via the Jetstream network and running a single-copy TCP/IP stack. The head server and client still run on one HP-9000/755 workstation. The frame rate degradation is less compared to the previous case where all the servers were running locally. Beyond a window slice size of 24 KB, the compute server saturates the CPU and the observed frame rate is almost the same as the local case. The benefit of performing the computation remotely significantly reduces the CPU usage on the display workstation leaving extra CPU cycles for other activities on the workstation. This is particularly important in typical multimedia sessions where often there are multiple multimedia applications running concurrently. For instance, a typical scenario involves a video conferencing or video-on-demand session in one window, simultaneous with a visualization session running in another window. In such application scenarios, it is therefore important that enough CPU cycles are available so that both applications run smoothly without degrading the performance of each other. This is why it is highly desirable to keep CPU usage to a minimum on uniprocessor conventional workstations, and to execute CPU-intensive tasks on other remote machines whenever possible. It has to be noted that in the past this was not really a high-performance approach because computer networks were slow. However, with new high-speed network technologies available today, the motivation for distributed computation on workstation clusters connected

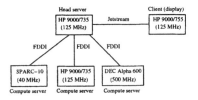

Figure 33.4 Three-machine configuration with the compute server running on different host architectures.

by high-bandwidth networks is becoming increasingly popular (Hansen and Tenbrink, 1993; Chowdhury *et al.*, 1997).

33.3.3 Impact of host architectures and processor performance

A key component of a true distributed system is that its clients and servers are *transparently* distributed over different machines (Tanenbaum and Renesse, 1985). In this section we demonstrate the impact of migrating the client, the head, and the compute servers on three different machines. The three-machine configuration used is shown in Figure 33.4. The client process executes on an HP 9000/755 workstation and performs the display of the rendered images. The head server runs on another workstation (HP 9000/735) where the image set for the rotating head is also stored. Both machines run HP-UX 9.01 and use single-copy TCP/IP stacks over the Jetstream network. The compute server runs on a third workstation which is connected by FDDI to the workstation running the head server. We experimented with the three-machine configuration by executing the compute server on different host architectures with different CPU performance: SPARCstation 10 (SunOS 4.1.3), HP 9000/735 (HP-UX 9.05), and DEC Alpha station 600 (Digital UNIX 4.0). Similar work has been described by other researchers (Carter *et al.*, 1996; Crandall *et al.*, 1996). (Carter *et al.*, 1996) have previously presented performance results for various workstation architectures. However, raw data measurements alone do not identify all the optimizations (*e.g.* display) needed to deliver high end-to-end application performance (Zeadally *et al.*, 1997). Our work differs from that of (Carter *et al.*, 1996) in that we investigate the performance of a real visualization application running over both gigabit networks and high-speed LAN technologies such as FDDI (compared to their work which uses only FDDI). In addition, we also exploit other performance enhancements such as single-copy protocol stacks to improve application performance. In other words, our work is demonstrating not only the impact of workstation architectures but also the benefit for real applications when exploiting gigabit networks, protocol optimizations, and application optimizations simultaneously.

The variation of frame rate performance with increasing window slice sizes are shown in Figure 33.5 (top). The maximum frame rate achieved with no slice window is 40 frames/second, which translates to 84 Mbits/s end-to-end user throughput. This is only about *one tenth* of the maximum data rate (*i.e.* 800 Mbits/s) that can be delivered by the underlying Jetstream network. The main reason for the difference is due to software overheads (*e.g.* device driver, operating system), especially those at the end

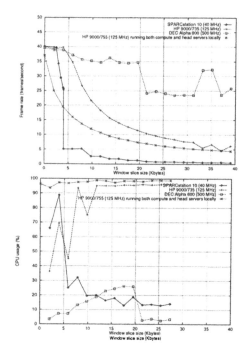

Figure 33.5   Frame rate performance with compute server running on different host architectures (top graph); CPU usage on different host architectures running the compute server
(bottom graph).

system which dominate communication performance. As a result, application end-
to-end performance is lagging behind improvements in network and processor speeds
(Damianakis *et al.*, 1996).

The top graph of Figure 33.5 shows that the frame rate performance decreases as the
windows slice size increases since more time is taken to compute the slice information
request. However, the rate of decrease for the different host architectures are different.
The frame rate performance with the compute server running on an HP 9000/735 is
far better than the case when all servers are local. In fact, at 20 frames/second, the
window slice size that can be obtained (12 KB) with the compute server running on an
HP 9000/735 is almost three times the window slice size (4 KB) that can be achieved
when all servers are running locally. As shown in Figure 33.5 (bottom), the CPU usage
is 100% beyond a 12 KB window slice size on the HP 9000/735 running the compute
server. Thus, CPU saturation accounts for the continuous degradation in frame rate
performance as the window slice size is increased.

In another experiment, the compute server was executed on a SPARCstation 10.
The variation of frame rate with window slice size is shown in Figure 33.5 (top). A
gradual decrease in frame rate can be observed for window slice sizes between 2 KB
and 4 KB. There is a corresponding increase in CPU usage on the Sun workstation as
the window slice size is increased until CPU saturation occurs at 4 KB, as shown in

Figure 33.5 (bottom). However, we observe a significant drop in the frame rate beyond a window slice of 4 KB giving performance that is worse than that obtained when all servers run locally on an HP workstation. Initially, we thought that the sudden drop is also due to CPU saturation on the Sun workstation. However, a look at the CPU usage beyond 4 KB revealed that was not the case since, as Figure 33.5 shows, the CPU usage is only around 10-30%. We could not find a satisfactory interpretation of this performance result. One plausible explanation is that beyond a window size of 4 KB, we are crossing a page boundary since the memory page size used by SunOS 4.1.3 on a SPARCstation 10 is 4 KB. Consequently, the overhead of writing to a new memory page with data sizes greater than 4 KB accounts for the sudden degradation in frame rate performance. Further experiments are required to confirm this explanation.

We also executed the compute server (with the three-machine configuration in Figure 33.4) on the DEC Alpha 600 workstation which is connected by FDDI to the workstation running the head server. Figure 33.5 (top) shows that the 500 MHz DEC Alpha station enables a high sustained frame rate for window slice sizes up to around 20 KB. In fact, the average frame rate obtained was around 35 frames/second and only a 12% frame rate degradation was observed for window slice sizes up to 20 KB. Moreover, the corresponding CPU usage on the DEC Alpha for window slices up to 20 KB was rather low (around 20-30% from Figure 33.5), as compared to the Sun and HP workstations which quickly reach CPU saturation at window slice sizes of 4 KB and 12 KB respectively. However, there is a sudden degradation from 35 to 25 frames/second for window slice sizes above 20 KB. The CPU on the DEC Alpha running the compute server was not the bottleneck since its observed usage was rather low. It was not possible to explain the 30% drop in frame rate performance and frame rate performance beyond 20 KB. Further investigation is required to explain this performance behavior. We also note from Figure 33.5 that for the same window slice size (*e.g.* 12 KB), the 500 MHz processor of the DEC Alpha station performs computations faster than the HP 9000/735 (125 MHz) workstation by a factor of four. In addition, the application frame rate is almost doubled (at 12 KB window slice size) when the compute server runs on the 500 MHz DEC Alpha, as compared to when running on the 125 MHz HP 9000/735 workstation.

33.4 PERFORMANCE OF THE VISUALIZATION APPLICATION ON PENTIUM WORKSTATIONS OVER ATM

The experiments described above were based on ring-based network architectures. We used a combination of high-speed networks such as Jetstream and 100 Mbits/s FDDI to achieve the best network performance[1]. In this section, we attempt to investigate the performance benefits of *low-cost* end systems running standard operating systems that are connected to a switch-based network architecture. Related efforts in this area include the Distributed Array of Inexpensive Systems (DAISy) project (Carter *et al.*, 1996) where the viability of commodity PC technology to the computation of sci-

---

[1]We have only two Jetstream adapters which explains why it was not possible to use a three-machine configuration on the Jetstream network.

entific and engineering problems is being explored using a PC cluster connected to switched Fast Ethernet (100 Mbits/s) and running the UNIX operating system. (Crandall *et al.*, 1996) compared the performance trade-offs between shared and distributed memory processing using desktop multiprocessing machines and workstation clusters connected to a high-speed switch. In all their experiments, (Crandall *et al.*, 1996) used Solaris 2.5 and measured execution times of some common algorithms at the core of many scientific applications. There is some overlap between these efforts and the work described in this section since we also investigate the potential of PC commodity technology. However, the work presented in this section also attempts to compare the performance of a "traditional" configuration (running UNIX on workstations connected to a shared-medium network such as FDDI) with a switch-based ATM[2] network consisting of PCs running Windows NT operating system. In this context, we further demonstrate the performance of Windows NT-based PCs in contrast to UNIX-based PCs which was reported in (Carter *et al.*, 1996). It must be stressed that we were not able to conduct a strict one-to-one performance comparison of our visualization application between HP workstations running HP-UX connected to FDDI and PCs running Windows NT connected to an ATM network. The reason for this was a lack of resources, such as having the same operating system running on different machine architectures, or having the same machine architectures running different operating systems. Moreover, we did not have network adapters for different hardware platforms for either FDDI or ATM. Despite the resource limitations explained above, we believe it is useful and interesting to investigate the performance of our visualization application on a switch-based architecture such as ATM with low-cost Pentium workstations running Windows NT 4.0.

In order to understand the bandwidth cost associated with FDDI and ATM, we calculated the cost per Mbits/s for both networks. In the simple cost calculation, only the networking hardware costs were included (this leaves out the cost of end-systems). With ATM, we calculated the total cost for DEC[3]ATM hardware. C_{ATM} is the cost per Mbits/s corresponding to MN machines on the network calculated as follows:

$$C_{MN} = \frac{(C_{bs} + C_{port} + C_{adapter})}{(NB \times MN)} \quad (33.1)$$

where C_{bs} is the cost of the basic switch, C_{port} is the cost per OC-3 port (one for each machine) at the ATM switch, $C_{adapter}$ is the cost of an ATM adapter (one per machine), and NB is the network bandwidth (155.52 Mbits/s). With FDDI, the cost per Mbits/s, C_{FDDI}, was calculated as follows:

$$C_{FDDI} = \frac{(C_{adapter} \times MN)}{NB} \quad (33.2)$$

where $C_{adapter}$ is the cost of a DEC FDDI adapter and NB is the FDDI bandwidth (100 Mbits/s). We plotted the variation of the cost per Mbits/s with the number of

---

[2]ATM is a communication architecture based on switching or relaying small fixed-length packets called cells (Vetter, 1995).

[3]Other network hardware vendors have been checked and prices were almost the same.

nodes in the network for both FDDI and ATM as shown in Figure 33.6. It is interesting to note that just beyond two machines, the fixed cost of an ATM switch (without any line cards) quickly becomes less significant compared to FDDI. With an increasing number of machines on the network, the cost per Mbits/s for FDDI is much higher than ATM.

We ported our visualization application to Windows NT 4.0. In our experiments, each Pentium workstation is equipped with an ATM adapter (ATMworks 351 from DEC) which can support connections up to 155.52 Mbits/s (*i.e.* OC-3 rate). As shown in Figure 33.7, the Pentium machines (running Windows NT 4.0) are connected to a DEC GIGAswitch/ATM switch (configurations b) and c)) compared to the ring-based FDDI architecture which uses three HP 9000 Series workstations running HP-UX 9.05 operating system (configurations a)).

The frame rate performance for the three configurations a), b), and c) are given in Figure 33.8 (top). The maximum frame rate achieved without slicing with the three-machine configuration a) is 25 frames/second. This translates to an end-to-end user throughput of 52.5 Mbits/s. This is slight more than half the performance obtained when a gigabit network was used between the head server and the head client. This is because the FDDI network becomes the bottleneck due to lack of network bandwidth. Previous raw data performance results have already demonstrated that maxi-

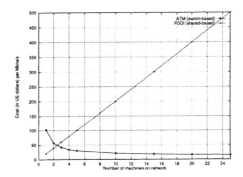

Figure 33.6 The figure shows the cost per megabit on ATM and FDDI networks as the number of machines connected to the network increases.

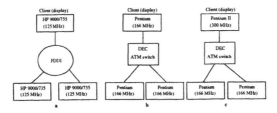

Figure 33.7 Distributed visualization over ATM and FDDI networks. The HP workstations in configuration a) run HP-UX 9.05. The Pentium machines, configurations b) and c), on the ATM network run Windows NT 4.0.

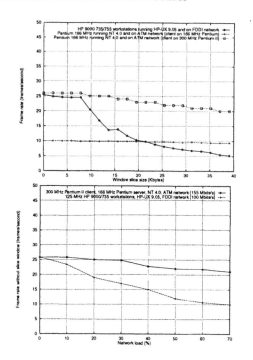

Figure 33.8 Variation of frame rate with window slice size on unloaded ATM and FDDI networks (top graph); variation of frame rate (without window slice) with network load on ATM and FDDI networks (bottom graph).

mum host memory to host memory TCP/IP throughput over a FDDI network is around 70 Mbits/s without display overhead (Zeadally *et al.*, 1997). With network data display, this drops to a lower throughput which is what is observed when the visualization application executes. Another reason for the lower throughput is that we have not used a single-copy TCP/IP stack with the FDDI implementation since it does not support it. There were also frequent pauses as the head rotates to show the sliced window at different angles with occasional image distortions. In other words, the "quality" of the image was much worse in this case and is the result of insufficient network bandwidth with FDDI. This was not the case when Jetstream was used where the head image was always consistent and the rotation proceeded smoothly without any intermittent pauses.

The decreasing costs and increasing performance of personal computers make them an attractive choice for low-cost high-performance distributed computing. With configuration b) in Figure 33.7, we achieved 10 frames/second (top graph in Figure 33.8) almost regardless of the window slice size. This is because the client machine performing the image display becomes the bottleneck and its CPU utilization of 100%. We therefore replaced the 166 MHz Pentium client (display) machine with a Pentium II 300 MHz workstation as shown in configuration c) of Figure 33.7. The improvement in frame rate was more than double as shown in Figure 33.8 (top). CPU usage on the Pentium II was around 60-70%. Furthermore, the frame rate decreases slowly as the window slice increases and remains slightly above 20 frames/second (top graph in Figure 33.8) for the range of window slice sizes tested. This is much better performance than that delivered by the HP 9000 Series workstations used on a FDDI network where the frame rate performance quickly degrades above 8 KB slice sizes. It is worth pointing out that the frame rate performance on the PCs has been achieved with only one Pentium II machine which executes the client and performs the display. The other machines used are cheap 166 MHz PCs. This result clearly shows the potential for low-cost PCs connected to a high-speed network such as ATM in meeting the requirements of high-performance distributed computing. Another important result we learn from these experiments is that it is not sufficient to achieve a high data transfer rate from network to main memory, it is also necessary to efficiently move the data from main memory to the graphics buffer for display[4]. Both of these operations need to be optimized for high-bandwidth networked applications in order to achieve high end-to-end user performance. Our results show that with an Intel 166 MHz processor, CPU saturation slows down the display path, and as a result affects the application performance—a low frame rate of 10 frames/second with our visualization application. With a more powerful Pentium II processor, the increased availability of CPU cycles enhances performance by allowing the display of network data to proceed at a faster rate. It is worthwhile noting that the display bottleneck is also partly due to the design of current conventional bus architectures which lack the capability to transfer data from a network adapter to main memory *concurrently* with data transfer from main memory to the graphics adapter for display. In other words, the Peripheral Component Interconnect (PCI) system bus can only handle one transaction at a time. However, the new Pentium II design exploits an Accelerated Graphics Port (AGP) architecture (Intel, 1997) where a dedicated high-bandwidth data path exists from main memory to the graphics adapter. With such an architecture, it should be possible to stream data from the network concurrently with data transfer to the graphics display via AGP at a high rate. However, on uniprocessor systems the CPU can still potentially become the bottleneck depending on how much CPU usage is consumed to move data between the different devices (*e.g.* network adapter, graphics display, and so on).

Due to space limitation, it is not possible to present a detailed study of the effects of loaded networks (*e.g.* with multiple users running multiple instances of the visualization application) on the performance of our distributed application. However, we briefly present in Figure 33.8 (bottom) the variation of the visualization application frame rate performance under various network loads (10 to 70%) for both FDDI and ATM. In both cases, four machines were used: one pair of machines was used to load the network (one generating, one receiving network traffic), the other pair of machines executes the visualization application with one running the head server and the other the head client. No slice window was chosen in this experimental set up and therefore only two machines were used by the application. With ATM, the same ATM link used by the head server to send the head images to the client was loaded with differ-

---

[4]Otherwise the display operation introduces a bottleneck which slows down overall end-to-end application performance.

ent network traffic loads. As shown in Figure 33.8 (bottom), performance degrades faster with FDDI than ATM when subject to the same network load because the available bandwidth becomes less with FDDI than ATM. Even when comparing the same available bandwidth on FDDI and ATM, the frame rate performance achieved over ATM is far superior than FDDI. For example, at 50% network load on FDDI (available bandwidth of 50 Mbits/s), and 70% network load on ATM (also with an available bandwidth of 50 Mbits/s), application frame rates obtained from Figure 33.8 (bottom) were 12.5 and 22 respectively. This result clearly shows that with competitive network loading, high-speed switching of ATM offers much better sustained performance than FDDI.

33.5 CONCLUSIONS AND FUTURE WORK

In this paper, we demonstrated the performance of current workstations, hardware, and high-speed network technologies using a high-performance distributed visualization application. We have particularly focussed on end-to-end user application performance using metrics such as frame rate performance and round-trip latency. The major research contributions, results, and observations of this work are summarized below.

For small message sizes (below 500 bytes), the observed latency was around 500 microseconds. This latency result is the *same* on both 100 Mbits/s FDDI and gigabit/s Jetstream networks. While gigabit networks can provide higher bandwidth capacity, the latency for small packets has not improved. Future gigabit networks need to be designed with low latency overheads, otherwise distributed multimedia applications, particularly those with high bandwidth and low latency requirements will suffer. One such network capable of providing multi-gigabit per second and low user-to-user latency is currently being developed as part of a Hewlett Packard/University of Southern California Parallel Optical Link Organization (POLO) project (Hahn and Dolfi, 1996). We also found that single-copy protocol stacks improve latency only for large message sizes. This is because with large messages, data copying overheads dominate communication costs and minimization of data copies therefore improves latency. For small messages, the inherent latency introduced by the end system due to application, operating system, device driver, and network software overheads dominate.

We have shown that distributing the server components of our visualization application yields better frame rate performance and significantly reduces the CPU usage compared to when the entire application runs on one machine only. This is especially crucial when executing multiple multimedia applications simultaneously. The availability of CPU left over will determine the quality of other applications running. The advent of high-speed networks such as ATM allows computationally-intensive tasks to be relegated to other machines on the network at a low-cost with minimal impact on performance. Our performance experiments also show that a 500 MHz DEC Alpha 600 workstation performs computations about four times faster than the 125 MHz HP 9000/735 workstation. In addition, we have also demonstrated that for networked applications displaying high-bandwidth images, a low-end PC such as a Pentium 166 MHz is not powerful enough since the CPU saturates quickly. With more advanced processors, such as a Pentium II, we have shown that very good performance is achieved.

Our performance results demonstrate the viability of low-cost, NT-based PCs and high-speed networks such as ATM in supporting distributed high-performance applications cost-effectively compared to expensive UNIX-based workstations and networks such as FDDI. We argue, like other researchers (Carter *et al.*, 1996), that the combination of low-cost commodity PC technology and high-speed switch-based LANs offers great potential for supporting high-performance distributed computing applications. However, unlike (Carter *et al.*, 1996) who exploit UNIX-based PCs, we promote commodity PC technology running Windows NT operating system which simplifies the integration of other industry-standard components available for this platform.

One area of future work we plan to investigate is methods to efficiently distribute the compute server operations to more than one node in a PC cluster. In this case, slice calculation information proceeds in parallel with small synchronization and command messages directing the computation.

Acknowledgments

The author thanks Ulrich Neumann and his students for their help in developing the visualization application. The author also thanks Tony Levi and the anonymous reviewers for their comments on early drafts of this paper and William Cui for his help in porting the application to Windows NT. This research been supported by the Integrated Media Systems Center NSF Grant EEC-9529-152 and the Intel Equipment Grant Program.

References

Carter, R., Laroco, J., and Armstrong, R. (1996). Commodity clusters: Performance comparison between PCs and workstations. In *Proceedings of High Performance Distributed Computing*.

Chowdhury, A., Nicklas, L., Setia, S., and White, E. (1997). Supporting dynamic space-sharing on non-dedicated clusters of workstations. In *17th IEEE International Conference on Distributed Computing Systems*.

Crandall, P., Sumithasri, E., and Clement, M. (1996). Performance comparison of desktop multiprocessing and workstation cluster computing. In *Proceedings of High Performance Distributed Computing*.

Dalton, C., Watson, G., Banks, D., Calamvokis, C., Edwards, A., and Lumley, J. (1993). Afterburner. *IEEE Network*, 7(4):36–43.

Damianakis, S., Bilas, A., Dubnicki, C., and Felten, E. (1996). Client-server computing on the shrimp multicomputer. Technical report, Computer Science Department, Princeton University.

Eicken, T., Basu, A., Buch, V., and Vogels, W. (1995). U-net: A user-level network interface for parallel and distributed computing. In *15th ACM Symposium on Operating System Principles*, pages 40–53.

Hahn, K. and Dolfi, D. (1996). Polo - a gigabyte/s parallel optical link. In *SPIE Optoelectronic Interconnects and Packaging*, volume CR62, pages 393–404.

Hansen, C. and Tenbrink, S. (1993). Impact of gigabit network research on scientific visualization. *IEEE Computer*.

Intel (1997). AGP tutorial. http://developer.intel.com/technology/agp.

Interphase (1996). *5515 ATM Adapter User Guide*. Interphase Corporation, Texas.

Jones, R. (1993). Netperf: A network performance benchmark. Technical report, Networks Division, Hewlett-Packard. http://www.cup.hp.com/netperf/-Netperf/Page.html.

Keeton, K., Anderson, T., and Patterson, D. (1995). LogP: The case for low overhead local area networks. In *Hot Interconnects III*.

Lin, M., Hsieh, J., Du, D., Thomas, J., and MacDonald, J. (1995). Distributed network computing over local ATM networks. *IEEE JSAC*, 13(4):733–747.

Nahrstedt, K. and Smith, J. (1994). Experimental study of end-to-end qos. Technical Report MS-CIS-94-08, University of Pennsylvania.

Sterling, T., Becker, D., Savarese, D., Berry, M., and Reschke, C. (1996). Achieving a balanced low-cost architecture for mass-storage management through multiple fast ethernet channels on the beowulf parallel workstation of workstations. In *International Symposium on Parallel Processing*.

Tanenbaum, A. and Renesse, R. (1985). Distributed operating systems. *ACM Computing Surveys*, 17(4).

Vetter, R. (1995). ATM concepts, architectures, and protocols. *Communications of the ACM*, 38(2):30–38.

Watson, G., Banks, D., Calamvokis, C., Dalton, C., Edwards, A., and Lumley, J. (1994). AAL5 at a gigabit for a kilobuck. *Journal of High Speed Networks*, 3(2):127–145.

Zeadally, S., Gheorghiu, G., and Levi, A. (1997). Improving end system performance for multimedia applications over high bandwidth networks. *International Journal of Multimedia Tools and Applications*, 5(3):307–322.

34   FAST AND EFFICIENT PARALLEL MATRIX COMPUTATIONS ON A LINEAR ARRAY WITH A RECONFIGURABLE PIPELINED OPTICAL BUS SYSTEM

Keqin Li[1], Yi Pan[2] and Si-Qing Zheng[3]

[1] Department of Mathematics and Computer Science,
State University of New York, U.S.A.
[2] Department of Computer Science,
University of Dayton, U.S.A.
[3] Department of Computer Science,
Louisiana State University, U.S.A.

li@mcs.newpaltz.edu
pan@cps.udayton.edu
zheng@bit.csc.lsu.edu

Abstract: We present fast and cost efficient parallel algorithms for a number of important and fundamental matrix computation problems on linear arrays with reconfigurable pipelined optical bus systems. These problems include computing the inverse, the characteristic polynomial, the determinant, the rank, the Nth power, an LU- and a QR-factorization of a matrix, and solving linear systems of equations. Our algorithms provide a wide range of performance-cost combinations. Compared with known results, the running time of parallel solutions to all these problems can be reduced by a factor of $O(\log N)$ while maintaining costs under $O(N^4)$.

Keywords: matrix computation, optical bus, processor array, time complexity.

34.1   INTRODUCTION

It has long been recognized that a number of important matrix problems are quite closely related, in the sense that up to polynomial changes in the size of the matrices, and constant factor changes in the running time of the algorithms, the amount of parallel time needed to solve these problems are the same. In particular, it is well known that finding the inverse, the characteristic polynomial, the determinant, the rank, the Nth power, and an LU-factorization of a matrix, and the product of a matrix chain, *etc.*, can be reduced to each other, such that all of them have the same parallel computing complexity (*c.f.* (Cook, 1985) and §2.4.5 of (Leighton, 1992) for proofs of the above claim). In addition to the parallel time complexity, the cost of an algorithm, which is the product of the execution time and the number of processors employed by the algorithm, is widely used to measure the efficiency of the algorithm. Currently the fastest parallel algorithms for these problems (pt LU-factorization) run in $O((\log N)^2)$ time using $O(N^4)$ processors connected by some commonly used static networks, such as meshes of trees and hypercubes (Leighton, 1992). Since the reduction among these problems may significantly increase the problem size, and hence, the number of processors, the costs are not preserved under these reductions. To keep the processor complexity at $O(N^4)$, the best algorithm for LU-factorization has time complexity $O((\log N)^3)$ on static networks.

While the problem of finding tight time and cost bounds of each of these problems remains open, answering the following two questions appears to be challenging:

Q1. Can the parallel time complexities of these problems (except LU-factoriza- tion) be reduced to $o((\log N)^2)$, and to $o((\log N)^3)$ for LU-factorization?

Q2. Can the costs of parallel algorithms for these problems be reduced to $o(N^4)$?

In answering question Q1, we notice that the major hurdle is that all algorithms for these problems ultimately use a matrix multiplication algorithm as a subroutine, and the best matrix multiplication algorithm has time complexity $O(\log N)$, even on theoretical models like parallel random access machines (PRAMs). The parallelism in matrix multiplication can be explored to such an extend that the N^2 vector products can be calculated simultaneously. The $O(\log N)$ bottleneck is due to the summation of the N values in an inner vector product, which cannot be done any faster on existing computing models. Theoretically, the answer to question Q2 is affirmative. It is well known that there exist matrix multiplication algorithms on PRAMs that run in $O(\log N)$ time using $O(N^\beta)$ processors, where $O(N^\beta)$, $2 < \beta < 3$, is the time complexity of the best sequential algorithm for matrix multiplication (Pan and Reif, 1985). The fact that the known smallest value of β is less than 2.3755 (Coppersmith and Winograd, 1990) implies that the costs of parallel PRAM algorithms for many matrix problems are less than $O(N^4)$. However, such a sequential algorithm is quite sophisticated, and its implementation is not considered practical even on sequential machines. Consequently, its parallelization on realistic parallel systems is far from feasible.

Recently, there have been significant advances in optical interconnections. Fiber optic communication technologies offer a combination of high bandwidth, predictable message delay, low interference and error probability, and gigabit transmission capacity. Based on the characteristics of fiber optical communications, a number of researchers have proposed using optical interconnections to connect processors in a parallel computer system (Benner *et al.*, 1991; Chiarulli *et al.*, 1987; Guo *et al.*, 1991; Levitan *et al.*, 1990; Pan and Li, 1998; Qiao and Melhem, 1993; Tocci and Caulfield, 1994; Zheng and Li, 1997). In such a system, messages can be transmitted concurrently on a pipelined optical bus by taking the advantages of unidirectional message transmission and predictable propagation delay. It is now feasible to integrate both optical message communication and electronic data computation in massively parallel processing systems. Many parallel algorithms from numerous application domains have been proposed for systems with optical interconnections (Hamdi and Pan, 1995; Li, 1997; Li *et al.*, 1996; Li *et al.*, 1998; Pan and Hamdi, 1996; Pan *et al.*, 1998a; Pan *et al.*, 1998b; Pavel and Akl, 1996; Rajasekaran and Sahni, 1997).

The advent of optical buses has created an entirely new parallel computing model, and opened up a broader avenue of parallel algorithm design. Consider a linear array with a reconfigurable pipelined bus system (LARPBS) (Li *et al.*, 1996; Pan and Li, 1998; Pan *et al.*, 1998b). First, pipelined optical buses can support concurrent accesses by many processors in a single bus cycle, and are able to transmit a large volume of data among processors simultaneously for various communication patterns. Second, a LARPBS can also be reconfigured into disjoint subsystems, each being an independent LARPBS of smaller size. This feature supports parallel implementation of divide-and-conquer computations. These capabilities have led to parallelization of nontrivial algorithms, especially the following result:

R0. Strassen's algorithm for multiplying two $N \times N$ matrices can be implemented on a LARPBS with $O(N^{2.8074})$ processors in $O(\log N)$ time (Strassen, 1969; Li *et al.*, 1996).

Such an implementation is very difficult (if not impossible) on all existing networks with electronic interconnections or systems with electronic buses. In addition to communication channels with tremendous capacity and flexibility, optical buses can also serve as active computing agents. One excellent example is that N values can be added up in a constant number of bus cycles on a LARPBS, where computations are essentially done by appropriate timing of optical signals. This implies that:

R1. Multiplying two $N \times N$ matrices can be done on a LARPBS in $O(1)$ time using N^3 processors (Li *et al.*, 1996).

Combining R0 and R1, we know that:

R2. Multiplying two $N \times N$ matrices can be done on a LARPBS in sublogarithmic time $O((\log N)^\delta)$, using $O(N^3/(\frac{8}{7})^{(\log N)^\delta})$ processors, where $0 \le \delta \le 1$. The cost is $o(N^3)$ for $0 < \delta \le 1$ (Li *et al.*, 1996).

With results R1 and R2, we are able to give affirmative answers to questions Q1 and Q2 on realistic parallel computing systems.

In this paper, we consider solving seventeen fundamental matrix problems on LARPBS. These problems and the time and processor complexities of our solutions are listed in Table 34.1. Problems (1)-(7) are basic matrix operations, whose algorithms are repeatedly used in other algorithms. Problems (8)-(17) include a number of fundamental matrix operations. Our results provide the following answers to questions Q1 and Q2:

A1. Problems (6)-(15) can all be solved in $O(\log N)$ time, and problems (16)-(17) in $O((\log N)^2)$ time, using no more than $O(N^4)$ processors. Thus, compared

Table 34.1   The problems and their time/processor complexities.

No.	Problem	Time Complexity	Processor Complexity
1	matrix transposition	$O(1)$	$O(N^2)$
2	vector chain addition	$O(1)$	$O(N^2)$
3	matrix chain addition	$O(1)$	$O(N^3)$
4	matrix-vector multiplication	$O(1)$	$O(N^2)$
5	matrix multiplication	$O((\log N)^\delta)$	$O(N^3/(\frac{8}{7})^{(\log N)^\delta})$
6	matrix powers	$O((\log N)^{1+\delta})$	$O(N^4/(\frac{8}{7})^{(\log N)^\delta})$
7	matrix chain product	$O((\log N)^{1+\delta})$	$O(N^4/(\frac{8}{7})^{(\log N)^\delta})$
8	lower triangular matrix inversion	$O((\log N)^{1+\delta})$	$O(N^3/(\frac{8}{7})^{(\log N)^\delta})$
9	upper triangular matrix inversion	$O((\log N)^{1+\delta})$	$O(N^3/(\frac{8}{7})^{(\log N)^\delta})$
10	the characteristic poly. of a matrix	$O((\log N)^{1+\delta})$	$O(N^4/(\frac{8}{7})^{(\log N)^\delta})$
11	the determinant of a matrix	$O((\log N)^{1+\delta})$	$O(N^4/(\frac{8}{7})^{(\log N)^\delta})$
12	the rank of a matrix	$O((\log N)^{1+\delta})$	$O(N^4/(\frac{8}{7})^{(\log N)^\delta})$
13	arbitrary matrix inversion	$O((\log N)^{1+\delta})$	$O(N^4/(\frac{8}{7})^{(\log N)^\delta})$
14	solving linear systems of equations	$O((\log N)^{1+\delta})$	$O(N^4/(\frac{8}{7})^{(\log N)^\delta})$
15	Krylov matrix	$O((\log N)^{1+\delta})$	$O(N^4/(\frac{8}{7})^{(\log N)^\delta})$
16	LU-factorization of a matrix	$O((\log N)^{2+\delta})$	$O(N^4/(\frac{8}{7})^{(\log N)^\delta})$
17	QR-factorization of a matrix	$O((\log N)^{2+\delta})$	$O(N^4/(\frac{8}{7})^{(\log N)^\delta})$

with known results, the running time of parallel solutions to all these problems mentioned in Table 34.1 can be reduced by a factor of $O(\log N)$ using these algorithms.

A2. Problems (6)-(15) can be solved in $o((\log N)^2)$ time, and problems (16)-(17) in $o((\log N)^3)$ time, with cost $o(N^4)$ on LARPBS.

34.2   THE LARPBS COMPUTING MODEL

A pipelined optical bus system uses optical waveguides instead of electrical wires to transfer messages among electronic processors. In addition to the high propagation speed of light, there are two important properties of optical signal (pulse) transmission on an optical bus, namely unidirectional propagation and predictable propagation delay. These advantages of using waveguides enable synchronized concurrent accesses of an optical bus in a pipelined fashion (Chiarulli *et al.*, 1987; Levitan *et al.*, 1990). Such pipelined optical bus systems can support a massive volume of communications simultaneously, and are particularly appropriate for applications that involve intensive communication operations such as broadcasting, one-to-one communication, multi-casting, global aggregation, and irregular communication patterns.

A linear array with a reconfigurable pipelined bus system (LARPBS) consists of N processors $P_1, P_2, ..., P_N$ connected by an optical bus. In addition to the tremendous communication capabilities, a LARPBS can also be partitioned into $k \geq 2$ independent subarrays LARPBS$_1$, LARPBS$_2$, ..., LARPBS$_k$, such that LARPBS$_j$ contains processors $P_{i_{j-1}+1}, P_{i_{j-1}+2}, ..., P_{i_j}$, where $0 = i_0 < i_1 < i_2 \cdots < i_k = N$. The subarrays can operate as regular linear arrays with pipelined optical bus systems, and all subarrays can be used independently for different computations without interference (see (Li et al., 1996; Pan and Li, 1998) for an elaborated exposition, and (Zheng and Li, 1997) for similar reconfigurable pipelined optical bus architectures).

As in many other synchronous parallel computing systems, a LARPBS computation is a sequence of alternate global communication and local computation steps. The time complexity of an algorithm is measured in terms of the total number of bus cycles in all the communication steps, as long as the time of the local computation steps between successive communication steps is bounded by a constant and independent of the problem size. This complexity measure implies that a bus cycle takes constant time, and this assumption has been adopted widely in the literature. (Remark: To avoid controversy, let us emphasize that in this paper, by "$O(f(p))$ time" we mean $O(f(p))$ bus cycles for global communication plus $O(f(p))$ number of local arithmetic/logic operations.)

For ease of algorithm development and specification, a number of basic communication, data movement, and global operations on the LARPBS model implemented using the coincident pulse processor addressing technique (Chiarulli et al., 1987; Levitan et al., 1990; Qiao and Melhem, 1993) have been developed (Li et al., 1996; Pan and Li, 1998). Each of these primitive operations can be performed in a constant number of bus cycles. These powerful primitives that support massive parallel communications, plus the reconfigurability of the LARPBS model, make the LARPBS very attractive in solving problems that are both computation and communication intensive, such as matrix manipulations. Optical buses are not only communication channels among the processors, but also active components and agents of certain computations, e.g. global data aggregations. The following primitive operations on LARPBS are used in this paper, and our algorithms are developed using these operations as building blocks.

One-to-One Communication. Assume that processors $P_{i_1}, P_{i_2}, ..., P_{i_m}$ are senders, and processors $P_{j_1}, P_{j_2}, ..., P_{j_m}$ are receivers. In particular, processor P_{i_k} sends a value in its register R(i_k) to the register R(j_k) in P_{j_k}. The operation is represented as:

for $1 \leq k \leq m$ do in parallel
$\qquad$ R$(j_k) \leftarrow$ R(i_k)
endfor

Note that we use R(i) to denote both the name and the content of register R(i).

Broadcasting. Here, we have a source processor P_i, who sends a value in its register R(i) to all the N processors:

$\qquad$ R$(1),$ R$(2),$ R$(3), ...,$ R$(N) \leftarrow$ R(i)

Multiple Multicasting. In a multicasting operation, we have a source processor P_i, who sends a value in its register $R(i)$ to a subset of the N processors $P_{j_1}, P_{j_2}, ..., P_{j_m}$:

$$R(j_1), R(j_2), ..., R(j_m) \leftarrow R(i)$$

Assume that we have g disjoint groups of destination processors, $G_k = \{P_{j_{k,1}}, P_{j_{k,2}}, P_{j_{k,3}}, ...\}$, $1 \leq k \leq g$, and there are g senders $P_{i_1}, P_{i_2}, ..., P_{i_g}$. Processor P_{i_k} has value $R(i_k)$ to be broadcast to all the processors in G_k, where $1 \leq k \leq g$. Since there are g simultaneous multicasting, we have a multiple multicasting operation, which is denoted as follows:

for $1 \leq k \leq g$ do in parallel
    $R(j_{k,1}), R(j_{k,2}), R(j_{k,3}), ... \leftarrow R(i_k)$
endfor

It is important to point out that the g groups of destination processors must be disjoint.

Element Pair-Wise Operations. Let $P_i, P_{i+1}, ..., P_j$ be a consecutive group of processors. We use $R[i..j]$ as an abbreviation of the registers $R(i), R(i+1), ..., R(j)$. $R[i..j]$ can be used to store a vector, or a matrix in the row-major or column-major order. Assume that A and B are two matrices of size N. All array elements are in a domain with a binary operator $\oplus$. Elements a_{ij} and b_{ij} are stored in $R[m + (i-1)N + j - 1]$ and $R[n + (i-1)N + j - 1]$, respectively. Then $C = A \oplus B$, where $c_{ij} = a_{ij} \oplus b_{ij}$, can be done as follows, where c_{ij} is found in $R[m + (i-1)N + j - 1]$:

for $1 \leq k \leq N^2$ do in parallel
    $R(m + k - 1) \leftarrow R(m + k - 1) \oplus R(n + k - 1)$
endfor

Global Aggregation. Assume that every processor P_i, where $1 \leq i \leq N$, has a register $R(i)$ which holds a value. We need to calculate $R(1) + R(2) + \cdots + R(N)$, and save the result in $R(1)$. The operation is represented as:

$$R(1) \leftarrow R(1) + R(2) + \cdots + R(N)$$

The following results have been proven in (Li et al., 1996; Pan and Li, 1998).

Theorem 1. One-to-one communication, broadcasting, multiple multicasting, element pair-wise operation, integer (of bounded magnitude) aggregation, and real value (of bounded precision and magnitude) aggregation all take $O(1)$ bus cycles in the LARPBS computing model. ∎

The extension of global aggregation to unbounded values have been discussed in (Li et al., 1996). In particular, it was shown that the summation of N integers can be calculated in $O(\log \log M)$ time, using $O(N \log M / \log \log M)$ processors in the LARPBS computing model, where M is the maximum magnitude; and the summation of N real values, with precision up to $2^{-(p+q-1)}$ and magnitude in the order of $M = 2^{2^q}$, can be calculated in $O(\log \log M + \log p)$ time, using $O(Np \log p)$ processors in the LARPBS computing model. Since all real machines have finite word length, which implies bounded magnitude and precision, we will not consider unbounded magnitude and precision in this paper.

The primitive operations can be directly used for some simple matrix manipulations. For instance, the one-to-one communication can be used for transposing a matrix. Assume that we have a LARPBS with N^2 processors P_1, P_2, ..., P_{N^2}, and each processor has a register $A(k)$, where $1 \leq k \leq N^2$. A matrix $A = (a_{ij})_{N \times N}$ is stored in the linear array in the row-major order. That is, $A((i-1)N + j) = a_{ij}$, for all $1 \leq i, j \leq N$. Then, our algorithm for matrix transposition simply does the following:

> for $1 \leq i, j \leq N$ do in parallel
> $\qquad A((i-1)N + j) \leftarrow A((j-1)N + i)$
> endfor

After the above one-to-one communication, which takes one bus cycle, we have $A((i-1)N + j) = a_{ji}$, for all $1 \leq i, j \leq N$.

Theorem 2. Matrix transposition can be done in $O(1)$ time using N^2 processors.   ∎

The aggregation operation can be used to calculate the summations of N vectors or N matrices. Given N N-dimensional vectors $\mathbf{v}_i = (v_{i1}, v_{i2}, ..., v_{iN})$, where $1 \leq i \leq N$, the vector chain addition problem is to calculate $\mathbf{v} = (v_1, v_2, ..., v_N)$, where $v_j = v_{1j} + v_{2j} + \cdots + v_{Nj}$, for all $1 \leq j \leq N$. Assume that the N vectors are stored in N^2 processors, where each processor has a register $V(k)$, $1 \leq k \leq N^2$, such that $\mathbf{v}_i$ is scattered in $V(i)$, $V(i+N)$, ..., $V(i+N^2-N)$. That is, $V(i+(j-1)N) = v_{ij}$, for all $1 \leq i, j \leq N$. We first configure the LARPBS into N subarrays LARPBS$_j$, such that LARPBS$_j$ contains processor $P_{(j-1)N+1}$, $P_{(j-1)N+2}$, ..., P_{jN}, where $1 \leq j \leq N$. The N processors in LARPBS$_j$ compute v_j by invoking the aggregation operation, and save the result v_j in $V((j-1)N + 1)$. Then, all the v_j's are moved to $V(1)$, $V(2)$, ..., $V(N)$ by a one-to-one communication operation. The complete vector chain addition algorithm is described below:

> for $1 \leq j \leq N$ do in parallel
> $\qquad V((j-1)N + 1) \leftarrow V((j-1)N + 1) + V((j-1)N + 2) + \cdots + V(jN)$
> endfor
> for $1 \leq j \leq N$ do in parallel
> $\qquad V(j) \leftarrow V((j-1)N + 1)$
> endfor

The above algorithm requires two bus cycles.

Theorem 3. Vector chain addition can be done in $O(1)$ time using N^2 processors.   ∎

Given N matrices A_1, A_2, ..., A_N, where $A_k = (a_{ij}^{(k)})_{N \times N}$, the matrix chain addition problem is to calculate $A = (a_{ij})_{N \times N}$, where $a_{ij} = a_{ij}^{(1)} + a_{ij}^{(2)} + \cdots + a_{ij}^{(N)}$, for all $1 \leq i, j \leq N$. Assume that we have N^3 processors, and matrix A_k is put into $P_{(k-1)N^2+1}$, $P_{(k-1)N^2+2}$, ..., P_{kN^2}, in the row-major order, where $1 \leq k \leq N$. That is, $A((k-1)N^2 + (i-1)N + j) = a_{ij}^{(k)}$, for all $1 \leq i, j, k \leq N$. We first rearrange the data via a one-to-one communication, such that all $a_{ij}^{(1)}$, $a_{ij}^{(2)}$, ..., $a_{ij}^{(N)}$ are packed together. We then reconfigure the LARPBS into N^2 subarrays LARPBS$_{ij}$, such that LARPBS$_{ij}$ contains processor $P_{((i-1)N+(j-1))N+k}$, $1 \leq k \leq N$, and that

$A(((i - 1)N + (j - 1))N + k) = a_{ij}^{(k)}$, where $1 \le i, j, k \le N$. LARPBS$_{ij}$ is used to calculate a_{ij} by aggregating $(a_{ij}^{(1)}, a_{ij}^{(2)}, ..., a_{ij}^{(N)})$. Finally, all the a_{ij}'s are moved to the first N^2 processors. The matrix chain addition algorithm is given as follows, which takes three bus cycles:

> for $1 \le i, j, k \le N$ do in parallel
> $\quad A(((i - 1)N + (j - 1))N + k) \leftarrow A((k - 1)N^2 + (i - 1)N + j)$
> endfor
> for $1 \le i, j \le N$ do in parallel
> $\quad A(((i - 1)N^2 + (j - 1))N + 1) \leftarrow \sum_{k=1}^{N} A(((i - 1)N^2 + (j - 1))N + k)$
> endfor
> for $1 \le j \le N$ do in parallel
> $\quad A((i - 1)N + j) \leftarrow A(((i - 1)N^2 + (j - 1))N + 1)$
> endfor

Theorem 4. Matrix chain addition can be done in $O(1)$ time using N^3 processors. ∎

In the sequel, we will show that many matrix problems can be solved using the primitive operations defined in this section and known algorithms for other matrix problems as subroutines. It seems that the description of these algorithms will be quite tedious if we insist on specifications down to the basic operation level, which is unnecessary. Thus, we will outline the algorithms, provide as much details as possible, and justify their time and processor complexities.

34.3   MATRIX-VECTOR AND MATRIX MULTIPLICATION

Assume that we have an $N \times N$ matrix $A = (a_{ij})$, and an N-dimensional vector $\mathbf{v} = (v_1, v_2, ..., v_N)$. Matrix A is stored in the row major order in processors P_1, P_2, ..., P_{N^2}, such that $A((i - 1)N + j) = a_{ij}$, for all $1 \le i, j \le N$. Vector $\mathbf{v}$ is put in processors $P_1, P_2, ..., P_N$, such that $V(j) = v_j$, for all $1 \le j \le N$. The N^2 processors will be reconfigured into N subarrays LARPBS$_i$, $1 \le i \le N$, which contains processors $P_{(i-1)N+j}$, $1 \le j \le N$. To calculate the matrix-vector product $A\mathbf{v} = \mathbf{x} = (x_1, x_2, ..., x_N)$, our algorithm performs the following steps:

1. The vector $\mathbf{v}$ is made available to all subarrays LARPBS$_i$, $1 \le i \le N$, via multiple multicasting in one bus cycle.

2. Processor $P_{(i-1)N+j}$ calculates $a_{ij}v_j$ locally, where $1 \le i, j \le N$, in constant time.

3. LARPBS$_i$, $1 \le i \le N$, aggregates $x_i = a_{i1}v_1 + a_{i2}v_2 + \cdots + a_{iN}v_N$ in one bus cycle.

4. The x_i's are moved to processors P_i, $1 \le i \le N$, using a one-to-one communication.

We can have the following claim.

Theorem 5. Matrix-vector product can be calculated in $O(1)$ time using N^2 processors. ∎

Matrix multiplication is perhaps the most important subproblem in many other matrix manipulations. A number of parallel matrix multiplication algorithms have been developed on a linear array with a reconfigurable pipelined bus system. In (Li *et al.*, 1996), it is shown that matrix multiplication can be performed:

- in $O(N)$ time using N^2 processors;

- in $O(1)$ time using N^3 processors;

- in $O(\log N)$ time using $O(N^{2.8074})$ processors;

- in $O(\log N + N^{1-\epsilon})$ time using $O(N^{2+0.8074\epsilon})$ processors, $0 \leq \epsilon \leq 1$;

and most noteworthy, the following result, which leads to performance enhancement for many matrix manipulation algorithms.

Theorem 6. There is a matrix multiplication algorithm MM_δ on LARPBS, which runs in $O((\log N)^\delta)$ time using $O(N^3/(\frac{8}{7})^{(\log N)^\delta})$ processors, where $0 \leq \delta \leq 1$. The cost of MM_δ is $o(N^3)$ for $0 < \delta \leq 1$. ∎

Now, let us consider how to calculate the powers A, A^2, A^3, ..., A^N of an $N \times N$ matrix A. We require a LARPBS with $N\rho(N, \delta)$ processors, where

$$\rho(N, \delta) = O\left(\frac{N^3}{1.1428^{(\log N)^\delta}}\right) \tag{34.1}$$

is the number of processors used in the matrix multiplication algorithm MM_δ. The LARPBS will be reconfigured into N subarrays $LARPBS_q$, $1 \leq q \leq N$, where $LARPBS_q$ consists of processors

$$P_{(q-1)\rho(N,\delta)+k}, \quad \text{for all } 1 \leq k \leq \rho(N, \delta).$$

Array A is initially held by the first N^2 processors of $LARPBS_1$. Without loss of generality, we assume that $N = 2^n$ is a power of two. Our algorithm, which calculates the first N powers of A, performs the following two major steps:

1. $LARPBS_1$ computes the powers A^{2^1}, A^{2^2}, ..., $A^{2^{n-1}}$, one by one, such that $A^{2^d} = A^{2^{d-1}} \times A^{2^{d-1}}$ is obtained by invoking algorithm MM_δ, where $1 \leq d \leq n - 1$. All these results are also sent to all $LARPBS_q$, $1 \leq q \leq N$, using the multiple multicasting operation.

2. If $q = c_{n-1}2^{n-1} + c_{n-2}2^{n-2} + \cdots + c_0 2^0$, then $LARPBS_q$ calculates $A^q = M_{n-1} \times M_{n-2} \times \cdots \times M_0$, where $M_d = A^{2^d}$ if $c_d \neq 0$, and $M_d = I_N$ if $c_d = 0$, and I_N is the $N \times N$ identity matrix.

Notice that both steps essentially perform $n - 1 = O(\log N)$ matrix multiplications. Thus, the overall time and processor complexities depend on those of matrix multiplication.

Theorem 7. The first N powers A, A^2, A^3, ..., A^N of a matrix A can be calculated in $O((\log N)^{1+\delta})$ time using $O(N^4/(\frac{8}{7})^{(\log N)^\delta})$ processors, where $0 \leq \delta \leq 1$. The cost is $o(N^4)$ for $0 < \delta \leq 1$. ∎

Given N matrices A_1, A_2, ..., A_N, where $A_k = (a_{ij}^{(k)})_{N \times N}$, the matrix chain product problem is to calculate $A = A_1 \times A_2 \times \cdots \times A_N$. Our algorithm for obtaining A is actually the standard binary tree method, and the overall running time is a factor of $O(\log N)$ more than that of matrix multiplication. Thus, similar to Theorem 7, we have:

Theorem 8. A matrix chain product $A = A_1 \times A_2 \times \cdots \times A_N$ can be obtained in $O((\log N)^{1+\delta})$ time, using $O(N^4 / (\frac{8}{7})^{(\log N)^\delta})$ processors, where $0 \le \delta \le 1$. The cost is $o(N^4)$ for $0 < \delta \le 1$.   ∎

34.4 INVERSION OF LOWER AND UPPER TRIANGULAR MATRICES

Let A be an $N \times N$ lower triangular matrix that is invertible, *i.e.* all the elements on A's main diagonal are nonzeros. We partition A into four submatrices of equal size $N/2 \times N/2$,

$$A = \begin{bmatrix} A_1 & 0 \\ A_3 & A_2 \end{bmatrix}.$$

Since A_1 and A_2 are also invertible lower triangular matrices, we have

$$A^{-1} = \begin{bmatrix} A_1^{-1} & 0 \\ -A_2^{-1} A_3 A_1^{-1} & A_2^{-1} \end{bmatrix}.$$

Therefore, A^{-1} can be obtained by inverting A_1 and A_2 recursively, and then multiplying A_1^{-1} and A_2^{-1} with A_3. Similarly, if A is an $N \times N$ upper triangular matrix that is invertible, we partition A into four submatrices of equal size $N/2 \times N/2$,

$$A = \begin{bmatrix} A_1 & A_3 \\ 0 & A_2 \end{bmatrix}.$$

Then, A_1 and A_2 are also invertible upper triangular matrices, and

$$A^{-1} = \begin{bmatrix} A_1^{-1} & -A_1^{-1} A_3 A_2^{-1} \\ 0 & A_2^{-1} \end{bmatrix}.$$

The above discussion yields the following method for inverting a lower (upper) triangular matrix.

A Method for Lower (Upper) Triangular Matrix Inversion.

1. Recursively calculate A_1^{-1} and A_2^{-1};

2. Compute $-A_2^{-1} A_3 A_1^{-1}$ ($-A_1^{-1} A_3 A_2^{-1}$) using the matrix multiplication algorithm MM.

This method reduces the lower/upper triangular matrix inversion problem to matrix multiplication. Without loss of generality, we assume that $N = 2^n$ is a power of two. The recursion can be unwound into $n + 1$ iterations, such that a sequence of matrices

$A_0, A_1, A_2, ..., A_n$ are calculated. Initially, A_0 is obtained from A by inverting the elements on the main diagonal. This is the base of the recursion. In general, A_k is obtained from A_{k-1} by further calculating 2^{n-k} submatrices of size 2^{k-1}. Finally, $A_n = A^{-1}$. To calculate A_k, a linear array is reconfigured into 2^{n-k} subarrays for 2^{k-2} simultaneous invocation of submatrix multiplications. To this end, certain data movements are required such that the entries of a submatrix are packed together. Fortunately, the communication patterns for this purpose are quite regular.

To calculate A_k, $1 \le k \le n$, there are 2^{n-k} submatrices of size 2^{k-1} that can be computed in parallel, and each requires only two matrix multiplications. The time complexity is:

$$O\left(\sum_{k=1}^{n}(k-1)^{\delta}\right) = O(n^{\delta+1}) = O((\log N)^{\delta+1}).$$

The total number of processors required is:

$$2^{n-k}\rho(2^{k-1}, \delta) = 2^{n-k}O\left(\frac{2^{3k-3}}{1.1428^{(k-1)^{\delta}}}\right) = O\left(\frac{N^3}{1.1428^{(\log N-1)^{\delta}}}\right),$$

where, $\rho(2^{k-1}, \delta)$ is defined in Equation (34.1). Thus, we have:

Theorem 9. The inverse of a lower triangular matrix can be obtained in $O((\log N)^{1+\delta})$ time, with number of processors $O(N^3/(\frac{8}{7})^{(\log N-1)^{\delta}})$, for $0 \le \delta \le 1$. The cost is $o(N^3)$ for $0 < \delta \le 1$. ∎

Similarly, we have:

Theorem 10. The inverse of an upper triangular matrix can be obtained in $O((\log N)^{1+\delta})$ time, with number of processors $O(N^3/(\frac{8}{7})^{(\log N-1)^{\delta}})$, for $0 \le \delta \le 1$. The cost is $o(N^3)$ for $0 < \delta \le 1$. ∎

34.5 DETERMINANTS, CHARACTERISTIC POLYNOMIALS, AND RANKS

We use $\det(A)$ to denote the determinant of a matrix $A = (a_{ij})_{N \times N}$. The characteristic polynomial of a matrix A is represented as

$$\phi_A(\lambda) = \det(\lambda I_N - A) = \lambda^N + c_1\lambda^{N-1} + c_2\lambda^{N-2} + \cdots + c_{N-1}\lambda + c_N,$$

where I_N is the $N \times N$ identity matrix. The trace $\text{tr}(A)$ of a matrix $A = (a_{ij})_{N \times N}$ is the sum of the entries on A's main diagonal, i.e. $\text{tr}(A) = a_{11} + a_{22} + \cdots + a_{NN}$.

The following classical result is the basis of a parallel algorithm for obtaining $\phi_A(\lambda)$ and $\det(A)$ (Verrier, 1840):

Leverier's Lemma. The coefficients $c_1, c_2, ..., c_N$ of the characteristic polynomial of a matrix A satisfy $S(c_1, c_2, ..., c_N)^T = (s_1, s_2, ..., s_N)^T$, where:

$$S = \begin{bmatrix} 1 & 0 & 0 & \cdots & 0 & 0 \\ s_1 & 2 & 0 & \cdots & 0 & 0 \\ s_2 & s_1 & 3 & \cdots & 0 & 0 \\ \vdots & \vdots & \vdots & \ddots & \vdots & \vdots \\ s_{N-2} & s_{N-3} & s_{N-4} & \cdots & N-1 & 0 \\ s_{N-1} & s_{N-2} & s_{N-3} & \cdots & s_1 & N \end{bmatrix},$$

and $s_k = \text{tr}(A^k)$, for all $1 \leq k \leq N$.   ∎

Based on Leverier's Lemma, (Csanky, 1976) devised the following method for calculating the characteristic polynomial of a matrix A. Since $\phi_A(0) = \det(-A) = c_N$, i.e. $\det(A) = (-1)^N c_N$, this algorithm can also be used to calculate $\det(A)$.

Csanky's Strategy for Characteristic Polynomial.

1. Calculate $A^2, A^3, A^4, ..., A^N$;

2. Calculate $s_k = \text{tr}(A^k)$, for all $1 \leq k \leq N$;

3. Find S^{-1}, where S is the lower triangular matrix in Leverier's Lemma;

4. Calculate the matrix-vector product $S^{-1}[s_1, s_2, ..., s_N]^T$ to obtain $c_1, c_2, ..., c_N$.

Step 1 involves the calculation of the first N powers of A (Theorem 7). Step 2 can be implemented using the aggregation operation, plus certain data movements. Step 3 invokes the lower triangular matrix inversion algorithm (Theorem 9). Finally, Step 4 is a matrix-vector multiplication (Theorem 5). The time and processor complexities of Csanky's method is dominated by Step 1.

Theorem 11.   The characteristic polynomial of a matrix can be obtained in $O((\log N)^{1+\delta})$ time, using $O(N^4/(\frac{8}{7})^{(\log N)^{\delta}})$ processors, where $0 \leq \delta \leq 1$. The cost is $o(N^4)$ for $0 < \delta \leq 1$.

    ∎

Theorem 12. The determinant of a matrix can be calculated in $O((\log N)^{1+\delta})$ time, and the number of processors is $O(N^4/(\frac{8}{7})^{(\log N)^{\delta}})$, where $0 \leq \delta \leq 1$. The cost is $o(N^4)$ for $0 < \delta \leq 1$.   ∎

The rank of a matrix A, rank(A), is the number of nonzero rows (or columns) in the row-reduced (or column-reduced) echelon form of A. It is well known that rank$(A) = $ rank$(A^T A)$, where A^T is the transpose of A (or conjugate transpose of A for complex matrices), and $A^T A$ is similar to a diagonal matrix whose elements are the roots of the characteristic polynomial $\phi_{A^T A}(\lambda)$. Therefore, rank(A) is the number of nonzero roots of $\phi_{A^T A}(\lambda)$. This leads to the following algorithm for finding rank(A) (Ibarra *et al.*, 1980):

An Algorithm for Calculating Matrix Rank.

1. Get the matrix $A^T A$;

2. Calculate $\phi_{A^T A}(\lambda) = c_0 \lambda^N + c_1 \lambda^{N-1} + c_2 \lambda^{N-2} + \cdots + c_{N-1}\lambda + c_N$;

3. Find $\text{rank}(A) = N - i$, where i, $0 \le i \le N$, is the largest integer such that $c_{N-i} \ne 0$, and $c_{N-i+1} = c_{N-i+2} = \cdots = c_N = 0$.

In the above algorithm, Step 1 performs a matrix transposition (Theorem 2) and a matrix multiplication (Theorem 6). Step 2 invokes Csanky's method for computing characteristic polynomial (Theorem 11). Step 3 can be implemented in a few bus cycles (*i.e.* constant time) by simple data testing, comparison, and movement.

Theorem 13. The rank of a matrix can be found in $O((\log N)^{1+\delta})$ time, and the number of processors is $O(N^4/(\frac{8}{7})^{(\log N)^\delta})$, where $0 \le \delta \le 1$. The cost is $o(N^4)$ for $0 < \delta \le 1$.   ∎

34.6 INVERSION OF ARBITRARY MATRICES

Inverting an arbitrary matrix A is closely related to the calculation of the characteristic polynomial $\phi_A(\lambda)$, as revealed by the following well-known theorem from linear algebra.

Cayley-Hamilton Theorem. Let $\phi_A(\lambda) = \lambda^N + c_1 \lambda^{N-1} + c_2 \lambda^{N-2} + \cdots + c_{N-1}\lambda + c_N$ be the characteristic polynomial of matrix A. Then

$$\phi_A(A) = A^N + c_1 A^{N-1} + c_2 A^{N-2} + \cdots + c_{N-1}A + c_N A^0$$

is the $N \times N$ zero matrix.   ∎

Cayley-Hamilton Theorem implies that

$$A(A^{N-1} + c_1 A^{N-2} + c_2 A^{N-3} + \cdots + c_{N-1}I_N) = -c_N I_N.$$

Hence, the inverse of a matrix A can be calculated using the following equation,

$$A^{-1} = -\frac{1}{c_N}\left(A^{N-1} + c_1 A^{N-2} + c_2 A^{N-3} + \cdots + c_{N-2}A + c_{N-1}I_N\right). \tag{34.2}$$

Csanky's method for calculating matrix inversion can be described as follows (Csanky, 1976):

Csanky's Strategy for Matrix Inversion.

1. Calculate the characteristic polynomial of A, that is, $c_1, c_2, ..., c_N$;

2. Compute A^{-1} using the identity in Equation (34.2).

The time and processor complexities of Step 1 are given in Theorem 11. Step 2 involves the computation of the first N powers of A (Theorem 7) and matrix chain addition (Theorem 4). Hence, we have the following result:

Theorem 14. The inverse of a matrix can be calculated in $O((\log N)^{1+\delta})$ time, and the number of processors is $O(N^4/(\frac{8}{7})^{(\log N)^\delta})$, where $0 \le \delta \le 1$. The cost is $o(N^4)$ for $0 < \delta \le 1$.

∎

34.7 LINEAR SYSTEMS OF EQUATIONS

Let A be a nonsingular $N \times N$ matrix $(a_{ij})_{N \times N}$, and $\mathbf{b} = (b_1, b_2, ..., b_N)^T$ be an N-dimensional vector. The problem of solving a linear system of equations is to find a vector $\mathbf{x} = (x_1, x_2, ..., x_N)^T$ such that $A\mathbf{x} = \mathbf{b}$. Hence $\mathbf{x} = A^{-1}\mathbf{b}$, *i.e.* solving the above linear system of equations can be accomplished by a matrix inversion (Theorem 14) and a matrix-vector multiplication (Theorem 5).

Theorem 15. A linear system of equations can be solved in $O((\log N)^{1+\delta})$ time, and the number of processors is $O(N^4/(\frac{8}{7})^{(\log N)^\delta})$, where $0 \le \delta \le 1$. The cost is $o(N^4)$ for $0 < \delta \le 1$.

∎

Let $\mathbf{k}_j = A^j\mathbf{b}$, where $0 \le j \le N - 1$. Then, the matrix $K(A, \mathbf{b}, N) = [\mathbf{k}_0, \mathbf{k}_1, \mathbf{k}_2, ..., \mathbf{k}_{N-1}] = [\mathbf{b}, A\mathbf{b}, A^2\mathbf{b}, ..., A^{N-1}\mathbf{b}]$ is called the Krylov matrix defined by the matrix A, the vector $\mathbf{b}$, and the integer N. By the Cayley-Hamilton Theorem, we have $A(A^{N-1}\mathbf{b} + c_1 A^{N-2}\mathbf{b} + c_2 A^{N-3}\mathbf{b} + \cdots + c_{N-1}\mathbf{b}) = -c_N \mathbf{b}$, which implies that:

$$\mathbf{x} = -\left(\frac{1}{c_N}\right)\mathbf{k}_{N-1} - \left(\frac{c_1}{c_N}\right)\mathbf{k}_{N-2} - \left(\frac{c_2}{c_N}\right)\mathbf{k}_{N-3} - \cdots - \left(\frac{c_{N-1}}{c_N}\right)\mathbf{k}_0.$$

In other words, $\mathbf{x}$ is a linear combination of the column vectors of the Krylov matrix $K(A, \mathbf{b}, N)$. Thus, we can first calculate the first N powers of A (Theorem 7), and then compute the Krylov matrix $K(A, \mathbf{b}, N)$ using matrix-vector multiplication (Theorem 5). Once $K(A, \mathbf{b}, N)$ is available, $\mathbf{x}$ can be obtained via vector chain addition in constant time (Theorem 3). This proves the following result:

Theorem 16. The Krylov matrix defined by the matrix A and the vector $\mathbf{b}$ can be calculated. Hence the linear system of equations $A\mathbf{x} = \mathbf{b}$ can be solved in $O((\log N)^{1+\delta})$ time using $O(N^4/(\frac{8}{7})^{(\log N)^\delta})$ processors, where $0 \le \delta \le 1$. The cost is $o(N^4)$ for $0 < \delta \le 1$.

∎

34.8 LU- AND QR-FACTORIZATIONS

The LU-factors of matrix A contain a nonsingular lower triangular matrix L and a nonsingular upper triangular matrix U such that $A = LU$. Suppose that the LU-factors of A exist. We divide A, L, and U into $N/2 \times N/2$ blocks as follows:

$$\begin{bmatrix} A_{11} & A_{12} \\ A_{21} & A_{22} \end{bmatrix} = \begin{bmatrix} L_{11} & 0 \\ L_{21} & L_{22} \end{bmatrix} \times \begin{bmatrix} U_{11} & U_{12} \\ 0 & U_{22} \end{bmatrix}.$$

The above identity implies that $A_{11} = L_{11}U_{11}$, $A_{12} = L_{11}U_{12}$, $A_{21} = L_{21}U_{11}$, and $A_{22} = L_{21}U_{12} + L_{22}U_{22}$. The following method for LU-factorization is due to (Pan, 1987):

Pan's Method for LU-Factorization.

1. Compute A_{11}^{-1};

2. Calculate $X_1 = A_{11}^{-1}A_{12}$, $X_2 = A_{21}A_{11}^{-1}$, and $X_3 = A_{21}A_{11}^{-1}A_{12}$;

3. Set $X_4 = A_{22} - X_3$; (It can be verified that $X_4 = L_{22}U_{22}$.)

4. Recursively LU-factorize A_{11} to get L_{11} and U_{11}, and recursively LU-factorize X_4 to get L_{22} and U_{22};

5. Calculate $L_{21} = X_2 L_{11}$, and $U_{12} = U_{11}X_1$.

Let $T(N)$ be the time complexity of the above algorithm. Then, we have

$$T(N) = T\left(\frac{N}{2}\right) + O((\log N)^{1+\delta}) = O\left(\sum_{l=0}^{\log N}(\log N - l)^{1+\delta}\right),$$

which gives $T(N) = O((\log N)^{2+\delta})$. As for number of processors, let us unwind the recursion, such that there are 2^k simultaneous LU-factorizations of matrices with size $N/2^k$. We notice that Step 1 requires $(N/2^k)\rho(N/2^k, \delta)$ processors. Each of Steps 2 and 5 involves at most two parallel matrix multiplications and, hence, $\rho(N/2^k, \delta)$ processors. Step 3 needs only $O(N^2)$ processors. Hence, the total number of processors is

$$2^k O\left(\frac{(N/2^k)^4}{1.1428^{(\log(N/2^k))^\delta}}\right) = O(N\rho(N, \delta)). \tag{34.3}$$

Theorem 17. LU-factorization of a matrix can be performed in $O((\log N)^{2+\delta})$ time using $O(N^4/(\frac{8}{7})^{(\log N)^\delta})$ processors, where $0 \leq \delta \leq 1$. The cost is $o(N^4)$ for $0 < \delta \leq 1$. ∎

The QR-factors of matrix A contain an orthogonal matrix Q (that is, Q satisfies $Q^T Q = I_N$) and a nonsingular upper triangular matrix R such that $A = QR$. If $A = QR$, then $A^T A = R^T Q^T QR = R^T R$. Since R is a nonsingular upper triangular matrix, R^T is a nonsingular lower triangular matrix. In other words, R^T and R are LU-factors of $A' = A^T A$. The above discussion suggests the following method:

A Method for QR-Factorization.

1. Calculate A^T;

2. Compute $A' = A^T A$;

3. LU-factorize A' to get R;

4. Find R^{-1};

5. Calculate AR^{-1} to obtain Q.

The time and processor complexities of Steps 1, 2, 4, and 5 are given by Theorems 2, 6, and 14 respectively. Step 3, whose complexities are given in Theorem 17, dominates the time and processor complexities of QR-factorization.

Theorem 18. QR-factorization of a matrix can be performed in $O((\log N)^{2+\delta})$ time, using $O(N^4/(\frac{8}{7})^{(\log N)^\delta})$ processors, where $0 \leq \delta \leq 1$. The cost is $o(N^4)$, for all $0 < \delta \leq 1$. ∎

34.9   FINAL REMARKS

In this paper we presented fast and cost-efficient parallel algorithms on linear arrays with reconfigurable pipelined optical bus systems for a number of important and fundamental matrix computation problems. We showed that our algorithms provide a wide range of performance-cost combinations which have not been achieved previously on existing parallel computation models, including theoretical PRAM models. Compared with previously known best parallel algorithms, our algorithms have an $O(\log N)$ reduction in time, while maintaining their cost below $o(N^4)$.

Our results demonstrate that the LARPBS is a very powerful model. In addition to its high-communication bandwidth, a LARPBS supports versatile communication patterns, and its communication reconfigurability constitutes an integral part of a parallel computation. These features allow for exploiting large degree of parallelism in a computational problem (with lower cost than in other systems) that most other machine models cannot achieve.

Acknowledgments

Keqin Li is partially supported by National Aeronautics and Space Administration and the Research Foundation of State University of New York through the NASA/University Joint Venture in Space Science Program under Grant NAG8-1313. Yi Pan is supported in part by the National Science Foundation under Grant CCR-9211621, the Air Force Avionics Laboratory, Wright Laboratory, Dayton, Ohio, under Grant F33615-C-2218, and an Ohio Board of Regents Investment Fund Competition Grant. Si-Qing Zheng is supported in part by the National Science Foundation under Grant ECS-9626215, and Louisiana Grant LEQSF (1996-99)-RD-A-16.

References

Benner, A., Jordan, H., and Heuring, V. (1991). Digital optical computing with optically switched directional couplers. *Optical Engineering*, 30:1936–1941.

Chiarulli, D., Melhem, R., and Levitan, S. (1987). Using coincident optical pulses for parallel memory addressing. *IEEE Computer*, 30:48–57.

Cook, S. (1985). A taxonomy of problems with fast parallel algorithms. *Information and Control*, 64:2–22.

Coppersmith, D. and Winograd, S. (1990). Matrix multiplication via arithmetic progressions. *Journal of Symbolic Computation*, 9:251–280.

Csanky, L. (1976). Fast parallel matrix inversion algorithms. *SIAM Journal on Computing*, 5:618–623.

Guo, Z., Melhem, R., Hall, R., Chiarulli, D., and Levitan, S. (1991). Pipelined communications in optically interconnected arrays. *Journal of Parallel and Distributed Computing*, 12:269–282.

Hamdi, M. and Pan, Y. (1995). Efficient parallel algorithms on optically interconnected arrays of processors. *IEEE Proceedings – Computers and Digital Techniques*, 142:87–92.

Ibarra, O., Moran, S., and Rosier, L. (1980). A note on the parallel complexity of computing the rank of order n matrices. *Information Processing Letters*, 11(4–5):162.

Leighton, T. (1992). *Introduction to Parallel Algorithms and Architectures: Arrays · Trees · Hypercubes*. Morgan Kaufmann, San Mateo, California.

Levitan, S., Chiarulli, D., and Melhem, R. (1990). Coincident pulse techniques for multiprocessor interconnection structures. *Applied Optics*, 29:2024–2039.

Li, K. (1997). Constant time Boolean matrix multiplication on a linear array with a reconfigurable pipelined bus system. *Journal of Supercomputing*, 11(4):391–403.

Li, K., Pan, Y., and Zheng, S.-Q. (1996). Fast and processor efficient parallel matrix multiplication algorithms on a linear array with a reconfigurable pipelined bus system. Technical Report #96-004, Department of Computer Science, Louisiana State University. Also to appear in *IEEE Transactions on Parallel and Distributed Systems*.

Li, K., Pan, Y., and Zheng, S.-Q., editors (1998). *Parallel Computing Using Optical Interconnections*. Kluwer Academic Publishers. To appear.

Pan, V. (1987). Complexity of parallel matrix computations. *Theoretical Computer Science*, 54:65–85.

Pan, V. and Reif, J. (1985). Efficient parallel solution of linear systems. In *7th ACM Symposium on Theory of Computing*, pages 143–152.

Pan, Y. and Hamdi, M. (1996). Efficient computation of singular value decomposition on arrays with pipelined optical buses. *Journal of Network and Computer Applications*, 19:235–248.

Pan, Y., Hamdi, M., and Li, K. (1998a). Efficient and scalable quicksort on a linear array with a reconfigurable pipelined bus system. *Future Generation Computer Systems*. To appear.

Pan, Y. and Li, K. (1998). Linear array with a reconfigurable pipelined bus system – concepts and applications. *Information Sciences – An International Journal*, 106(3–4):237–258.

Pan, Y., Li, K., and Zheng, S.-Q. (1998b). Fast nearest neighbor algorithms on a linear array with a reconfigurable pipelined bus system. *Journal of Parallel Algorithms and Applications*. To appear.

Pavel, S. and Akl, S. (1996). Matrix operations using arrays with reconfigurable optical buses. *Journal of Parallel Algorithms and Applications*, 8:223–242.

Qiao, C. and Melhem, R. (1993). Time-division optical communications in multiprocessor arrays. *IEEE Trans. on Computers*, 42:577–590.

Rajasekaran, S. and Sahni, S. (1997). Sorting, selection, and routing on the array with reconfigurable optical buses. *IEEE Transactions on Parallel and Distributed Systems*, 8(11):1123–1131.

Strassen, V. (1969). Gaussian elimination is not optimal. *Numerische Mathematik*, 13:354–356.

Tocci, C. and Caulfield, H. (1994). *Optical Interconnection – Foundations and Applications*. Artech Nouce, Inc.

Verrier, U. L. (1840). Sur les variations seculaires des elementes elliptiques des sept planets principales. *J. Math. Pures Appl.*, 5:220–254.

Zheng, S.-Q. and Li, Y. (1997). Pipelined asynchronous time-division multiplexing optical bus. *Optical Engineering*, 36(12):3392–3400.

35 DEVELOPMENT OF PARALLEL, ADAPTIVE FINITE ELEMENT IMPLEMENTATIONS FOR MODELING TWO-PHASE OIL/WATER FLOW

Donald J. Morton

Department of Computing Science,
University of Montana, U.S.A.

morton@cs.umt.edu

Abstract: An adaptive finite element code is implemented for the parallel solution of multiphase fluid flow problems. This modular code, although not optimal, provides reasonable performance gains while maintaining flexibility for future developments. Issues of dynamic mesh modification and load balancing are encapsulated within highly cohesive units which communicate minimally with other units for the parallel solution of the differential equations over a global domain.

Portability is emphasized, so PVM is used for interprocessor communication. The code has been tested on the Arctic Region Supercomputing Center's Cray T3E, and on a cluster of Linux Pentiums at the University of Montana.

Keywords: adaptive finite element methods, multiphase flow, parallel computing.

35.1 INTRODUCTION

Finite element methods have been proven useful in the simulation of complex science and engineering problems described by partial differential equations. One of the greatest advantages of the finite element method over other traditional approaches, such as finite difference methods, is that irregular and heterogeneous regions can be incorporated accurately in models through the use of elements that vary in shape, size and material properties. When designed carefully, no matter what shape, size, or type of element being used, the same basic operations will be performed on the element to incorporate it into a global system of equations and arrive at a solution. Thus, from a programming point of view, the low-level details of element shape functions, Gauss points, and even coordinates can be encapsulated beneath an appropriate interface, allowing modelers to focus more on the governing equations and their numerical solutions rather than tedious details of mesh implementation.

Because of the flexibility in shape, size, and type, elements may be arranged in a mesh so that smaller, more accurate elements are placed in regions of high activity while larger elements are used to capture the relatively "quiet" aspects of the problem. This allows for a generic approach for performing accurate, yet efficient, computations within a problem domain. Furthermore, *adaptive* finite element methods extend this concept by dynamically modifying the finite element mesh according to the current state of the solution variables. Using a predetermined metric for accuracy or activity, an adaptive finite element approach will determine where smaller elements need to be, and modify the mesh to facilitate this, often by *refining* elements through the partitioning of a single element into several smaller ones. Likewise, a "calm" region will require less accuracy, so a group of smaller elements may be combined into a single larger element.

Section 35.2 describes the adaptive finite element method as applied to a fluid flow problem, and identifies its parallel performance bottlenecks. In Section 35.3, a key parallel performance issue, load balancing, is addressed. Section 35.4 presents some experimental data, using Linux PCs and Cray supercomputers, that show how the adaptive methods can improve performance. Finally, Section 35.5 outlines future research.

35.2   APPLICATION - TWO-PHASE FLOW THROUGH POROUS MEDIA

Over the past several years, we have constructed experimental software for the investigation of adaptive finite element techniques used in the solution of science and engineering problems. The application area has been petroleum engineering, specifically the modeling of oil/water flow through porous media. The theoretical equations describing these processes are provided in Equations 35.1 and 35.2 (Peaceman, 1977):

$$\nabla \cdot \hat{\mathbf{M}}_{\mathbf{w}} \nabla p_n - \nabla \cdot \hat{\mathbf{M}}_{\mathbf{w}} \frac{dp_c}{dS_n} \nabla S_n \quad -$$

$$\nabla \cdot \hat{\mathbf{M}}_{\mathbf{w}} \nabla (\rho_w g z) \quad - \quad \frac{q_w}{\rho_w} =$$

$$-\phi \frac{\partial S_n}{\partial t} + c_w \phi (1 - S_n) \frac{\partial p_n}{\partial t}. \tag{35.1}$$

$$\nabla \cdot \hat{\mathbf{M}}_{\mathbf{n}} \nabla p_n - \nabla \cdot \hat{\mathbf{M}}_{\mathbf{n}} \nabla (\rho_n g z) \quad - \quad \frac{q_n}{\rho_n} =$$

$$\phi \frac{\partial S_n}{\partial t} + c_n \phi S_n \frac{\partial p_n}{\partial t}. \tag{35.2}$$

In general, these equations are highly nonlinear, often describing the transition of a sharp oil/water interface through the problem domain. This interface represents a near-discontinuity as the concentration of oil (or water) varies from 0% to 100%. Our initial

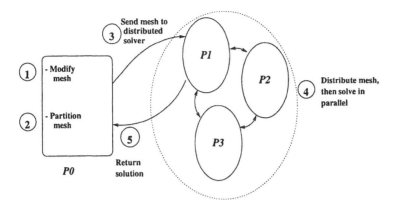

Figure 35.1 Flow diagram of heterogeneous, distributed adaptive finite element solution during a single time step. Flow sequence is depicted by the circled numbers. Step 2, the mesh partitioning, was recently incorporated, as described in this paper.

work (Morton *et al.*, 1994) solved this problem on single-processor systems through the use of adaptive finite element methods. A crude, initial mesh was constructed by hand. Then, at the beginning of each timestep in a simulation, the mesh would be evaluated and dynamically modified if necessary to achieve high resolution near the oil/water interface, with coarser resolution elsewhere. Within a timestep, the equations were linearized through iteration, using a standard Galerkin approach for constructing a global system of equations, then solving with direct techniques.

This work has evolved from the investigation of serial software components to the integration of serial components into a modular, heterogeneous software system for multiple processors. Naturally, computational efficiency is a driving motivation behind the work, but not at the expense of modifiability. Our efforts are focused more on developing a software system which can be easily used by finite element modelers to construct simulation software for a wide range of problems in numerous disciplines. The intent of this software is that scientists and engineers, knowledgeable in the basic concepts of finite element methods, will be able to rapidly construct highly-modifiable models.

Prior to the work described in this paper, we had developed code which made it possible to run two-dimensional adaptive finite element models on parallel architectures ranging from Linux PCs to Cray MPP platforms by concentrating mesh modifications in one module (on one processor), then distributing newly modified meshes to distributed processes for the solution phase. As described in (Morton and Tyler, 1996), parallelization was accomplished through sub-structuring techniques, whereby each processor would construct local systems of equations coupled to other processors' equations through nodes lying on interprocessor boundaries. As illustrated in Figure 35.1, at each time step, or less frequently if desired, the "driving process" (*P*0, in Figure 35.1) would periodically evaluate the current solution over the mesh. The mesh would be dynamically adjusted so that smaller elements were placed in

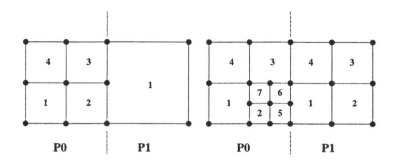

Figure 35.2   Before and after views of mesh refinement along an interprocessor boundary.

regions of activity with larger elements elsewhere. Where needed, smaller elements were created by taking an existing element and partitioning it into several similar, but smaller elements in the same location. Likewise, where necessary (for example, where things were getting "calmer"), previously refined groups of small elements would be re-unified to form a single, larger element.

Once a new mesh had been obtained, it would be sent to a single "master" process of the parallel finite element solver which, upon reception, processed the mesh as if it were static. In this way, numerous issues of dynamic mesh modification were kept separate from issues of parallel solutions. After the timestep solution had been obtained (which often required several iterations due to nonlinearities), it would be sent back to the "driving" process for the next round of mesh modification and subsequent parallel solution.

This approach delivered respectable performance gains, but the issue of load-balancing had never been considered. Test problems were chosen in which it was known beforehand that even as an evolving solution resulted in a changed mesh, processors would store roughly the same number of unknowns. Clearly, "interesting" problems would not exhibit such facilitating behavior, so load-balancing was recently addressed to broaden the scope of solvable problems with our technique.

Although various approaches for fully parallelized mesh adaptation and subsequent load balancing have been explored with success (Devine *et al.*, 1993; Biswas *et al.*, 1996), we have been intent on minimizing the interdependencies with a parallel finite element solution over a given mesh. A fully parallelized approach requires that modifications made in one processor along a processor boundary be "signaled" to the adjacent processor. The adjacent processor in turn will need to adjust numerous data structures reflecting added or deleted nodes, along with any information reflecting conditions on the other side of the border. Consider the finite element meshes of Figure 35.2, in which an interprocessor boundary is illustrated. To maintain inter-element continuity, the adaptive mesh procedures we utilize require that adjacent element refinement levels differ by no more than one. So, if we were to refine $P0$'s element 2 in the left hand mesh of Figure 35.2, we would first need to refine $P1$'s element 1 to produce the valid mesh shown on the right in Figure 35.2. Doing so across interprocessor boundaries added communication and synchronization requirements, in addition to numerous complexities. Once an element is modified, any processors that possess a node on that element must play a role in updating data structures and modifying systems of equations to account for new unknowns. In fact, before *any* boundary element could be modified, adjacent elements in other processors would need to be checked for possible violation of constraints. Even the modification of "interior" elements can become cumbersome, since sometimes it is necessary to first modify neighboring elements, and their neighboring elements, and so on, out to the processor boundary. As nodes and elements within a processor are modified, the structure of the linear systems of equations must also be modified, leading to an increased coupling of the mesh and solution modules, which would make future modifications difficult to maintain.

35.3   IMPLEMENTATION OF LOAD BALANCING

A load balancing implementation in the context of a fully parallelized mesh modification would add even more complexity. As new elements are created in a processor, it begins to inherit a heavier load than other processors and must eventually "give away" some of its load to others. This leads to frequent re-arrangement and re-numbering of nodes and elements in each processor. Although, as stated earlier, fully parallelized methods have been implemented successfully, we believe that modifiability of parallel, adaptive finite element code can be better facilitated by adhering to loosely coupled, cohesive modules.

Therefore, to incorporate load balancing into our approach, we continue to separate the "mesh maintenance" activities from the general finite element solution routines. Our implementation introduces load balancing immediately after a new mesh has been dynamically constructed, maintaining a weighted graph representation of the mesh for input to a graph partitioning algorithm. The graph partitioning results are then utilized to make processor assignments for each element. Then the mesh data is sent to the distributed finite element solver (where it is viewed as a static mesh) and partitioned among the processors according to the assignments made after the graph partitioning in the serial, driving process. Thus, from the serial, driving process' point of view, mesh data is simply being sent to a solver which will eventually return a solution. Other than the load balancing activities, which are incorporated in an added module, the serial process has no knowledge of a parallel solution; it simply sends a mesh out and gets a solution for the timestep. Likewise, the distributed solution portion of the system has no knowledge that it is working with a dynamically adaptive mesh. From its perspective, it receives a mesh which will remain static while a solution is computed, and it will return the solution to the driving, serial process. With such an implementation, we believe modification and experimentation will be less tedious.

The actual incorporation of load balancing in the mesh modification module centers about the maintenance of a weighted graph which is subsequently partitioned to obtain processor assignments. To reduce the cost of constructing, maintaining, and partitioning huge graphs every time the mesh is modified, our approach constructs a graph whose overall structure remains unchanged during the course of a simulation. However, weights on graph vertices are changed after mesh modifications to reflect the current level of refinement in a particular region. For example, referring to the top of Figure 35.3, an initial mesh with four elements is illustrated, from which a static

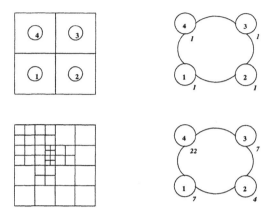

Figure 35.3   Mapping of adaptive finite element mesh to a weighted graph.

graph structure is created where each graph vertex represents one of the initial mesh elements, and each graph edge represents an adjacency between two elements, A weight is assigned to each graph vertex, representing the number of elements located in that region of the mesh. Thus, as an element is refined into smaller elements, the weight on its corresponding graph vertex increases, as in the bottom of Figure 35.3. So, upon completion of any mesh modification, there exists a weighted, undirected graph representing the number of elements underlying each of the original elements. This graph is subsequently presented to a graph partitioning package (*METIS*) (Karypis and Kumar, 1995), which additionally seeks to minimize the edge-cut between partitions.

35.4   TEST CASES

Finite element methods are difficult to simulate accurately, and adaptive finite element methods (in a single-processor environment) had been previously developed to simulate flows in real world laboratory experiments (Morton *et al.*, 1994). The test problems presented here were chosen not to match a specific real-world occurrence, but to demonstrate the type of problems that adaptive finite element methods might handle, and to illustrate the degree of load-balancing offered by our implementation.

Two similar problems were chosen, representing the injection of water into an oil-saturated region. Boundaries of the two-dimension domain were maintained at no-flux conditions, with the exception that water was injected at one corner, and fluids were removed from the opposite corner. The expected behavior was the movement of an oil/water front starting from the injection corner and migrating towards the production corner, "flooding" the problem domain with water.

The square problem domain, shown in Figure 35.4, was partitioned in both test cases into 16 × 16 quadrilateral elements. In the first test case, the problem domain was homogeneous, so all elements had the same material properties. In the second test case, elements which lay roughly along a diagonal between the injection and production nodes had material properties such that the absolute permeability of the medium

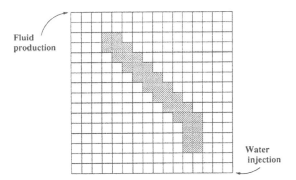

Figure 35.4 16 × 16 element initial mesh. In first test case, all elements had the same material properties; in the second test case, the shaded elements had an absolute permeability and porosity that was significantly greater than the other elements.

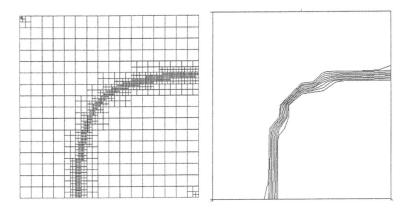

Figure 35.5 Finite element mesh and contour plot during homogeneous property simulation.

in these regions was an order of magnitude greater than it was in the other elements. In other words, fluid would flow more easily through the elements along the diagonal. Figures 35.5 and 35.6 show the adaptive finite element mesh and contour plots approximately midway through the simulations.

Test cases were first run on the Cray T3E at the Arctic Region Supercomputing Center. Timing and load balancing data for 4, 8, 16, and 32 processors is provided in Table 35.1. In each case, one processor served as the driving serial process for mesh maintenance while the rest were utilized for the actual parallel finite element solution. The first two columns of the table show the times required for the serial process to modify the mesh at the beginning of a timestep, then transfer the new mesh data to the parallel processes. The third column shows the time required for the parallel processes to obtain a distributed solution. Note that the distributed solution time when increasing

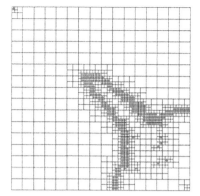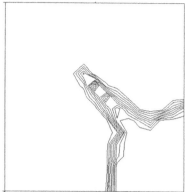

Figure 35.6 Finite element mesh and contour plot during heterogeneous property simulation.

Table 35.1 Cray T3E wall time (in seconds) and load balancing for a single timestep approximately midway through two-phase flow simulation. At this time, there were 4,548 degrees of freedom, and 1981 elements. The last column indicates the load balancing, showing the minimum and maximum number of elements in the "solution" processors.

Processors	Mesh Modification	Mesh Transfer	Distributed Solution	Min/Max Elements
4	0.96	0.24	165.74	657/664
8	0.96	0.24	26.06	248/315
16	0.96	0.24	4.17	108/142
32	0.97	0.23	4.51	40/80

from 4 to 8 processors, and 8 to 16 processors, exhibits more than the expected two-fold performance gain. These "superlinear" speedups were a result of performing $O(N^3)$ matrix operations on matrices which became progressively smaller as they were partitioned among increasing numbers of processors (Morton and Tyler, 1996). Finally, the last column in the table illustrates the load balance, showing the number of elements in the processors that were loaded the least and the most, respectively.

For portability testing and for inexpensive design and development, the code was also implemented on a cluster of nine Linux workstations at the University of Montana's Scientific Computing Laboratory. Timing and load balancing data for 4 and 8 processors is provided in Table 35.2.

35.5 CONCLUSIONS

The purpose of this work is to explore methods for the construction of parallel, adaptive finite element models which are accessible to the general scientist. Although computational efficiency is certainly not ignored, the primary emphasis is on the devel-

Table 35.2 Linux wall time (in seconds) and load balancing for a single timestep approximately midway through two-phase flow simulation. At this time, there were 4,548 degrees of freedom, and 1,981 elements. The last column indicates the load balancing, showing the minimum and maximum number of elements in the "solution" processors.

Processors	Mesh Modification	Mesh Transfer	Distributed Solution	Min/Max Elements
4	14.60	14.34	1130.50	656/663
8	14.75	10.98	209.71	270/297

opment of modular tools which will enable scientists to construct reasonably efficient models for a wide range of problems in a relatively short period of time. Further, to encourage experimentation and exploration of various techniques, these models should be easily modifiable.

The system discussed in this paper represents our current progress to date. We have constructed components which allow for the implementation of two-dimensional adaptive finite element methods in a portable fashion, from Linux PCs to Cray supercomputers. We have also performed preliminary work on extending these concepts to three dimensions (Morton *et al.*, 1995). Although "square" elements have been used up to now, there is nothing in our methods that prevents us from utilizing irregular shaped elements for more flexible modeling, and these will be incorporated in the future.

So far, all code has been written in Fortran 77 for portability. Although it is a language that scientists are typically comfortable with, it clearly lacks many facilities that would provide robust and flexible encapsulation of many tedious details. An effort is now being made to begin a comprehensive analysis and design on the construction of a full object-oriented system.

Finally, because computational efficiency is of great importance to all, we constantly seek approaches for making our code more efficient, while retaining properties of loosely-coupled, cohesive modules. Numerous enhancements are possible in the distributed finite element solution modules, particularly in the implementation of efficient, sparse, iterative solvers for parallel environments. Although we have limited mesh maintenance activities to a serial environment, there are possibilities for improvement through the adoption of various parallel techniques which would distribute mesh maintenance and load balancing across several processors, still separate from the actual distributed finite element solution. With properly designed components, future optimizations will be constructed below the programmer's interface so that the general finite element modeler need not be concerned with such details. This also opens the door to making architecture-specific optimizations.

Acknowledgments

The support of this work by the National Science Foundation, Cray Research, Inc., and the University of Alaska's Arctic Region Supercomputing Center is gratefully acknowledged.

References

Biswas, R., Oliker, L., and Sohn, A. (1996). Global load balancing with parallel mesh adaptation on distributed-memory systems. In *Supercomputing'96*, Pittsburgh, PA. On CD-ROM.

Devine, K. D., Flaherty, J. E., Wheat, S. R., and Maccabe, A. B. (1993). A massively parallel adaptive finite element method with dynamic load balancing. In *Supercomputing'93*, pages 2–11, Portland, Oregon. IEEE Computer Society Press.

Karypis, G. and Kumar, V. (1995). *METIS: Unstructured Graph Partitioning and Sparse Matrix Ordering System.* http://www.cs.umn.edu/~karypis/metis/metis.html.

Morton, D. and Tyler, J. (1996). Minimizing development overhead with partial parallelization. *IEEE Parallel & Distributed Technology*, 4(3):15–24.

Morton, D., Tyler, J., Bourgoyne, A., and Schenewerk, P. (1994). An adaptive finite element methodology for 2d simulation of two-phase flow through porous media. In *ACM Symposium on Applied Computing*, pages 357–362, Phoenix, Arizona.

Morton, D. J., Tyler, J., and Dorroh, J. R. (1995). A new 3d finite element for adaptive *h*-refinement in 1-irregular meshes. *International Journal For Numerical Methods in Engineering*, 38:3989–4008.

Peaceman, D. W. (1977). *Fundamentals of Numerical Reservoir Simulation*. Elsevier, Amsterdam.

36 PARALLEL SOLUTION ALTERNATIVES FOR SPARSE TRIANGULAR SYSTEMS IN INTERIOR POINT METHODS

Huseyin Simitci

Department of Computer Science,
University of Illinois at Urbana-Champaign, U.S.A

simitci@cs.uiuc.edu

Abstract: The repeated solution of large symmetric sets of linear equations, which constitutes the major computational effort in interior point algorithms, requires efficient implementations of triangular system solvers. In this study, we present some parallel solution alternatives for these sparse triangular linear systems. Several forward and backward solution algorithms are tested, and a scalable buffered backward solution algorithm, which outperforms the other back substitution algorithms, is developed. Performance results are presented from a number of real application problems from the NETLIB suite.
Keywords: sparse triangular systems, linear programming, interior point algorithms, forward substitution, backward substitution.

36.1 INTRODUCTION

There is increasing interest in exploiting the power of parallel computers to solve difficult optimization problems arising in operations research (OR), especially linear programming (Vanderbei, 1996) which is central to the practice of this technical area. A linear program is an optimization problem in which a linear function (of a large number of variables) is minimized or maximized, subject to a large number of linear equality and inequality constraints. Besides many scientific applications, it is being used by the U.S Air Force Military Airlift Command (MAC) to solve critical logistics problems and by commercial air-lines to solve scheduling problems, such as crew planning (Chian Cheng *et al.*, 1989; Eckstein, 1993).

During the last decade, interest in linear programming methods has been revived by the discovery of the projective method of Karmarkar (Karmarkar, 1984), and the claims that variants of this method are much faster than the simplex method in practice (Adler *et al.*, 1989; de Ghellinck and Vial, 1986; Lustig *et al.*, 1994).

Even though there is considerable interest in parallel implementations of interior point methods (Housos *et al.*, 1989; Karypis *et al.*, 1994; Simitci, 1994), the lack of sufficiently general, highly-parallel sparse linear algebra primitives constitutes a technical barrier to easy application of parallel computing technology to large-scale optimization models. If we examine the "standard" algorithms of numerical optimization, we will see that several critical operations appear repeatedly. Perhaps the most common is the solution of large sparse linear systems. This operation forms the bulk of the work in most Newton-related methods, including interior point methods for linear programming. Without an efficient parallel version of sparse linear system solver, it will not be practical to simply "port" existing general-purpose sparse optimization codes to parallel architectures.

Most formulations of interior point algorithms require the formation and factorization of a matrix

$$K = ADA^T$$

in every iteration, where A is a constant constraint matrix and the diagonal matrix D changes between the iterations. This implies that the sparsity pattern of K remains the same from iteration to iteration. This factorization is used to solve a system of linear equations

$$Kx = b,$$

where K is an $n \times n$ symmetric positive definite matrix, b is a known vector, and x is the unknown solution vector to be computed. One way to solve the linear system is first to compute the Cholesky factorization

$$K = LL^T,$$

where the Cholesky factor L is a lower triangular matrix with positive diagonal elements. Then the solution vector x can be computed by successive forward and backward substitutions to solve the triangular systems

$$Ly = b, \quad L^T x = y.$$

These operations are generally done in four steps: ordering, symbolic factorization, numeric factorization, and triangular solution. Detailed explanations of these steps and exposition of the graph theoretical notions used in sparse linear systems can be found in (George and Liu, 1981).

Parallel implementations of Cholesky factorization also make use of these four solution steps (Heath *et al.*, 1991; Ng and Peyton, 1993). Generally, some variant of the minimum degree ordering (George and Liu, 1989), or a nested dissection method is used. Alternatives for parallel numeric factorization include block and panel approaches, and multi-frontal methods (Gupta and Kumar, 1994; Rothberg, 1996).

The structure of the forward and backward substitution algorithms is more or less dictated by the sparse data structure used to store the triangular Cholesky factor L and the data distributions used in the earlier solution steps. It is very difficult to achieve high computational rates with parallel algorithms for forward and backward triangular solutions. Data dependencies and a paucity of work to distribute among the processors make this problem harder, even for the dense case. When solving the sparse problems, due to preservation of sparsity in the factor matrix, there is usually far less work to distribute among the processors. The number of required floating-point operations is approximately two times the number of non-zeros in the factor matrix rather than the $n(n-1)$ floating-point operations available in the dense case.

Though these computations take much less time compared to sequential Cholesky factorization, there are several reasons to solve them in parallel using interior point methods. Interior point implementations for linear programming generally run between 20 to 80 iterations and this range does not change much with the problem size (Lustig *et al.*, 1994; Vanderbei, 1996). Every iteration in these methods requires a triangular solve. Predictor-corrector formulations require the solution of a triangular system with two different right-hand side vectors in every iteration. So, it is imperative to have a good parallel implementation of the triangular solvers, if we want reasonable performance from interior point algorithms. Between iterations, only the numerical value of the factor matrix changes. So, any symbolic analysis done on the sparsity pattern of the factor matrix will be valid during all subsequent iterations of the interior point algorithm. This amortizes the cost of symbolic analysis across several iterations. Furthermore, the Cholesky factor is usually available across the processors as a result of parallel factorization. In such a case, the parallel solution of triangular systems avoids the overhead of collecting the factor matrix into a single processor.

Several previous studies explored the ways to obtain "optimal" parallel implementations of triangular solvers (Alvarado and Schreiber, 1993). Earlier studies assumed column distribution of the triangular system among the processors (George *et al.*, 1989; Heath and Romine, 1988). More recent studies (Kumar *et al.*, 1993; Rothberg, 1995) argue that block and panel distributions give much higher performance than the column distribution. In the experiments we have conducted for this study, we have seen that good parallel performance can be obtained even with the column distribution, which is much easier to implement, if some symbolic analysis is done prior to the computation. Current parallel architectures have more favorable communication latencies compared to previous generations, which remedies the high communication needs of column distributions.

In the following section, we will describe several forward and backward solution algorithms besides our optimized backward solution algorithm. Then, we will discuss their performance on a multicomputer with real application data.

36.2 PARALLEL TRIANGULAR SOLUTION

We introduce some notation that will be used to describe the algorithms. We use $ColStruct(\mathbf{M}, k)$ to denote the set of row indices of the nonzero entries in column k of the lower triangular part of the matrix $\mathbf{M}$. Similarly, $RowStruct(\mathbf{M}, k)$ denotes the set of column indices of the nonzero entries in row k of the lower triangular part of the matrix $\mathbf{M}$.

$k = \text{mynode}()$
$u = 0$
if v_k is not a leaf
    while $smod(k) > 0$
        receive update vector u' from a child v_i
        for $j \in \text{Struct}(u')$ do
            $u_j = u_j + u'_j$
        decrement $smod(k)$
$y_k = (b_k - u_k)/l_{kk}$
for $j \in \text{ColStruct}(L, k)$ do
    $u_j = u_j + l_{jk} * y_k$
send $u(j \mid j \in \text{ColStruct}(L, k))$
    to $\text{map}(L, \text{parent}(k))$

Figure 36.1   Elimination Tree Based FS (EBFS) algorithm.

In (George and Liu, 1981), a *parent*() relationship is defined for the columns of **L** such that when there is at least one off-diagonal nonzero in column j of **L**, *parent*(j) is the row index of the first off-diagonal nonzero in that column, and it is j otherwise.

K and **b** will be assumed to be ordered previously, and permutation matrix **P** will be implicit. The elimination tree $T(\mathbf{K})$ associated with the Cholesky factor **L** of a given matrix **K** has $\{v_1, v_2, \ldots, v_n\}$ as its node set, and has an edge between two vertices v_i and v_j, with $i > j$, if $i = parent(j)$. In this case, node v_i is said to be the *parent* of node v_j, and node v_j is a *child* of node v_i. The elimination tree is fixed for a given ordering and is a heap ordered tree with v_n as its root. It captures all the dependencies between the columns of **K** in the following sense. If there is an edge (v_i, v_j), $i < j$, in the elimination tree, then the factorization of column j cannot be completed unless that of column i is completed.

We will assume that columns of the factor matrix **L** are distributed among processors using the subtree-to-subcube mapping (George *et al.*, 1989), which is generalized by (Geist and Ng, 1989) to arbitrarily unbalanced elimination trees.

Each column, l_{*k}, is stored on one and only one of $\mathcal{P}$ available processors. An n-vector map is required to record the distribution of columns to processors: if column k is stored on processor p, then $map(\mathbf{L}, k) = p$. We use $mycols(\mathbf{L})$ to denote the set of columns owned by a processor. Finally, **FS** and **BS** will be used as shorthands for forward solution and backward solution, respectively.

36.2.1   Elimination Tree Based FS (EBFS)

(Kumar *et al.*, 1993) exploit the elimination tree concept to develop forward and backward triangular solvers. They consider the situation when the number of processors is same as the number of columns of **L**, and column i along with b_i are assigned to processor i. Their parallel algorithm for FS phase, which we call EBFS, starts from the leaves of the elimination tree. An update vector **u** of size n is associated with each processor. An internal vertex v_i waits till it receives update vectors from all of its children and adds each of them to its update vector. The value of y_i is calculated as

$$y_i \leftarrow (b_i - u_i)/l_{ii}.$$

The update vector is then modified as follows:

$$\forall j \in ColStruct(\mathbf{L}, i), u_j \leftarrow u_j + l_{ji}y_i.$$

The modified update vector is sent to the parent of v_i. Here, only the relevant elements of the update vector are sent. When the computation terminates, processor i contains the value y_i. The algorithm is given in Figure 36.1. In the algorithm, $smod(k)$ initially contains the number of children of v_k.

(Kumar et al., 1993) later consider the case where the columns are mapped to processors according to the $map(\cdot)$ function. They modify the algorithm to remove the redundancy in the messages arising in this situation.

36.2.2   Fan-in FS (FIFS)

The sparse forward solution algorithm proposed by (George et al., 1989) is an adaptation of the fan-in algorithm for factorization. The columns of $\mathbf{L}$ and the corresponding elements of the right hand vector $\mathbf{b}$ are distributed among the processors according to the $map(\mathbf{L}, \cdot)$ function. To compute the value of y_k, the processor $map(\mathbf{L}, k)$ needs the inner product of the k-th row of $\mathbf{L}$ and $(y_1, y_2, \dots, y_{k-1})$. This computation is partitioned among the processors. Each processor computes the products of the elements of the k-th row it contains with the corresponding elements of the solution vector and sends their sum to processor $map(\mathbf{L}, k)$. On receiving the contributions of all the processors, processor $map(\mathbf{L}, k)$ computes the value of y_k.

36.2.3   Elimination Tree Based BS (EBBS)

Again consider the situation when the number of processors is same as the the number of columns of $\mathbf{L}$. The computation starts from the root of the elimination tree. The processor containing column n of $\mathbf{L}$ computes the value of $x_n \leftarrow y_n/l_{nn}$. It sends x_n to processors that contain children of v_n. Each processor i receives a message from processor $parent(i)$, and computes x_i as follows:

$$x_i \leftarrow \left(y_i - \sum_{j \in ColStruct(\mathbf{L}, i)} l_{ji} * x_j \right).$$

It then appends x_i to the received message and sends it to processors that contain the children of v_i in the elimination tree. If v_i is a leaf in the tree, then no message is sent from processor i. The algorithm is given in Figure 36.2.

As in EBFS, (Kumar et al., 1993) consider the case where the columns are mapped to processors according to $map(\cdot)$ function and exploit $map(\cdot)$ to reduce the size of the messages.

$i = \text{mynode}()$
if $(i = n)$
$\qquad x_n = y_n / l_{nn}$
$\qquad$ for each child v_k of v_n do
$\qquad\qquad$ send x_n to P_k
else
$\qquad$ receive $x_k, k \in \text{anc}(i)$ from $P_{parent(i)}$
$\qquad$ u = 0
$\qquad$ for $k \in \text{ColStruct}(L, i)$ do
$\qquad\qquad u = u + x_k * l_{ki}$
$\qquad x_i = (y_i - u)/l_{ii}$
$\qquad$ for each child v_k of v_i do
$\qquad\qquad$ send $x_j, j \in \{i\} \bigcup \{\text{anc}(i)\}$ to P_k

Figure 36.2   Elimination Tree Based BS (EBBS) algorithm.

36.2.4 Send-Forward BS (SFBS)

The sparse backward solution discussed in (George *et al.*, 1989) is based on the dense backward algorithm in which the value of x_k is broadcast to all the processors as soon as it is computed. All the processors update elements of **y** vector in parallel as follows: $y_j \leftarrow y_j - l_{jk} x_k$, $j < k$. Next, the processor containing the $(k - 1)$th row of $\mathbf{L}^T$ computes x_{k-1} and broadcasts the same, and so on.

In the sparse case, we can avoid broadcasting x_k, if we know the structure of the k-th column of $\mathbf{L}^T$. We only need to send x_k to all the processors which contain rows of $\mathbf{L}^T$ that have nonzero entries in column k. But, since $\mathbf{L}^T$ is stored row-wise, the column structure is not readily available. This problem can be solved in the preprocessing phase. After finding the nonzero structure of $\mathbf{L}^T$ column-wise (*ColStruct*), a data structure is formed such that, *sendset*(k) contains the processors which need x_k. This new algorithm, which we name send-forward BS (SFBS), is shown in Figure 36.3.

36.2.5 Buffered BS (BFBS)

The Send-forward BS algorithm can be improved radically. In SFBS, the value of x_k is sent to processors in *sendset*(k) as soon as it is computed. This process causes the sending processor to do early work and to delay the necessary work, because most of the processors in *sendset*(k) need x_k only many iterations later. Furthermore, sending x_k values individually incurs a high message count. So, if we buffer x values that will be sent to a processor and, in subsequent iterations, send them in a combined message, we will reduce the number of messages and prevent the early sends. These ideas are used to develop the buffered BS algorithm (BFBS) given in Figure 36.4.

Here, *sendset*(k) is modified to contain the tuples (s, t) such that processor s contains rows of $\mathbf{L}^T$ that have non-zeros in column k, and t is the highest index among these columns. In other words, the tuple (s, t) denotes that the earliest iteration in which x_k is needed by processor s is t. The cost of constructing the *sendset* will be

{ *initialize the* sendset() }
for $i = 1$ to n do
   sendset(i) = $\emptyset$
    for $k \in$ ColStruct(L^T, i) do
      sendset(i) = sendset(i) $\bigcup$ {map(L, k)}

{ *BS using* sendset() }
for $j = 1$ to n do
   valid(j) = false;
for $j = n$ downto 1 do
   if $j \in$ mycols(L)
    u = 0
    for $k \in$ RowStruct(L^T, j) do
      if not valid(k)
        receive x_k
        valid(k) = true
      $u = u + x_k * l_{kj}$
    $x_j = (y_j - u)/l_{jj}$
    valid(j) = true
    for $i \in$ sendset(j) do
      send x_j to processor i

Figure 36.3   Send-Forward BS (SFBS) algorithm.

amortized during many iterations in the interior point algorithms. Since the nonzero structure of the factor matrix will remain the same, so will the *sendset* be the same.

Additionally, we introduce buffers, where $buf(i)$ is used to combine x_k values to be sent to processor i. Associated with $buf(i)$, there is a value $maxj(i)$, which denotes the earliest iteration in which the current contents of $buf(i)$ needs be sent.

In the k-th iteration of the algorithm, the value of x_k is computed and added to buffers denoted in *sendset(k)* by the processor having column k. Other processors control the buffer $buf(map(L, k))$ and, if the current iteration value exceeds $maxj(map(L, k))$, the buffer is sent to processor $map(L, k)$.

The BFBS algorithm approximates a fan-in algorithm by by buffering the messages to individual processors. This is done without the replication of the column structure of L^T, which is not available to the processors because we assumed a column distribution of L.

36.3   COMPUTATIONAL RESULTS AND DISCUSSIONS

All algorithms have been implemented on an Intel Paragon multicomputer with 31 i860XP nodes each with 16 MB. Performance results were obtained from actual runs on linear programming problems from the NETLIB suite (Gay, 1985). These data sets represent realistic problems in industry applications ranging from small-scale to

{ *initialize the* sendset() }
for $i = 1$ to n do
   sendset$(i) = \emptyset$
      for $k \in$ ColStruct(L^T, i) do
         $p = $ map(L, k)
         if $\exists (p, t) \in$ sendset(i)
            if $k > t$
               sendset$(i) = $
                  (sendset(i) - $\{(p, t)\}) \bigcup \{(p, k)\}$
         else
            sendset$(i) = $ sendset$(i) \bigcup \{(p, k)\}$

{*BS using* sendset() }
for $j = 1$ to n do
   valid$(j) = $ false;
for $i = 1$ to $\mathcal{P}$ do
   buf$(i) = \emptyset$
   maxj$(i) = 0$
for $j = n$ downto 1 do
   if $j \in$ mycols(L)
      u = 0
      for $k \in$ RowStruct(L^T, j) do
         while not valid(k)
            receive next message W
            for $(v, t) \in W$
               $x_t = $ v
               valid$(t) = $ true
         $u = u + x_k * l_{kj}$
      $x_j = (y_j - u)/l_{jj}$
      valid$(j) = $ true
      for $(i, t) \in$ sendset(j) do
         buf$(i) = $ buf$(i) \bigcup \{(x_j, j)\}$
         if $t > $ maxj(i)
            maxj$(i) = t$
   else
      $i = $ map(L, j)
      if maxj$(i) \geq j$
         send $(v, t) \in$ buf(i) to processor i
         buf$(i) = \emptyset$
         maxj$(i) = 0$

Figure 36.4   Buffered BS (BFBS) algorithm.

Table 36.1   Statistics for the NETLIB problems used.

Problem Name	Constraints (m)	Variables (n)	Non-zeros in Constraints	Non-zeros in L	Operation Count (Factorization)
80bau3b	2262	9799	21002	42510	2832016
cycle	1903	2857	20720	73427	5434413
pilot87	2030	4883	73152	444543	189269384
woodw	1098	8405	37474	47757	3280086
cre-b	9648	72447	256095	955473	380780740
ken-11	14694	21349	49058	118801	4174850
ken-13	28632	42659	97246	325506	17101012
pds-06	9881	28655	62524	540199	182137657

Table 36.2   Total number of messages sent by each algorithm for the NETLIB problems.

Problem Name	Total Number of Messages				
	FIFS	EBFS	SFBS	EBBS	BFBS
80bau3b	5082	492	5082	492	4802
cycle	11636	728	11636	728	10147
pilot87	24514	1268	24514	1268	22596
woodw	7189	522	7189	522	5772
cre-b	40663	2714	40663	2714	35159
ken-11	1301	179	1301	179	1295
ken-13	1998	300	1998	300	1962
pds-06	21943	1327	21943	1327	21050

large-scale. Statistics regarding the NETLIB problems used in this study are shown in Table 36.1.

The total number of messages exchanged in each algorithm is shown in Table 36.2 for each data set. As expected, the number of messages in the elimination tree based algorithms (EBFS and EBBS) are exactly the same, which is equal to the number of edges in the elimination tree connecting the nodes which are mapped to different processors. It is interesting to note that the number of messages in FIFS and SFBS are the same. A closer look at the SFBS results reveals that it is mimicking a fan-out algorithm, which performs the same message exchanges with a fan-in algorithm but at different times. Comparing the number of messages sent in SFBS and BFBS, we see that buffering in the BFBS algorithm also helps to reduce the number of messages.

Space-time diagrams and utilization count graphs for the substitution phases of the algorithms are shown in Figure 36.5. For space considerations, only the graphs for the 80bau3b problem are shown. The space-time diagrams illustrate snapshots of the execution of the algorithms, with time on the horizontal axis. Processor activity is shown by horizontal lines and interprocessor communication by slanted lines. The

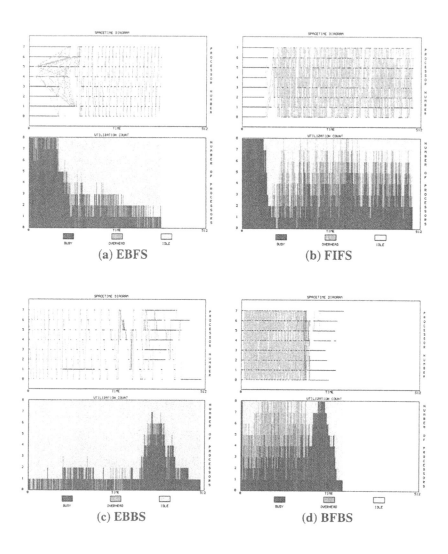

Figure 36.5   Space-time diagrams and utilization counts for the FS and BS algorithms. Results are for the 80bau3b problem on an Intel Paragon multicomputer.

horizontal line corresponding to each processor is either solid or blank, depending on whether the processor is busy or idle, respectively. Each message sent between processors is shown by a line drawn from the sending processor at the time of transmission to the receiving processor at the time of reception of the message. Utilization count diagrams show the total number of processors in each of three states — busy, overhead, and idle — as a function of time. Each processor is categorized as *idle* if it has suspended execution awaiting a message that has not yet arrived, *overhead* if it is executing in the communication subsystem, and *busy* if it is executing some portion of the program other than the communication subsystem. These diagrams were produced using a package developed at Oak Ridge National Laboratory for visualizing the behavior of parallel algorithms (Heath, 1990).

Comparison of the diagrams for FIFS and EBFS, which have the same time scale, reveals that the elimination tree based algorithms have a low message count[1] and it takes less time to complete the FS in this setting. Similarly, comparing space-time diagrams of BFBS and EBBS algorithms shows that even though elimination tree based algorithm have a low message count, it takes much more time than BFBS because of idle waiting.

Figure 36.6 contains the performance comparison of the five BS and FS algorithms discussed in the previous sections. The performance is computed as the megaflops achieved, which is a function of the non-zeros in the factor matrix and the execution time. All the plots in this figure were drawn using the same axis scales to see the effect of different data sets on the performance. It is apparent that the bigger the problem size is, the higher the performance because there is more work to distribute among multiple processors.

These plots reveal that all the algorithms have extremely bad performance on a single processor. This is more apparent for bigger problems, which suggests that the memory on a single processor is not sufficient and the virtual memory activity is degrading the performance. This is another argument in favor of parallel implementations.

Performance of the EBFS algorithm is better than the performance of FIFS for most of the settings, except ken-11 and ken-13 which have huge but very sparse constraint matrices. Looking at the performance of the backward solution algorithms we see that, even though EBBS and SFBS have some mixed performance figures, BFBS has superior performance than the two others. For larger problem sizes, BFBS displays a promising scalability.

As explained in Section 36.2.5, SFBS causes the processors to do early message sends. This causes the performance degradation of SFBS for large problems and as the number of processors increases.

36.4   CONCLUSIONS

We have examined several parallel sparse triangular system solution algorithms which are essential to get reasonable performance from most numerical algorithms like interior point methods. For the forward solution component, we compared elimination tree based and fan-in algorithms. For the backward solution component, we compared send-forward, elimination tree based, and buffered backward solution algorithms.

Our studies show that column distribution of the factor matrix gives reasonable parallel performance. This is despite the arguments in the recent literature in favor of block distributions, which are much more complicated to implement and will require data redistribution if the previous phases used column distribution. Elimination

---

[1]In fact, communication count of fan-in forward solution is equal to communication count of fan-in factorization.

tree based solution algorithms reduce the communication count, but they cause idle waiting. Though they gave better performance values for small number of processors, they did not scale well when larger number of processors were used. A study of the space-time diagrams of the fan-in forward solution shows that this algorithm requires a huge amount of communication, but it does not cause idle waiting as much as the elimination tree based forward solution. Hence, in some cases the fan-in forward solution scaled well for a larger number of processors. The situation is the same between the elimination tree based backward solution and the buffered backward solution. The buffered backward solution scales better as the number of processors increases for large size problems. We plan to explore the generalization of our buffered backward solution algorithm to other data distributions.

References

Adler, I., Karmarkar, N., Resende, M. G. C., and Veiga, G. (1989). Data structures and programming techniques for the implementation of Karmarkar's algorithm. *ORSA Journal on Computing*, 1(2):84–106.

Alvarado, F. L. and Schreiber, R. (1993). Optimal parallel solution of sparse triangular systems. *SIAM J. SCI. COMPUT.*, 14(2):446–460.

Chian Cheng, Y., Houck, D. J., *et al.*(1989). The AT&T KORBX© system. *AT&T Technical Journal*, pages 7–19.

de Ghellinck, G. and Vial, J.-P. (1986). A polynomial Newton method for linear programming. *Algorithmica*, 1:425–453.

Eckstein, J. (1993). Large-scale parallel computing, optimization, and operations research: A survey. *ORSA CSTS Newsletter*, 14(2):1,8–12,25–28.

Gay, D. M. (1985). Electronic mail distribution of linear programming test problems. *Mathematical Programming Society COAL Newsletter*.

Geist, G. A. and Ng, E. (1989). Task scheduling for parallel sparse Cholesky factorization. *International J. of Parallel Programming*, 18(4):291–314.

George, A., Heath, M. T., Liu, J., and Ng, E. (1989). Solution of sparse positive definite systems on a hypercube. *J. of Comp. and Applied Math.*, 27:129–156.

George, A. and Liu, J. (1981). *Computer Solution of Large Sparse Positive Definite Systems*. Prentice-Hall, Englewood Cliffs, NJ.

George, A. and Liu, J. W. H. (1989). The evolution of the minimum degree ordering algorithm. *SIAM REVIEW*, 31(1):1–19.

Gupta, A. and Kumar, V. (1994). A scalable parallel algorithm for sparse Cholesky factorization. In *Supercomputing'94*. IEEE Computer Society.

Heath, M. (1990). Visual animation of parallel algorithms for matrix computations. In *5th Distributed Memory Computing Conference*.

Heath, M. T., Ng, E., and Peyton, B. W. (1991). Parallel algorithms for sparse linear systems. *SIAM REVIEW*, 33(3):420–460.

Heath, M. T. and Romine, C. H. (1988). Parallel solution of triangular systems on distributed memory multiprocessors. *SIAM Journal on Scientific and Statistical Computing*, 9:558–588.

Housos, E. C., Huang, C. C., and Min Liu, J. (1989). Parallel algorithms for the AT&T KORBX© system. *AT&T Technical Journal*, pages 37–47.

Karmarkar, N. (1984). A new polynomial-time algorithm for linear programming. *Combinatorica*, 4(4):373–395.

Karypis, G., Gupta, A., and Kumar, V. (1994). A parallel formulation of interior point algorithms. In *Supercomputing '94*. IEEE Computer Society.

Kumar, P. S., Kumar, M. K., and Basu, A. (1993). Parallel algorithms for sparse triangular system solution. *Parallel Computing*, 19:187–196.

Lustig, I. J., Marsten, R. E., and Shanno, D. F. (1994). Interior point methods for linear programming: Computational state of the art. *ORSA Journal on Computing*

Ng, E. G. and Peyton, B. W. (1993). A supernodal Cholesky factorization algorithm for shared-memory multiprocessors. *SIAM J. SCI. COMPUT.*, 14(4):761–769.

Rothberg, E. (1995). Alternatives for solving sparse triangular systems on distributed-memory multiprocessors. *Parallel Computing*, 21(7):1121–1136.

Rothberg, E. (1996). Performance of panel and block approaches to sparse Cholesky factorization on the iPSC/860 and Paragon multicomputers. *SIAM Journal on Scientific Computing*, 17(3):699–713.

Simitci, H. (1994). Parallelization of an interior point algorithm for linear programming. Technical Report bu-ceis-9428, Bilkent University, Ankara.

Vanderbei, R. J. (1996). *Linear Programming, Foundations and Extensions*, chapter 17, pages 269–283. Kluwer Academic Pub.

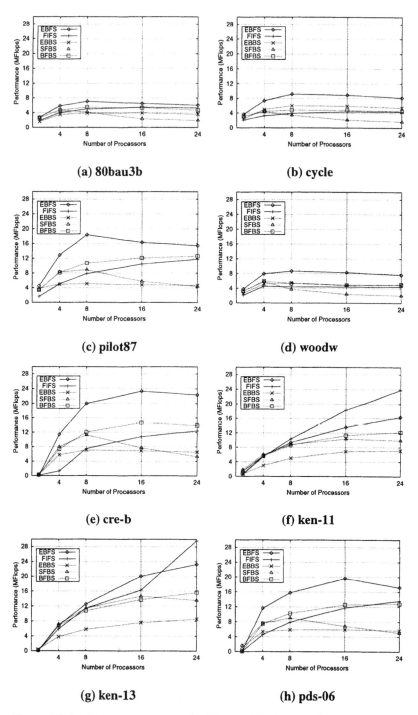

(a) 80bau3b

(b) cycle

(c) pilot87

(d) woodw

(e) cre-b

(f) ken-11

(g) ken-13

(h) pds-06

Figure 36.6 Performance curves for FS and BS algorithms. Performance given using megaflops on an Intel Paragon multicomputer.

X General Session

37 MANAGING LONG LINKED LISTS USING LOCK-FREE TECHNIQUES

Mohammad Farook and Peter Graham

Department of Computer Science,
University of Manitoba, Canada

{ farook,pgraham } @cs.umanitoba.ca

Abstract: When writing parallel programs for shared memory machines, it is common to use shared data structures (linked lists, queues, trees, *etc.*). Concurrency control for such data structures may be implemented using either blocking or non-blocking/lock-free synchronization. Blocking synchronization is popular because it is familiar and also amenable to worst-case performance analysis (important for certain real-time applications). Unfortunately it also suffers from some undesirable characteristics. These include the possibility of deadlock, limited concurrency since processes are blocked, and certain scheduling anomalies. Non-blocking techniques, although less prevalent, do not suffer from these problems and may be well suited to many parallel programming problems.

Despite increased interest in non-blocking techniques, even simple data structures such as singly linked lists often have annoying limitations. This paper presents a new implementation of non-blocking linked lists which overcomes some common problems with existing implementations. In addition to offering greater overall concurrency, the presented technique also significantly increases performance for long linked lists. We provide algorithms for list traversal, insertion and removal, and show their correctness. We also analyze the results of a simulation study which compares the performance of our implementation against two existing techniques under varying load conditions and access patterns and for various list lengths.

Keywords: concurrency control, lock free data structures, non-blocking synchronization, parallel programming.

37.1 INTRODUCTION

In shared memory parallel programming, concurrent processes/threads operate on shared data structures thereby cooperating to solve a problem. Conventionally, the consistency of such data structures is maintained by ensuring mutually exclusive access to them using locks or other blocking synchronization primitives. Blocking techniques, however, have several drawbacks including decreased concurrency, the poten-tial for deadlock, and certain undesirable scheduling side effects (*i.e.* priority inver-sion and convoying). An alternative approach is to use non-blocking synchronization techniques (*e.g.* (Barnes, 1993), (Greenwald and Cheriton, 1996), (Herlihy, 1988), (Herlihy, 1991), (Massalin and Pu, 1991), (Prakash *et al.*, 1991),(Valois, 1995)). Non-blocking techniques do not suffer from these problems since processes optimistically execute concurrently and therefore never wait on one another. As such, they offer dis-tinct advantages for certain parallel applications. Non-blocking techniques that make hard guarantees concerning completion of processes are often referred to as being "wait free".

Existing work on non-blocking data structures has been fruitful, yielding non-blocking implementations of a wide variety of conventional data structures including linked lists, queues, and trees. Nevertheless, even for such simple structures as singly linked lists, problems persist. Among the limitations of current implementations are high overhead, restrictions on forms of concurrency, and high re-execution costs in the face of conflicting concurrent operations. This paper addresses these problems for singly linked lists. Through simple changes to existing techniques for non-blocking linked lists, we derive a new implementation that significantly improves on existing techniques and that offers increasing performance improvement as the size of the list grows.

The rest of this paper is organized as follows. Section 37.2 presents related work. Section 37.3 discusses the limitations of existing non-blocking linked list designs. Section 37.4 proposes a new design that addresses the problems, describes the algo-rithms, and then illustrates their correctness. Simulation results comparing the per-formance of our design to two other designs are presented in Section 37.5. Finally, Section 37.6 concludes and discusses possible future research.

37.2   RELATED WORK

Work related to that presented in this paper falls into three categories: non-blocking synchronization primitives, generic techniques for implementing non-blocking data structures, and specific techniques for implementing non-blocking linked lists. Work already done on other non-blocking data structures is beyond the scope of this paper and is therefore not discussed.

37.2.1   Non-Blocking Synchronization Primitives

Shared, non-blocking data structures are built using different synchronization prim-itives than those from which shared, blocking data structures are implemented. The most discussed of these synchronization primitives is the "Compare And Swap" (CAS) instruction[1].

When using the CAS instruction, a programmer records the state of a shared vari-able (commonly a pointer for shared data structure implementations) at the beginning

---

[1]Most modern machine architectures provide support for both blocking and non-blocking synchronization in hardware via specific machine instructions.

```
CAS(A,B,C)
Begin Atomic
  if (A==B){
    A=C;
    return True;
  }
  else
    return False;
End Atomic
```

Figure 37.1   Implementation of CAS.

of a computation. A new value for the shared variable is then computed and the CAS instruction is used to atomically update the shared variable with that value only if it has not changed value from the beginning of the computation. If the shared variable cannot be updated, then an error indication is returned and the computation must be re-executed[2] using the new value for the shared variable. This technique may be used to "swing" a pointer in a linked list to point to a new node. This use of CAS to effect a "conditional store" is the basis for both insertions into and deletions from lists.

CAS is a three-operand atomic operation (pseudo-code for the CAS operation is shown in Figure 37.1). The operands A, B, and C are single word variables. B is a shared variable and A is a private copy of B made prior to a computation involving B. C is the newly computed value for B which the process wishes to write into B. The process is allowed to write C into B only if B is unchanged since it was last read (*i.e.* only if A==B).

The use of CAS suffers from what is known as the "A-B-A problem". This occurs when the value of B is the same as A even though B has been modified at least once since the process read the value of B into A. In this situation, the CAS instruction succeeds even though the value C was potentially computed using stale data. In certain applications this may lead to erroneous results.

A variant of CAS is the "Double Compare And Swap" (DCAS) instruction. This form of CAS exchanges not one, but two values indivisibly. If the second value is used as a "version number" for the first (which is incremented on every update to the first shared variable) then the A-B-A problem may be avoided. The pseudo-code for the DCAS instruction is shown in Figure 37.2.

Recently a number of processors (*e.g.* the MIPS family of machines) have implemented a more general non-blocking synchronization primitive in the form of the "Load Linked" (LL) and "Store Conditional" (SC) instruction pair. Our new algorithms for insertion and deletion into a shared linked list, which are presented in Section 37.4, utilize only the CAS and DCAS instructions.

---

[2]Such techniques, utilizing re-execution, are familiar to the database community as *optimistic* concurrency control (as used in database management systems).

```
DCAS(A1,A2,B1,B2,C1,C2)
Begin Atomic
  if ((A1==B1) && (A2==B2)){
    A1=C1;
    A2=C2;
    return True;
  }
  else
    return False;
End Atomic
```

Figure 37.2   Implementation of DCAS.

37.2.2   Generic Non-Blocking Data Structures

A number of so-called "universal methods" (*e.g.* (Barnes, 1993), (Herlihy, 1993), (Herlihy and Moss, 1993), (Herlihy and Wing, 1987), (Prakash *et al.*, 1991)) for constructing non-blocking data structures of any type have been discussed in the literature. (Lamport, 1983) described the first lock-free algorithm for the problem of managing a single-writer, multiple-reader shared variable. Such variables may be used as a basis for developing a wide variety of non-blocking data structures but Lamport's work is not generally considered to be a universal method.

(Herlihy, 1993) presented the first widely accepted universal method. He maintained a write-ahead log of operations for each shared data structure. The order of entries in the log determines a serialization order. Private copies of structures, built by applying a sequential reconstruction algorithm to the operations in the log, are updated and then finally added to the log itself. Check-pointing the log is used to decrease the cost of state reconstruction.

(Herlihy and Wing, 1987) also proposed a method for constructing non-blocking data structures using "copying". Utilizing the "Load Linked" (LL) instruction, a copy of the data structure (or, in certain situations, a subset of it) may be made. Changes are made to the local copy and then the "Store Conditional" (SC) instruction is used to update the shared structure. This method has two drawbacks. First, it can be very expensive in terms of copying overhead if the data structure is large. Second, it permits only a single process to successfully update the structure at a time.

A third universal method has been proposed by (Barnes, 1993). In this technique, private copies of individual cells/nodes are made (*i.e.* they are "cached") using LL. Updates are made to the cells and then they are validated and returned to the shared structure using SC. An interesting feature of Barnes' work is that it supports "cooperative completion" whereby one process may complete the work of another process rather than having that process subsequently fail its update. To do this, each process records its operation list in a global work queue. If a process is interrupted before completing its work, the remaining operations will be recorded in the work queue and may be completed by a later executing process when it accesses the shared data structure.

37.2.3   Non-Blocking Linked Lists

Non-blocking algorithms that are specific to particular data structures are typically more efficient than universal methods. In this paper we focus on algorithms for insertions and removals on non-blocking singly linked lists. We now review existing approaches that will be used for comparison purposes with our new technique described later in Section 37.4.

(Valois, 1995) describes a non-blocking linked list where every list node has an "auxiliary" node that is used during concurrent insertions and deletions to help maintain the consistency of the list. Each auxiliary node consists of a single pointer to the next regular node in the list. An empty linked list consists of two dummy nodes (first and last) with an auxiliary node between them.

A structure known as a cursor is defined which contains three pointers to nodes (dummy or regular) which surround the point of insertion/deletion. The first pointer, target, points to the node being visited. The second pointer, pre-aux, points to the auxiliary node immediately preceding target. The final pointer, pre-cell, points to the last normal node preceding target in the list.

Referring to Figure 37.3, suppose that a process wants to insert a node new (together with the required auxiliary node new.aux) between nodes 'B' and 'C' in the shared linked list. The cursor c maintains pointers, c.pre-aux pointing to the auxiliary node before node 'C' and c.target pointing to node 'C'. The new node and following auxiliary node are created and initialized so that node new points to its auxiliary and its auxiliary points to node 'C'. CAS is then used to atomically check that c.pre-aux still points to c.target. If the pointers have not changed, then c.pre-aux is set to point to new and True is returned. If they have changed, then False is returned and the process must attempt the insertion again.

When a process wants to do a deletion, the node to be deleted is pointed to by the target node in the cursor c. CAS is used to atomically set c.pre-aux to c.target->next. The result of a deletion is a pair of adjacent auxiliary nodes. As deletions continue, long chains of contiguous auxiliary nodes may develop. As described, Valois' algorithm also suffers the A-B-A problem since it uses only CAS.

In a recent paper, (Greenwald and Cheriton, 1996) propose the use of non-blocking data structures in the design of a multiprocessor operating system. They describe a non-blocking linked list approach that uses the DCAS instruction and a version number for each list to avoid the A-B-A problem.

During deletion, a process traverses the linked list until it finds the node to be deleted. If the node is not found, the version number is checked. If the version number has changed, then the desired node may have just been inserted and the list must be scanned again. Assuming the node has been found, DCAS is used to atomically swing the next pointer of the node preceding the target to point to the node following it and to increment the version number. This may fail or succeed depending on whether the list has been changed or not.

Inserting an element is done by traversing the linked list to locate the node after which the new node should be inserted. DCAS is then used to increment the version number and swing the next pointer of the target node to point to the new node (in

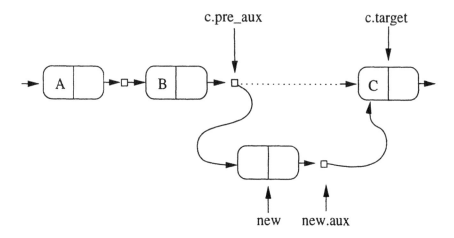

Figure 37.3   Insertion under Valois' implementation.

much the same way as in Valois' algorithm). Again the DCAS may succeed or fail depending on the state of the list.

While Greenwald and Cheriton's algorithm is simple and intuitive, the use of a single version number is too coarse grained to be effective. This is particularly true as the size of the linked list grows.

Finally, (Massalin and Pu, 1991) have also described a non-blocking implementation of a shared linked list using the DCAS instruction for deletion. Deletion of nodes from within the linked list is done in two steps. First, the nodes that are to be deleted are marked but left in the list. Then, in the second step, if the node preceding the one marked for deletion is not itself marked for deletion the marked node may be safely removed. This implementation, however, limits concurrency to a single process.

37.3   PROBLEMS WITH EXISTING NON-BLOCKING LINKED LISTS

There are a number of problems with existing non-blocking implementations of linked lists. Many of these problems have already been illustrated in the implementations discussed in the preceding section.

First, successful concurrency is seldom attained. This is true of the universal methods as well as many of the linked list specific approaches. By successful concurrency we mean concurrent operations on the list, which do not cause re-execution of one another. For example, insertions into distinct parts of a linked list should not affect one another yet, for example, in Greenwald and Cheriton's algorithm and in Massalin and Pu's algorithm they do. In the former, this is due to the coarse granularity at which the version number is applied. As there is a single version number for the entire linked list, even a change to an unrelated part of the list results in the re-execution of other concurrent updates.

The A-B-A problem is also a concern for those techniques that do not use an explicit version number (*e.g.* Valois' algorithm). For some applications of linked lists, the A-B-A problem may not be tolerated and thus it must be addressed.

While Valois' algorithm does permit concurrent updates and may be modified to manage the A-B-A problem, it introduces execution overhead in two ways. First, long chains of auxiliary nodes may develop after many deletions have taken place. This introduces overhead either in traversing them (Valois' algorithm suffers 100% overhead in list traversal even when there are no adjacent auxiliary nodes) or to delete the excess nodes. Second, when re-execution is necessary due to two adjacent update operations in the list, the list must always be re-scanned from the start of the list. Particularly for long linked lists, this represents high overhead in an environment where such conflicts are relatively common. Massalin and Pu's algorithm also suffers from this latter form of overhead.

37.4 AN IMPROVED ALGORITHM FOR NON-BLOCKING LINKED LISTS

The desirable characteristics of an effective non-blocking implementation of linked lists are:

- As much successful concurrency as possible,

- Effective handling of the A-B-A problem,

- Minimal re-execution costs after conflicting operations, and

- Minimal overhead relative to a conventional linked list implementation (time overhead primarily but also space overhead as a secondary concern).

We now present an implementation that achieves most of these goals. It does provide a high degree of successful concurrency, handles the A-B-A problem, has minimal re-execution costs in all but one case, and has minimal time overhead compared to a conventional linked list implementation. It does achieve these goals at the expense of increased space requirements but we consider the additional space overhead to be negligible for all linked lists except those storing the simplest (and hence, smallest) data items.

37.4.1 Structure of the Linked List

The structure of the linked list in our design includes two dummy nodes head (pointing to the first node in the linked list) and tail (pointing to the last). An empty list consists of these two nodes pointing to one another.

Each process manipulating a particular node, target, maintains three pointers; prev, target, and next. The prev pointer points to the node preceding target and the next pointer points to the node following target. These three pointers are obtained by traversing the linked list.

Each node in the linked list consists of four fields; a data field, two pointer fields, and a count field. One of the pointers, next, is used to maintain an up-to-date linked list. These pointers never point to a node that has been deleted or one that has been marked for deletion. The other pointer, traverse, is used to facilitate concurrent

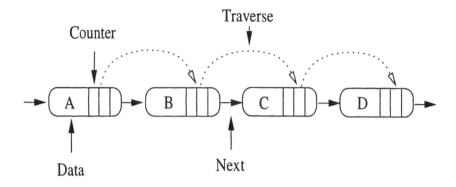

Figure 37.4   Structure of the linked list.

processes traversing the list despite the fact that some nodes may have been concurrently marked for deletion. This situation occurs when a process is at a particular node when another process marks that node for deletion (by making its pointer field Null). The first process may still traverse the list using the spare pointer thereby increasing overall concurrency. Actual deletions within the list are more difficult because a node may be deleted and de-allocated while another process is traversing the list. If a deleted node is then re-used for some other purpose, its new pointer values may lead to invalid memory references. To avoid this problem, the counter field in each list node is used. Whenever a process is visiting the node, it increments the counter field and decrements it only when it is leaving. A process attempting a node deletion must first verify that its counter field is zero before proceeding.

An example of the proposed linked list structure is shown in Figure 37.4.

37.4.2 Linked List Traversal

Algorithm Cursor (see Figure 37.5) returns the head and tail nodes as prev and target if the list is empty (lines 2–5). When a process finds the key by traversing the list using the next pointer, it returns pointers to the prev, target (matching the key), and next nodes (lines 6–11). If the key is not found, the algorithm returns NotFound. If the next field of any node is Null and the node is currently being visited by another process, then traverse is used to chain along the list. While traversing the list, each process decrements the counter of a node it is leaving and increments the counter of a node it is about to traverse (lines 14 and 16). If a process visiting a node which has a Null pointer field finds the following node contains the key it returns TryAgain (lines 22–27) which will force the caller to re-traverse the list. This is necessary because the process must re-read its pointers to get a non-stale prev pointer. Finally, if the next field is not Null and the following node does not contain the key then the search continues (lines 28–33).

Algorithm Cursor(key) // 'prev', 'target', & 'next' are globally visible

```
1.   prev=head;
2.   if (prev->next==tail) { // empty list
3.      target=tail;
4.      return NotFound;
5.   }
6.   while (prev->next!=tail) { // traverse the list
7.      if ((prev->next!=Null) && (prev->next->data==key)) {
8.         target=prev->next;
9.         next=target->next;
10.        return Found;
11.     }
12.     if ((prev->next==Null)&&(prev->traverse->data!=key)) {
13.        // marked for deletion so use 'traverse' pointer
14.        prev->counter--;
15.        temp=prev;
16.        prev->traverse->counter++;
17.        prev=prev->traverse;
18.        if (temp->counter==0)
19.           release(temp);
20.        continue;
21.     }
22.     if ((prev->next==Null)&&(prev->traverse->data==key)) {
23.        // check if 'prev' points to 'key' via 'traverse' pointer
24.        prev->counter--;
25.        if (prev->counter==0)
26.           release(prev);
27.        return TryAgain;
28.     } else { /* follow normal ('next') field to next node */
29.        if (prev!=head)
30.           prev->counter--;
31.        prev->next->counter++;
32.        prev=prev->next;
33.     }
34.  }
35. return NotFound;
End Cursor
```

Figure 37.5 Linked list traversal.

```
Algorithm Delete(key)
1.   flag=True;
2.   while (flag==True) {
3.      repeat {
4.         r=Cursor(key);    // - delete position returned in 'prev',
5.      } until (r!=TryAgain);   // 'target', and 'next'
6.      if (r==False)
7.         return NotFound;
8.      repeat {
9.         t=TryDelete(prev,target,next); // attempt deletion
10.     } until (t!=DeleteAgain);
11.     if (t!=TryAgain)
12.        flag==False; // deletion successful
13.  }
14.  return t;
End Delete
```

Figure 37.6   Deletion from the linked list.

37.4.3 Deletion

The algorithm for deleting a node from the linked list is shown in Figure 37.6. It depends on another algorithm called TryDelete which is shown in Figure 37.7.

Algorithm Delete repeatedly tries to delete a node with the given key. Repetition is necessary because of potentially concurrent insertions/deletions of the target node. The algorithm first uses algorithm Cursor to determine the location of the target node (lines 3–5). If a node with the required key is not found, then NotFound is returned (lines 6–7). If the target node exists, algorithm TryDelete is then called to do the deletion (lines 8–10). TryDelete may return TryAgain causing the repeat loop to re-execute.

Algorithm TryDelete (Figure 37.7) uses DCAS to attempt the deletion of the target node. If the DCAS is successful, it sets the next field of node prev to point to node next. The code then decrements the counter of node prev (as it is no longer being used) and, if possible, releases the deleted node (line 4) to reclaim space. If the DCAS instruction fails it is because one or more of the pointers have changed. If the next pointer has changed, the next pointer is updated and DeleteAgain is returned (line 9). In this case, algorithm Delete can call algorithm TryDelete again without re-traversing the list (a significant saving over Valois' algorithm). If the prev pointer has changed, however, the algorithm does not know what the new prev pointer should be so it returns TryAgain.

37.4.4 Insertion

The algorithm for inserting an element into the linked list is shown in Figure 37.8.

Algorithm Insert first uses algorithm Cursor to locate the correct insertion point in the list (lines 1–3). As with the deletion case, there is a second, companion

Algorithm TryDelete(prev,target,next)

```
1.   r=DCAS(prev->next,target->next,target,next,next,Null);
2.   if (r==True) {
3.       prev->counter--;
4.       if (target->counter==0)
5.           release(target);
6.   } else {
7.       if ((target->next!=next)&&(target->next!=Null)) {
8.           next=target->next;
9.           return DeleteAgain;
10.      }
11.      prev->counter--;
12.      if ((prev->counter==0)&&(prev->next==Null))
13.          release(prev);
14.      return TryAgain;
15.  }
16.  if (r==True)
17.      return Found;
18.  else
19.      return NotFound;
```

End TryDelete

Figure 37.7   Delete an existing node from the linked list.

Algorithm Insert(key,value)

```
1.   repeat {
2.       r=Cursor(key);
3.   } until (r!=TryAgain);
4.   if (r==Found) {
5.       repeat {
6.           t=TryInsert(prev,target,value);
7.       } until (t!=Retry);
8.   }
9.   return t;
```

End Insert

Figure 37.8   Insertion into the linked list.

Algorithm TryInsert(prev,target,value)
```
1.   node=new node();
2.   node->data=value;
3.   node->counter=0;
4.   node->next=target;
5.   node->traverse=target;
6.   r=CAS(prev->next,target,node); // try to insert the node
7.   if (r==False) {
8.       release(node);
9.       if (prev->next!=Null) { // check for retry
10.          target=prev->next;
11.          return Retry;
12.      }
13.  }
14.  prev->counter--;
15.  return Found;
End TryInsert
```

Figure 37.9   Insertion into the linked list at a known position.

algorithm TryInsert (shown in Figure 37.9) which is used to perform the actual insertion once the insertion point is known.

The TryInsert algorithm starts by allocating a new node to store the new value. It initializes the node to point to the following node in the list (target) and then uses CAS (line 6) to attempt to insert the newly created node. If CAS fails, the new node is released and a check is made to see if the target node has been marked for deletion (line 9). If so, the target pointer is updated and Retry is returned (lines 10–11) so that the Insert algorithm will call TryInsert again. Otherwise, the prev node's counter field is decremented (line 14) and the appropriate result status is returned (line 15).

Note that due to space limitations, the insertion and deletion algorithms presented do not attempt to handle duplicate key values in the list.

37.4.5   Correctness

Space does not permit a formal proof of correctness of the algorithms presented. We sketch a proof and refer the reader to (Farook, 1998) for details.

We begin by noting that non-adjacent updates do not affect one another since synchronization is localized (i.e. fine granularity concurrency control). The correctness of such inserts and deletes are guaranteed simply through the specified use of CAS and DCAS (respectively). There are three cases to consider:

1. adjacent inserts,

2. adjacent deletes,

3. a delete followed by an adjacent insert or vice versa.

Theorem 1: *When processes try to insert nodes between the same set of adjacent nodes* prev *and* target, *only one of the processes succeeds while the others fail (and re-execute).*

In this case, two or more processes have read the same set of nodes prev and target to insert a new node between them. Only that process, from among the processes competing to insert the nodes, which finds the pointer of the node prev.next and the nodes prev and target unchanged will succeed (see algorithm Insert). All other processes will fail and later retry their insertions.

Theorem 2: *When processes try to delete adjacent nodes, some of the processes succeed while the others fail.*

In this case, two or more processes are trying to delete adjacent nodes. Consider two such processes. One process will succeed while the other will fail (because the first process will have updated one of the pointers checked by DCAS for the second). This can happen due to a change either to the node pointing to the deleted node or due to a change of the node the deleted node is pointing at. Similarly, when more than two processes are attempting adjacent deletions some of the processes will succeed (those that find their pointers unchanged) while others will fail (those that find modified pointers).

Theorem 3: *When processes try to insert and/or delete adjacent nodes then some of the processes succeed while others fail.*

As with the case for concurrent deletes of adjacent nodes, some processes will execute their CAS or DCAS operations successfully because no other process has yet updated any relevant pointers. These processes will update pointers that some other processes require to remain unchanged in order to succeed. Those processes will then fail and re-execute. Which processes succeed and which fail cannot be determined *a priori* as this is dependent on the run-time scheduling order. It should, however, be noted that in a parallel programming environment, the order in which unrelated but potentially concurrent events on shared data structures takes place is insignificant.

We omit discussion of the pathological cases dealing with empty lists.

37.5 PERFORMANCE OF THE NEW ALGORITHM

To assess the performance of our new implementation of non-blocking linked lists, we conducted a simulation study. We compared the behavior of our algorithm with the algorithms of (Valois, 1995) and (Greenwald and Cheriton, 1996). We varied both the length of the linked list and the relative frequencies of inserts and deletes performed on the linked list. The number of concurrent processes was held constant at 20. For each experiment, a linked list having a fixed number of nodes was constructed. Each process was given a random insert or delete to perform and was then allowed to traverse the list for its time slice. The time slice was generated randomly and determines the number of nodes a process may visit. Once a process' time slice expires, other processes are allowed to operate on the linked list. Scheduling was performed in a

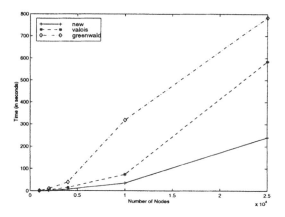

Figure 37.10   100% deletions.

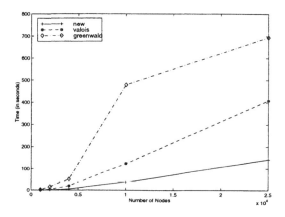

Figure 37.11   100% insertions.

round robin fashion. This approach models a parallel environment in which each process executes alone on a processor (or where each process on a machine shares the processor equally).

Figure 37.10 shows the performance of the three algorithms with 100% deletes while Figure 37.11 shows the performance of the three algorithms with 100% inserts. Finally, Figure 37.12 shows the performance of the three algorithms when there are 50% inserts and 50% deletes.

In considering these graphs, it is clear that our algorithm is relatively insensitive to the operation mix. From the data we also see that the performance of our new algorithm is consistently better than both Valois' and Greenwald & Cheriton's algorithms, and the performance differential improves with the size of the list. Further, although data is not provided in this paper, our algorithm also outperforms their algorithms for smaller link lists albeit less significantly.

There are three reasons why our algorithm performs better than the others. First, and foremost, unlike Valois' algorithm, we require only n nodes to represent a list

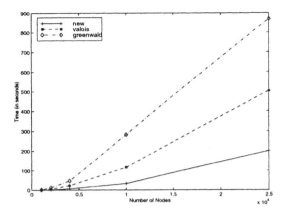

Figure 37.12 50% inserts, 50% deletions.

containing n distinct key values. This offers significant savings during list traversal (especially when considering the effects of spatial locality on memory caching performance). The algorithm also provides significantly enhanced success in concurrent reading and writing. Concurrency control is fine-grained (at the level of adjacent node triples rather than the entire list as with Greenwald & Cheriton's approach). Finally, if, during a deletion, a process fails to delete a node because its next pointer has changed, the updated next pointer is available without re-reading the list up to the deletion point.

37.6   CONCLUSIONS AND FUTURE WORK

In this paper we have described a new implementation for non-blocking linked lists. The presented algorithms (list traversal, insertion, and deletion) exploit a fine granularity synchronization strategy to significantly outperform existing algorithms, particularly for long linked lists.

Possible areas of future research include extensions to other linked data structures. The idea of localizing concurrency control can also be applied to structures such as trees and graphs. It remains to carefully develop the necessary algorithms. Application of non-blocking structures in new areas (*e.g.* parallel databases) is also of interest as is the idea of applying Barnes' "operation completion" concept to our algorithm to decrease the number of conflicts.

Acknowledgments

This research was supported in part by the Natural Sciences and Engineering Research Council of Canada under grant number OGP-0194227.

References

Barnes, G. (1993). A method for implementing lock-free shared data structures. In *5th International Symposium on Parallel Algorithms and Architectures*, pages 261–270.

Farook, M. (1998). Fast lock-free link lists in distributed shared memory. MSc Thesis, University of Manitoba.

Greenwald, M. and Cheriton, D. (1996). The synergy between non-blocking synchronization and operating system structure. In *2nd Symposium on Operating System Design and Implementation*, pages 123–130.

Herlihy, M. (1988). Impossibility and universality results for wait-free synchronization. In *7th International Symposium on Principles of Distributed Computing*, pages 276–290.

Herlihy, M. (1991). Wait-free synchronization. *ACM Transactions on Programming Languages and Systems*, 13(1):123–149.

Herlihy, M. (1993). A methodology for implementing highly concurrent data objects. *ACM Transactions on Programming Languages and Systems*, 15(5):745–770.

Herlihy, M. and Moss, J. (1993). Transactional memory: Architectural support for lock-free data structures. In *20th Annual International Symposium on Computer Architecture*, pages 289–301.

Herlihy, M. and Wing, J. (1987). Axioms for concurrent objects. In *14th International Symposium On Principles of Programming Languages*, pages 13–26.

Lamport, L. (1983). Specifying concurrent program modules. *ACM Transactions on Programming Languages and Systems*, 5(2):190–222.

Massalin, H. and Pu, C. (1991). A lock-free multiprocessor OS kernel. Technical Report CUCS-005-91, Columbia University.

Prakash, S., Lee, Y., and Johnson, T. (1991). Non-blocking algorithms for concurrent data structures. Technical Report TR91-002, University of Florida.

Valois, J. (1995). Lock-free linked lists using compare-and-swap. In *14th International Symposium on Principles of Distributed Computing*, pages 214–222.

38 SINGLE STEPPING IN CONCURRENT EXECUTIONS: DEALING WITH PARTIAL ORDERS AND CLUSTERING

Thomas Kunz and Marc Khouzam

Systems and Computer Engineering,
Carleton University, Canada

tkunz@scs.carleton.ca

Abstract: Event visualization tools are commonly used to facilitate the debugging of parallel or distributed applications, but they are insufficient for full debugging purposes. The need for traditional debugging operations, such as single stepping, is often overlooked in these tools. When integrating such operations, the issue of concurrency (executions are only partially ordered, unlike totally-ordered sequential executions) needs to be addressed. This paper justifies and describes three single-stepping operations that we found suitable for partially-ordered executions: *global-step*, *step-over* and *step-in*.

Abstraction techniques are often used to reduce the overwhelming amount of detail presented to the user when visualizing non-trivial executions, such as grouping processes into clusters. These abstraction operations introduce additional problems for single stepping. The paper identifies two problems introduced by process clustering and describes how to extend the single-stepping operations to address these issues.

Keywords: concurrency, debugging, single stepping, PVM, Poet, visualization, process clustering.

38.1 INTRODUCTION

To take advantage of available computer resources in the most efficient fashion, one must look to parallel and distributed applications instead of traditional sequential ones. However, those applications involve all the complexity of sequential ones, plus problems specific to the parallel or distributed environment. Hence, tools that help understand the behavior of such applications are becoming more common. A useful approach for describing and reasoning about a parallel or distributed execution is the

use of event-visualization tools, which give the user a graphical view of the execution (Eick and Ward, 1996; Kohl and Geist, 1995; Kranzlmüller *et al.*, 1996; Kunz *et
al.*, 1998; Zernik *et al.*, 1992).

These tools aim to help the user understand and debug a parallel or distributed
application. Although visualization is very useful for this purpose, it is often not
sufficient. Traditional debugging operations are also needed in such tools, so a user
can study a faulty program more efficiently. One of the most frequently used functions
of a traditional debugger is single stepping. It allows the user to study the program in
an incremental manner, while examining its behavior. Such a facility would be a
valuable addition to any event-visualization tool. However, defining the meaning of a
single step within a non-sequential execution, where events are only partially ordered,
is a non-trivial task. This paper justifies and describes single-stepping operations for
partially-ordered executions. These operations here have been implemented in Poet,
a Partial Order Event Tracer, developed at the University of Waterloo (Kunz *et al.*,
1998).

An example of using Poet during the visualization of a PVM (Geist *et al.*, 1994)
application is given in Figure 38.1. Within Poet, at the simplest level, each entity
exhibiting sequential behavior is represented by a horizontal trace line. These entities
can be processes, tasks, threads, objects, semaphores, and so on. In the remainder
of this paper, for the sake of simplicity, the word *process* will be used to refer to
any one of these entities. On each trace line, *events* are positioned in their order of
occurrence to describe the behavior of a process over time. An event represents some
computation and is considered to occur instantaneously. *Primitive events* constitute
the lowest level of observable behavior in a parallel or distributed execution, such as
the sending or receiving of a message, or the creation or termination of a process. On
the display, pairwise related events are joined by arrows, such as sending a message
from one process to another. At a higher level of complexity, *process clustering* and
event abstraction are used to simplify the view of the execution (Kunz *et al.*, 1998).

A basic difference between traditional source-code debuggers and event-
visualization tools is that, unlike code statements, the events of an execution are not
known *a priori*. Because of this limitation, the research discussed here deals with
post mortem single stepping, in which the events have been recorded during normal
execution. The event traces can then be visualized by the user to understand the execution after the fact or serve as a reference point to deterministically *replay* the execution (LeBlanc and Mellor-Crummey, 1987). In the former case, a user is limited to
understanding the execution behavior by going through it incrementally at a controlled

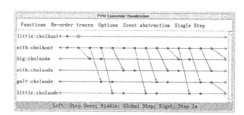

Figure 38.1   Poet, the Partial Order Event Tracer.

pace. In the latter case, traditional source-code debuggers can be attached to the individual processes, collecting detailed information about each process during execution, such as variable content, code path taken, *etc.*

The paper is organized as follows. Section 38.2 reviews related work. Section 38.3 gives a short introduction to a few basic theoretical concepts, in particular the definition of the *past* of an event. This concept is fundamental for the remaining sections. Sections 38.4 and 38.5 are the core of the paper. Section 38.4 describes three single-stepping operations for partially-ordered executions. Section 38.5 introduces the notion of process clustering, points out the problems that arise when single-stepping in the context of process clustering, and presents our solutions. Section 38.6 concludes the paper with the description of future work.

38.2   RELATED WORK

Distributed applications are inherently complex and large. Accordingly, graphical visualizations can play a significant role in understanding their behavior. These visualizations are often based on process-time representations of events which are available to the user after the execution has taken place. While many visualization tools have been proposed in the literature, only a few address the need to combine graphical visualizations and traditional debugging operations.

ATEMPT (Kranzlmüller *et al.*, 1996) is a *post mortem* event-visualization tool. The authors mention the use of consistent cuts as a way of breakpointing the execution. However, as a replay mechanism is not implemented, such a facility is presumably not yet available. No single-stepping facility seems to be provided in ATEMPT.

Hicks and Berman (Hicks and Berman, 1996) describe Pangaea, an event-visualization tool allowing both post-mortem debugging and program replay. Pangaea combines the visualization and trace facilities of XPVM (Kohl and Geist, 1995) with a new event-graph view based on process-time diagrams. No single-stepping or breakpointing facilities are available in this tool.

Paragraph (Heath and Etheridge, 1991) is a post-mortem event-debugger that provides 25 different displays to describe the collected event data. Despite the difficulties inherent in establishing a global clock, Paragraph attempts to use such a mechanism, creating a total serialization of events. The tool adopts a dynamical visualization approach which re-enacts the original live action of the execution by sequentially showing the occurrence of the events. For a more detailed study, the tool provides a sequential single-stepping mode which processes the collected data according to user requests.

Enterprise (Iglinski *et al.*, 1996) is a parallel programming system that supports debugging operations. A user can set breakpoints and single-step through the execution at the level of events. Each single step will cause the "next" event to occur, so single-stepping is reduced to a total ordering of events that is consistent with the underlying partial order of event occurrence. In partially-ordered executions, additional single-stepping operations are required, as discussed below.

Parade (Kraemer and Stasko, 1998) is a parallel program animation development environment, supporting reorderable, synchronous, and independent execution visualizations. The authors claim that synchronous displays enable a user to step through an execution one program event at a time (which, again, ignores the issue of concurrency) or by stepping through the execution of a group of concurrent events at a time. This latter option is similar to one of the single-stepping operations proposed here. However, as the discussion will show, other single-stepping operations are feasible and desirable when dealing with partially-ordered concurrent executions.

(Zernik *et al.*, 1992) discuss a post-mortem visualization prototype based on process-time representations. By performing different operations on these representations, it is possible to get the equivalent of the debugging operations of a sequential debugger. The authors describe an approach that allows for single stepping in terms of events. The events of the graph are partitioned into three sets, *past*, *present*, and *future* and operations are defined to enable the user to move events from one set to another. Single stepping is performed by repeatedly moving an event from *future* to *present*, and updating the sets accordingly. To select the event that changes sets, a total ordering of events is necessary. Different methods are proposed to restrict the partial order of the events to a total order, such as a global-clock ordering or a user-defined ordering.

In summary, even those tools that provide single-stepping as integrated part of their event-based display fail to address the issue of concurrency. They fall back on imposing a total order on the sequence of event occurrence, "executing" one event at a time. In our opinion, this does not deal adequately with inherently partially-ordered executions.

38.3   THEORETICAL BACKGROUND

The model of a partially ordered execution that we deal with is based on the *happened-before* relation defined by Lamport (Lamport, 1978). Unlike his model, however, we deal not only with asynchronous messages but also with synchronous communication. We therefore use an extension of the relation, which is defined in (Black *et al.*, 1993). With this relation, any two events can be compared to see which one occurs first in the partial order. When event a precedes event b we write $a \to b$. When $a \nrightarrow b$ and $b \nrightarrow a$, the two events are said to be *concurrent*, which means that their order of occurrence is not important. When a particular event is of interest, it is often useful to examine the set of events which influence it. The past of an event is used for that purpose.

Definition 1 (Past of an event)
The past of an event a *is the set of events that precede it:* $past(a) = \{b \mid b \to a\}$.

The event-based view of program execution was originally applied to message-passing parallel and distributed applications and programming environments. However, the same partial-order model can be applied to shared-memory systems as well. In a nutshell, each entity that exhibits sequential behavior (process, thread, object, semaphore, monitor, *etc.*) is modeled, and events capture how these entities relate to each other (a process enters a monitor, a thread acquires a semaphore, *etc.*). These event pairings can be synchronous or asynchronous in nature. The work reported here applies equally well to the more general case, where a concurrent application consists of active or passive sequential entities, whose synchronous or asynchronous interactions can be modeled by pairs of events. Note the limitation to event pairs, which excludes broadcasts or barrier synchronizations (unless these are modeled at a lower level as sets of pairwise interactions). To accurately model distributed executions with interactions that are one-to-many, many-to-one, or many-to-many, extensions of this basic model are needed. For the purpose of this paper, however, we will focus on the basic model only, which characterizes the partially-ordered execution of many message-passing and shared-memory applications.

38.4 SINGLE-STEPPING OPERATIONS

This section introduces and justifies three single-stepping operations for partially-ordered executions. These operations can be incorporated in any event visualization tool that captures and displays the partial order of event occurrence.

Our operations are based on three sets, the *execution sets*, which characterize the execution state as follows. The *executed* set contains all events that already occurred. The *ready* set is the set of events that could potentially happen next. Since events can be concurrent, at any given point in time multiple events can be in this set. In fact, this is the major difference to totally-ordered executions. Finally, the *non-ready* set contains all events that cannot occur next (and have not yet occurred). Since event occurrences are unique, an event belongs to exactly one set at a time, making the three sets exhaustive (they cover all events in an execution) and mutually distinct.

For the execution sets to be consistent, two conditions must hold: any event preceding an executed or ready event must be executed, and any event preceded by a ready or non-ready event must be non-ready. The term "preceded" is used with respect to the precedence relation characterizing the ordering of events. By enforcing these conditions, the execution is guaranteed to progress in an orderly fashion, and in particular, to adhere to the total order within each process. An alternate description is that the cut defined by the executed and ready sets is always a consistent cut (Chandy and Lamport, 1985). Let E be the set of all events that occurred during a distributed execution. We define the ready and non-ready sets in terms of a given executed set as follows:

Definition 2 (Ready and non-ready sets)
$ready = \{e \in E \mid e \notin executed \land past(e) \subseteq executed\}$
$non\text{-}ready = \{e \in E \mid e \notin executed \land past(e) \not\subseteq executed\}.$

For an event to be ready, its entire *past* must be in the executed set. Due to the total order of events in the same process, this implies that each process will have at most one ready event. Nevertheless, there will be situations where a process will have no ready events; this occurs when the event that should be ready is preceded by an event not in executed from another process.

Our single-stepping operations can best be expressed in terms of manipulations on the set membership of events: moving events from non-ready to ready or executed, and moving events from ready to executed. All three operations try to advance the computation by the smallest logical amount. Single stepping should allow a user to progress through an execution in small steps. In a sequential program, this corresponds to the execution of a single assembler or code statement. In concurrent applications, establishing a smallest amount of execution is less straightforward. A first approach is to allow the execution of only one event that is ready. However, transparently selecting one of many concurrent events in the *ready* set will often seem obscure to the user, since the motivation behind the choice may not be clear or even known. Asking the user to choose which ready event to execute leads to a very slow and cumbersome interface when stepping through a large execution. Both options also suffer from the problem that they impose a total order on the event occurrence, obscuring the fact that the user is dealing with a concurrent execution. A second approach, which we prefer, avoids choosing between concurrent events as much as possible. Instead, *multiple* ready events are added to the executed set. The exact number of ready events depends on the specific operation, but the selection is logical and can easily be understood by the user. So in our approach the *smallest* amount of execution is not always a single event, and a single step will often cause multiple events to be executed at once.

Before single stepping begins, the execution sets must be initialized. At the start, when no event has been executed, it is clear that the executed set is empty. Hence, the ready events will be ones with an empty past. The remaining events are part of the non-ready set. An *update_execution_sets* function computes the ready and non-ready sets according to a given executed set, as expressed in Definition 2.

In some situations it will not be desirable to start the single-stepping session from the very beginning of the execution. To deal with this issue, the executed set can be initialized with any arbitrary but consistent event set. For example, a user can set a breakpoint at a specific event. When the breakpoint is hit, the execution is stopped in a consistent cut. The executed set is then the set of all events that occurred so far and single stepping can resume from this point on.

We identified three single-stepping operations, which are described in detail below: *global-step*, *step-over* and *step-in*. Each of these operations supports particular scenarios encountered during a single-stepping session.

38.4.1   Global-Step

A simple approach at single stepping in a parallel or distributed execution allows the user to request a single step for the entire execution. A *global-step* causes every ready event to be added to the executed set:

> executed := executed $\cup$ ready
> update_execution_sets

This operation is simple to understand for the user: it allows for an incremental progression through the execution. The global-step operation allows the user to request a small step in the entire execution, without focusing on a particular event or trace. There are, however, situations where a global-step is too general; one can imagine a situation where a specific process is of interest and the user wishes to focus on that process.

38.4.2   Step-Over

Imagine a debugging scenario where a message is never sent; to determine the cause of the error, the user will want to focus on the particular process where the send event should have occurred and its interactions with other processes. This type of situation calls for single stepping *in the particular process* up to where the message should have been sent. Other processes are permitted to execute only when it is necessary for

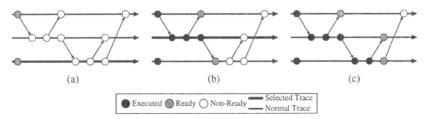

(a)   (b)   (c)

Executed ● Ready ◉ Non-Ready ○   ▬▬ Selected Trace   ▬▬ Normal Trace

Figure 38.2   A single-stepping session using step-over.

allowing the process in focus to execute; in particular, a process fully concurrent to
the one in focus will not execute at all.

Once a process has been specified for such an operation, establishing which event or
events should execute is essential. Intuitively, since the operation focuses on a specific
process, a single step should cause execution towards the *next event* of the process.
Because events in the same process are totally ordered, a simple definition for the next
event of a process can be given. In this definition (and all the ones following later),
we do not distinguish between a process P and the totally ordered set of events in this
process.

Definition 3
$\mathcal{N}_P$ *is the first non-ready event of P:*
$$\mathcal{N}_P = \{e \in \textit{non-ready} \cap P \mid \forall e' \in P : e' \to e \Rightarrow e' \notin \textit{non-ready}\}$$

This definition is given in set notation to allow for generalization in the presence
of clusters, discussed later. In the current situation however, since the definition is
restricted to a particular process, $\mathcal{N}_P$ always consists of at most one event. In the
particular case where all the events of P are either executed or ready, $\mathcal{N}_P$ is empty.

With this definition, we define a step-over operation as follows. A process P is
selected by the user, then $\mathcal{N}_P$ is determined and made ready by adding the events in
its past to the executed set regardless of the processes they belong to. This approach
is similar to stepping over a function in a sequential debugger. The execution sets are
then updated to reflect the new state of the execution:

  executed := executed $\cup$ past($\mathcal{N}_P$)
  update_execution_sets

Figure 38.2 shows an example of a small debugging session using step-over. In the
example, the process selected for the operation is represented by a bold line. To reach
the state of Figure 38.2(b) from (a), a step-over was performed on the lowest process
of the diagram: it forced the Next event to become ready by adding its past to the
executed set. A step-over is then requested in the middle process. The top process is
not affected by the operation since it has no events in the past of the Next event which
are not already in the executed set. The example also demonstrates that a step-over
operation can be performed on a process that has a ready event (38.2(a)) or does not
(38.2(b)).

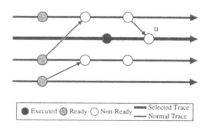

Figure 38.3   Motivating the step-in operation.

38.4.3 Step-In

A third debugging scenario is depicted in Figure 38.3. Imagine that the receive event u is known to be wrong. To determine the cause of the error, the user needs to examine the events in the past of u. Global-step does not allow one to focus on a process and therefore cannot serve our purpose. On the other hand, although step-over focuses on a process, requesting such a step would cause event u to immediately become ready. What the user needs is a method of single stepping *towards* event u, in such a fashion that it will be possible to find the event(s) causing u to be wrong. We can think of this requirement as a need to step in the past of u, instead of stepping over it. This approach is much like a sequential step-in operation which jumps to the beginning of a function instead of automatically executing it. The step-in function should cause the smallest amount of execution leading to the eventual occurrence of the Next event of the selected process. If the selected process has a ready event, the step-in operation will cause only this event to execute. If the process has no ready event, its Next event, $\mathcal{N}_P$, is determined and all ready events in its past are added to the executed set. Doing so allows the execution to progress towards $\mathcal{N}_P$:

> if (P has a ready event e)
> > then executed := executed $\cup$ e
> > else executed := executed $\cup$ (past($\mathcal{N}_P$)$\cap$ ready)
> > update_execution_sets

The step-in operation is interesting for two main reasons. First, it allows the user to cause a single ready event to execute by requesting a step-in for the process containing the specific ready event. A step-over would not only execute the ready event but also everything in the past of the Next event of the trace. Second is the ability to gradually execute events in the past of a specific event, in such a way as to make the execution progress towards this event. A step-over would immediately cause the event to become ready instead of gradually getting closer to it.

Both step-in and step-over cause execution only in the past of $\mathcal{N}_P$, consequently, other events are unaffected. Furthermore, as multiple step-in operations add ready events in the past of $\mathcal{N}_P$ to the executed set, the execution will progress within this past until it contains no more ready events. When this situation occurs, $\mathcal{N}_P$ is itself ready, which is what is accomplished directly by a single step-over. It follows that step-in performs small execution increments within a step-over.

In addition, it is worth noting that performing a step-in within every process containing a ready event will result in the same overall execution state as a global-step.

Although this shows that a global-step can be emulated with multiple step-in operations, providing the user with the capability to do a global-step simplifies stepping through large executions with many processes and a high degree of concurrency.

38.5 PROCESS CLUSTERING

Due to the overwhelming amount of information that can be collected by an event-visualization tool, reducing the quantity of detail presented to the user is necessary. Most graphical visualization tools provide for abstraction operations to reduce the display complexity. One such abstraction operation is *process clustering*, supported by, among other tools, Poet (Kunz *et al.*, 1998), ATEMPT (Kranzlmüller *et al.*, 1996), and SeeSeq (Eick and Ward, 1996). Process clustering attempts to reduce the number of processes presented to the user. In many situations, certain processes, as well as the interactions between them, are of no interest. If such processes could be grouped, or clustered together, the user would be presented only with a single, higher-level, process-like entity, instead of the many processes hidden within.

The discussion in this section is based on the cluster representation used by Poet, but is easily adapted to other cluster representations which share the problems identified below. To understand these problems, the cluster representation in Poet is briefly reviewed first. In the simplest of situations, when a set of processes is clustered, the cluster is displayed as a single trace line and all events within it are hidden unless they involve interactions with processes outside the cluster. However, it is important that visualizations containing process clusters correctly reflect the partial order of the execution. Otherwise, a clustered display might appear to contradict the original execution. In some cases, a cluster will contain concurrent events which involve communication outside of the cluster. Displaying these events on a single trace line would imply that these events were precedence-related. Therefore, a cluster must be displayed using multiple trace lines to reflect concurrency, one trace for each concurrent event. There may also exist precedence between events of different trace lines which are not reflected by the current cluster representation; such precedence may be caused by messages internal to the cluster, which are, therefore, not displayed. To visually indicate these precedence relations, *linking arrows* are used to connect separate lines whose events are precedence-related. An example of a partially-ordered cluster is shown in Figure 38.4. In (a), the first two processes, P1 and P2, are clustered to lead to the representation shown in (b). Two trace lines are needed to represent the concurrency between events a and b and a linking arrow is used to show that event a precedes event c.

Process clustering can be organized in a hierarchical fashion, so that clusters may contain other clusters as well as processes. Given a cluster hierarchy, Poet's *debugging focus* describes the set of processes and clusters the user is currently interested in. Changing the debugging focus can be achieved through a high-level graphical interface. At a lower level, all changes can be characterized by a sequence of two simple operations: *open* and *close*. Opening a cluster replaces it with its constituent processes and sub-clusters, while closing it hides the components within the cluster in question.

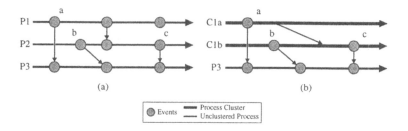

Figure 38.4   A process cluster example.

38.5.1   *Problems due to Clusters*

Until now, the single-stepping operations dealt with totally-ordered, individual traces, describing the behavior of a single process and its interactions with other processes. All events collected by the visualization tool are displayed to the user and are a visible part of the execution. On the other hand, the use of process clusters requires that the single-stepping operations behave according to the new display presented to the user, where not all events are displayed and clusters, occupying more than one trace line, are treated as single entities. These two issues are orthogonal and can be addressed separately.

Event hiding prevents the clustered view of the execution from corresponding directly to the non-clustered view. To respect the current display, single stepping must deal with hidden events transparently. Hidden events must not inhibit the execution of visible events. Consequently, hidden events that are found to be ready should be added to the executed set. If not, visible events will wait for the execution of hidden ones, a situation which is counterintuitive to the user and not reflected in the current display.

A *process cluster* can be represented visually by one or more trace lines. As these traces are only used to express concurrency, a cluster is treated as a single entity, independently of the number of traces needed to represent it. Consequently, a single-step operation dealing with process clusters must behave as if the cluster was an indivisible entity. This observation complicates single stepping because the smallest entity is no longer characterized by a total order. Step-over and step-in, as defined so far, rely on the total ordering of events within individual processes. Specifically, the behavior of these two operations was based on the concept of the Next event of a process. With a partial ordering, there is not necessarily a single Next event, but instead, there may be multiple ones, a possibility which forces us to deal with sets of Next events. To determine the set of Next events in the execution order of a cluster, we look for the *first* non-ready events in the cluster. To fall in that category, an event must be non-ready and have no non-ready events of the cluster in its past. Since we deal with a set of events, we refer to this set as the *non-ready boundary* of a cluster. Finally, to respect the view presented to the user, hidden events must be ignored, and therefore, only visible events are considered when defining the non-ready boundary of a cluster.

Definition 4 (Non-ready boundary of a cluster C)
The non-ready boundary *of C, β_C, is the set of non-ready visible events whose visible predecessors belonging to C are ready or executed, i.e. :*
$$\beta_C = \{e \in \textit{non-ready} \cap C \mid$$
$$\forall e' \in C : e' \rightarrow e \wedge e \textit{ visible} \wedge e' \textit{ visible} \Rightarrow e' \notin \textit{non-ready}\}$$

38.5.2   The Single-Stepping Operations

In the presence of process clusters, the definitions of the execution sets remain the same but the algorithms for single stepping must be extended. A modification of the *update_execution_sets* algorithm takes care of hidden events. When determining if an event is ready in a process, the algorithm also checks if the event is visible. If the event is ready but hidden, it is moved to the executed set and another potential ready event for the process is examined. This must be repeated until the process is found to have no ready event, or that the ready event found is visible. When a ready event that is hidden is executed automatically, non-ready events that depended on it may become ready. Moreover, the new ready events may also be hidden and thus automatically executed, possibly creating more new ready events.

Figure 38.5 gives an example of how the modified *update_execution_sets* algorithm works. In the figure, processes P2 and P3 are part of a cluster, which implies that the only visible events of these processes are d and f. Figure 38.5 (a) shows the original state of the execution sets, before the ready set is computed. The *update_execution_sets* algorithm will iterate through each process to determine which events are ready. First, process P1 is examined and it is determined that event g is ready and visible. Process P2 is examined next and its potential ready event b is found to be non-ready, waiting on event a, which is ready. This state is shown in Figure 38.5 (b). Since a is hidden, the algorithm adds both a and b (which are hidden) to the executed set and continues in process P2. Event c is found to be ready but hidden so it is added to the executed set. Finally, event d is found ready and visible, terminating the examination of P2. Next, P3 is examined and event f is found to be non-ready. The final state, showing hidden events, is depicted in Figure 38.5 (d).

The *global-step* operation described in the previous section caused every ready event in the execution to be added to the executed set. In the presence of process clusters, ready events are all visible since hidden events cannot be ready. Therefore, a global-step remains the same. The only difference lies in the *update_execution_sets* step described above.

Originally, the *step-over* operation required the user to select a process in which the step should be performed. The Next event of that process was then made ready through the execution of its entire past. In the presence of process clusters, the operation should be similar. Accordingly, when a step-over is requested, the user must select a specific cluster. The non-ready boundary of this cluster C is then computed and made ready by adding the entire past of this set to executed:

> executed = executed $\cup$ past(β_C)
> update_execution_sets

The *step-in* function is also performed on a specific cluster. Similarly to its previous definition, a step-in executes all ready events within the chosen cluster. If there

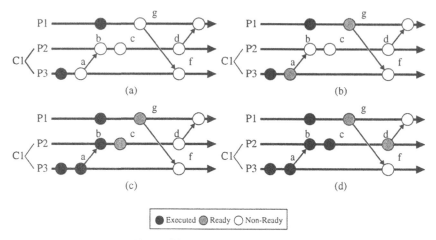

Figure 38.5   Updating the ready set.

are no ready events, the algorithm executes all ready events leading to its non-ready boundary:

$$\text{if } (\text{ready} \cap C \neq \emptyset)$$
$$\text{then executed} := \text{executed} \cup (\text{ready} \cap C)$$
$$\text{else executed} := \text{executed} \cup (\text{past}(\beta_C) \cap \text{ready})$$
$$\text{update\_execution\_sets}$$

To treat clusters as indivisible entities it is important that all ready events in a cluster be treated as a unit; differentiating between these events is not justifiable as the traces in a cluster have no meaning beyond the expression of concurrency. Consequently, the step-in operation will not always allow the user to execute a single event. To force the execution of a single event, the user has to open up the cluster until the process that contains that event is displayed and then use the step-In operation on that process.

38.5.3 Debugging Focus Changes

To enable a user to examine an execution at different abstraction levels, the debugging focus can be changed dynamically. Therefore, clusters can be opened or closed at any time. When opening a cluster, hidden events may become visible; closing a cluster may cause visible events to become hidden.

Opening a cluster requires no action to be taken with respect to single stepping. Since the algorithms used to determine the execution sets deal with the execution at its lowest level of abstraction, hidden events that become visible are already assigned to the correct sets. On the other hand, closing a cluster might force some events to change sets. Since the change of focus may cause ready events to become hidden, these events must be added to the executed set and the ready set must then be recomputed. This can be achieved by running the *update_execution_sets* algorithm after each change of focus.

Although dealing with a change of focus is simple, the forced execution of newly-hidden ready events can have noticeable consequences. Figure 38.6 shows a situation

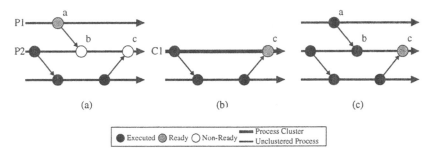

Figure 38.6   Changing the debugging focus.

where changing the abstraction focus will cause execution visible to the user. The original state of the execution is given in (a) in which event c is non-ready, waiting for events a and b to become executed. The user then changes the focus by clustering processes P1 and P2 into cluster C1 as seen in (b). Events a and b are internal to C1 and are now hidden. Since ready hidden events are automatically added to the executed set, both a and b are now considered executed (after a is added to the executed set since it was ready, b becomes ready, and since it is hidden, it is also added to executed). At this point event c does not depend on any events which are not executed and is added to the ready set. Unfortunately, if the user were to re-open the cluster, the execution would now be in the state shown in (c) where events a and b are executed although no stepping operation was performed.

Nevertheless, the ability to change the debugging focus during a single-stepping session is an important feature in our visualization tool. The problem described above could be circumvented with an automatic undo operation which would revert to the state of the execution before the clustering operations. However, since a user can perform many focus-update operations in any order and because of the existence of cluster hierarchies which complicate the possible sequence of operations, reverting to the correct previous state is difficult if not impossible. Moreover, such an automatic undo operation could not be extended to replay mode, since an actual execution cannot be undone. Therefore, although the execution should not be influenced by anything other than an actual single step, some focus-update operations may cause execution.

38.6   CONCLUSIONS AND FUTURE WORK

This paper discusses the fundamental issues related to single stepping in event-visualization tools. Three single-stepping operations have been defined to deal with partially-ordered parallel and distributed executions. These operations were implemented in Poet, a visualization tool developed at the University of Waterloo. Executions can be visualized post mortem, using the single-stepping operations to incrementally step through and understand the monitored execution. Alternatively, Poet can deterministically replay the execution, using the recorded event data as a reference to break potential (non-deterministic) races (Kunz *et al.*, 1998). During *replay*, sequential debuggers can be attached to the processes of the application. In such an environment, single stepping allows the user to examine the internal state of each process while incrementally progressing through the execution.

The three single-stepping operations were extended to address the problems arising from process clustering as well. Process clustering is a common abstraction operation to reduce the display complexity in a number of graphical visualization tools. Clustering introduces problems of event hiding and the need to deal with partially-ordered entities. The algorithms described in Section 38.5 address these problems.

An open problem is how to define and implement single stepping in the absence of event data. As mentioned before, our approach assumes that we know about the occurrence of events *a priori*, either because we visualize the execution post mortem or because we have access to the reference event data during deterministic replay. Without this information, it is, for example, not possible to define a step-over operation, which requires knowledge about the past of an event that has not yet occurred. It therefore becomes impossible to decide which processes should be blocked and which processes should be allowed to execute and by how much.

Acknowledgments

The work described here has been supported by the Natural Sciences and Engineering Research Council of Canada (NSERC). It benefited from various discussions with the members of the Shoshin research group at Waterloo, in particular Professor David Taylor.

References

Black, J., Coffin, M., Taylor, D., Kunz, T., and Basten, T. (1993). Linking specification, abstraction, and debugging. CCNG E–232, University of Waterloo.

Chandy, M. and Lamport, L. (1985). Distributed snapshots: Determining global states of distributed systems. *ACM Trans. on Comp. Sys.*, 3(1):63–75.

Eick, S. and Ward, A. (1996). An interactive visualization for message sequence charts. In *4th Workshop on Program Comprehension*, pages 2–8.

Geist, A., Beguelin, A., Dongarra, J., Jiang, W., Manchek, R., and Sunderam, V. (1994). *PVM: Parallel Virtual Machine. A User's Guide and Tutorial for Networked Parallel Computing*. MIT Press.

Heath, M. and Etheridge, J. (1991). Visualizing the performance of parallel programs. *IEEE Software*, 8(5):29–39.

Hicks, L. and Berman, F. (1996). Debugging heterogeneous applications with Pangaea. In *SPDT'96: Symposium on Parallel and Distributed Tools*, pages 41–50, Philadelphia, PA, USA.

Iglinski, P., Kazouris, N., MacDonald, S., Novillo, D., Parsons, I., Schaeffer, J., Szafron, D., and Woloschuk, D. (1996). Using a template-based parallel programming environment to eliminate errors. In *10th International Symposium on High-Performance Computing Systems*, Ottawa, Ontario, Canada. On CD-ROM.

Kohl, J. and Geist, A. (1995). The PVM 3.4 tracing facility and XPVM 1.1. Technical report, Computer Science & Mathematics Division, Oak Ridge National Lab., TN, USA.

Kraemer, E. and Stasko, J. (1998). Creating an accurate portrayal of concurrent executions. *IEEE Concurrency*, 6(1):36–46.

Kranzlmüller, D., Grabner, S., and Volkert, J. (1996). Event graph visualization for debugging large applications. In *SPDT'96: Symposium on Parallel and Distributed Tools*, pages 108–117, Philadelphia, PA, USA.

Kunz, T., Black, J., Taylor, D., and Basten, T. (1998). Poet: Target-system-independent visualizations of complex distributed applications. *Computer Journal*, 40(8):499–512.

Lamport, L. (1978). Time, clocks, and the ordering of events in a distributed system. *Communications of the ACM*, 21:558–565.

LeBlanc, T. and Mellor-Crummey, J. (1987). Debugging parallel programs with instant replay. *IEEE Trans. on Comp.*, 36(4):471–482.

Zernik, D., Snir, M., and Malki, D. (1992). Using visualization tools to understand concurrency. *IEEE Software*, 9(3):87–92.

39 ADAPTIVE PARALLELISM ON A NETWORK OF WORKSTATIONS

Mohan V. Nibhanupudi and Boleslaw K. Szymanski

Department of Computer Science,
Rensselaer Polytechnic Institute, U.S.A

{ nibhanum, szymansk } @cs.rpi.edu

Abstract: Several computing environments, including wide area networks and non-dedicated networks of workstations, are characterized by frequent unavailability of the participating machines. Parallel computations, with interdependencies among their component processes, cannot make progress if some of the participating machines become unavailable during the computation. As a result, to deliver acceptable performance, the set of participating processors must be dynamically adjusted following the changes in computing environment. In this paper, we discuss the design of a run-time system to support a Virtual BSP Computer that allows BSP programmers to treat a network of transient processors as a dedicated network. The Virtual BSP Computer enables parallel applications to remove computations from processors that become unavailable and thereby adapt to the changing computing environment. The run-time system, which we refer to as *adaptive replication system* (ARS), uses replication of data and computations to keep current a mapping of a set of virtual processors to a subset of the available machines. ARS has been implemented and integrated with a message-passing library for the Bulk-Synchronous Parallel (BSP) model. The extended library has been applied to two parallel applications with the aim of using idle machines in a network of workstations (NOW) for parallel computations. We present the performance results of ARS for these applications.
Keywords: networks of non-dedicated workstations, bulk-synchronous parallel model, adaptive parallel computing, transient processors, virtual BSP computer.

39.1 INTRODUCTION

Several computing environments are characterized by the frequent unavailability of the participating machines. Machines that are available for use only part of the time are referred to as *transient processors* (Kleinrock and Korfhage, 1993). A transition of the host machine from an *available* to a *non-available* state is considered a *transient failure.* Such a model of a network of transient processors applies to several computing paradigms, including wide area networks such as the Internet and local networks of non-dedicated workstations (NOWs). In the latter case, a workstation is available for the parallel computation only when it is *idle*; that is, when it is not being used by its owner; a part of the parallel computation running on a particular workstation must be suspended when its owner activity resumes. Use of workstations in this manner allows additional sequential programs to accumulate work and thereby improve overall workstation utilization (Kleinrock and Korfhage, 1993). However parallel programs, with interdependencies among their component processes, cannot make progress if some of the participating workstations become unavailable during the computation. Parallel computations in such environments must adapt to the changing computing environment to deliver acceptable performance.

The Bulk-Synchronous Parallel (BSP) model (Valiant, 1990) is a universal abstraction of a parallel computer. By providing an intermediate level of abstraction between hardware and software, BSP offers a model for general-purpose, architecture-independent parallel programming. However the standard libraries for parallel programming using the BSP model offer only static process management (the initial allocation of processors cannot be changed while the parallel computation is in progress) and thus cannot adapt to changing computing environments, such as the ones described above.

In this paper we discuss the design of run-time support for Virtual BSP Computer to enable parallel applications to adapt to the changing computing environment. We refer to the run-time system as the *adaptive replication system* (ARS). We describe our approach to adaptive parallel computations in Section 39.2 and compare it to related work in Section 39.3. Section 39.4 describes adaptive parallelism in the BSP model. In Section 39.5 we discuss the design and implementation of the adaptive replication system. The performance of the adaptive replication system is evaluated using two applications in Section 39.6. Finally, we summarize our work and conclude in Section 39.7.

39.2   PARALLEL COMPUTING ON NON-DEDICATED WORKSTATIONS

39.2.1   *Bulk-Synchronous Parallel model*

Our view of the parallel computation is based on the Bulk-Synchronous Parallel (BSP) model (Valiant, 1990), in which computation proceeds as a series of *super-steps* comprising of computation and communication operations. All the participating processors synchronize at the end of the super-step. By providing an intermediate level of abstraction between hardware and software, BSP provides a model for general-purpose, architecture-independent parallel programming. The BSP model has been used to implement a wide variety of scientific applications including numerical algorithms (Bisseling and McColl, 1993), combinatorial algorithms (Gerbessiotis and Siniolakis, 1996) and several other applications (Calinescu, 1995). We used the model to build an efficient implementation of plasma simulation on a network of workstations (Nibhanupudi *et al.*, 1995) as well as a partial differential equation solver for modeling of bioartificial artery (Barocas and Tranquillo, 1994).

Although the barrier synchronization at the end of each BSP super-step can be expensive, its cost can often be reduced by overlapping communication with local computation and enforcing only logical but not physical synchronization. Barriers have desirable features for parallel system design. By making circular data dependencies impossible, they avoid potential deadlocks and live locks, eliminating the need for their costly detection and recovery (Skillicorn *et al.*, 1997). Barriers ensure that all processes reach a globally consistent state, which allows for novel forms of fault tolerance (Skillicorn *et al.*, 1997). In our model of parallel computation based on BSP, the participating processors are all in a globally consistent state at the beginning of each computation step, which becomes a point of a consistent checkpoint. The synchronization at the end of a super-step also provides a convenient point for checking transient process failures. Should one or more processes fail, the surviving processes can start the recovery of the failed processes at this point.

39.2.2   *Adaptive replication of computations to tolerate transient failures*

Our approach relies on executing (replicating) the computations of a failed process on another participating processor to allow the parallel computation to proceed. Note that in the BSP computation, the computation states of the participating processes are consistent with each other at the point of synchronization. By starting with the state of a failed process at the most recent synchronization point and executing its computations on another available participating workstation, we are able to recover the computations of the failed process. This allows the parallel computation to proceed without waiting for the failed process. To enable recovery of the computation of a failed process, the computation state of each process, C_s, is saved at every synchronization point on a peer process. Thus, our approach uses *eager replication of computation state* and *lazy replication of computations*.

39.3   RELATED WORK

Piranha (Carriero *et al.*, 1995) is a system for adaptive parallelism built on top of the tuple-space based coordination language *Linda*. Piranha implements master-worker parallelism. The master process is assumed to be persistent. The worker processes are created on idle machines and destroyed when the machine becomes busy. Due to master-worker parallelism, Piranha is applicable to only coarse-grained parallel applications involving independent tasks. Synchronous parallel computations with the computation state distributed among the component processes cannot be modeled with master-worker parallelism.

A limited form of adaptive parallelism can be achieved by dynamically balancing the load on the participating workstations. Parform (Cap and Strumpen, 1993) is a system for providing such capability to parallel applications. Parform is based on the strategy of initial heterogeneous partitioning of the task according to actual loads on workstations, followed by dynamic load balancing. To benefit from Parform, parallel applications must be written in such a way that they can handle a changing communication topology for an arbitrary number of tasks.

There have been several efforts to implement transparent recovery from processor failures for parallel applications. (Leon *et al.*, 1993) discuss the implementation of a consistent checkpointing and roll-back mechanism to transparently recover from individual processor failures. The consistent checkpoint is obtained by forcing a global synchronization before allowing a checkpoint to proceed. CoCheck (Stellner, 1996) tries to blend the resource management capabilities of systems like Condor (Litzkow *et al.*, 1988) with parallel programming libraries such as PVM (Sunderam, 1990) and MPI (Snir *et al.*, 1996; Gropp *et al.*, 1994). It provides consistent checkpointing and process migration mechanism for MPI and PVM applications.

Stardust (Cabillic and Puaut, 1997) is a system for parallel computations on a network of heterogeneous workstations. Stardust uses techniques similar to ours. It captures the state of the computation at the barrier synchronization points in the parallel program. In saving the application state, only the architecture-independent data is saved on disk and transferred to other nodes, which allows for the migration of the application between heterogeneous hosts. A major restriction of Stardust's mechanism of using naturally occurring synchronization barriers is that it limits the number of places where an application can be stopped and migrated. Our approach allows for a component process executing on a user's machine to be stopped at any point during the computation. This makes our approach much less intrusive to the individual owners of the workstations and encourages them to contribute their workstations for additional parallel computations during their idle times.

For synchronous parallel applications, our approach provides a less expensive alternative to checkpointing on the disk by replicating the computation state of component processes of the parallel computation on peer processes, which can be considered a form of diskless checkpointing. Our approach offers ways to reduce the amount of data to be replicated on other processes. For example part of the computation state that is common across the component processes need not be replicated on a peer process. As the speed of networks increases relative to the speed of disks, diskless checkpointing should become more attractive for synchronous parallel computations. In addition, replicating computations of a failed process can easily be extended to work across heterogeneous architectures by providing automatic conversion of data representations. Replicating data on peer processes to enable replication of computations in case of failures fits well with emerging architectures like the network computers (OpenGroup, 1997) which may not come with a local disk.

39.4   ADAPTIVE PARALLELISM IN BULK-SYNCHRONOUS PARALLEL MODEL

39.4.1   *Protocol for replication and recovery*

The adaptive replication scheme assumes that one of the processes is on a host owned by the user and hence this process is immune to transient failures. We refer to this reliable process as the *master process*. The master process coordinates recovery from transient failures without replicating for any of the failed processes. Figure 39.1 il-

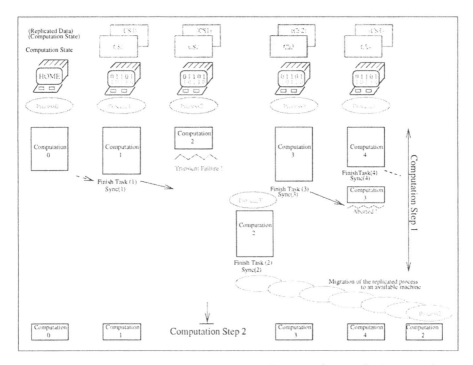

Figure 39.1   Protocol for replication and recovery illustrated for a replication level of one.

lustrates the protocol. The participating processes other than the master process are organized into a logical ring topology in which each process has a predecessor and a successor. At the beginning of each computation step, each process in the ring communicates its computation state C_s to one or more of its successors, called *backup processes*, before starting its own computations. Each process also receives the computation state from one or more of its predecessors.

When a process finishes with its computations, it sends a message indicating successful completion to each of its backup processes. The process then checks to see if it has received a message of completion from each of its predecessors whose computation state is replicated at this process. Not receiving a message in a short timeout period is interpreted as the failure of the predecessor. The process then creates new processes, one for each of the failed predecessors. The computation state of each new process is restored to that of the corresponding failed predecessor at the beginning of the computation step using the computation state received from that predecessor. Each of the newly created processes performs the computations on behalf of a failed predecessor and performs synchronization on its behalf to complete the computation step. In general, such a newly created process assumes the identity of the predecessor and can continue participating in the parallel computation as a legitimate member. However, for the sake of better performance, this new process is migrated to a new host if one is available. For more details on the protocol, refer to (Nibhanupudi and Szymanski, 1996). It should be noted that the assumption of existence of a master process is not necessary for the correctness of the protocol. Using the standard techniques from distributed algorithms, synchronization can be achieved over the virtual ring regardless of transient failures. However, the master process is a convenient solution for a majority of applications, so we used it in the prototype implementation of the system discussed in this paper.

Systems that intend to provide fault tolerance by transparently recovering from failures rely on some form of replication of resources. In general, we can replicate either data (computation state) or computations or both. When each parallel task is replicated on a group of several processors, it is too costly to update replicas with the status of the fastest processor in each group. On the other hand, migrating a process after discovering the failure intrudes upon the primary user. These considerations led us to the approach in which data are replicated on all processors but the computations are replicated only when needed. In our approach, the recovery of the failed computations and subsequent migration to a new available host are performed on an available host, which is much less intrusive.

39.4.2   Tolerating multiple failures

The number of successors at which the computation state of a process is replicated is referred to as the *replication level*, denoted by R. R is also the number of predecessors from which a process will receive the computation state. A process can therefore act as a backup to any of the R predecessors from which it receives the computation state. It is easy to see that the replication level defines the maximum number of consecutive process failures in the logical ring topology that the system can tolerate. Failure of more than R consecutive processes within the same computation step will force the processes to wait till one of the host processors recovers. A higher level of replication increases the probability of recovery from failures, but it also increases the overhead during normal (failure free) execution. The probability of failure of R consecutive processes is P_f^R, where P_f is the probability of failure of a workstation, that is, the probability of a workstation becoming unavailable for computation during a computation step. Since the duration of the computation step is small compared to the mean available and non-available periods, the probability of failure is small ($P_f \ll 1$). Thus the probability of unrecoverable failures decreases exponentially with the replication level R. The required level of replication to avoid unrecoverable failures is small for the degrees of parallelism used in practice.

39.4.3   Performance of adaptive replication

The cost of data replication includes the additional memory required for the replicated data and the cost of transferring the computation state to the successors. The additional memory needed for data replication is proportional to the level of replication, R, and the size of the computation state, C_s. The cost of communicating the computation state depends on replication level R, the size of the computation state C_s and the underlying communication network. A communication network that scales with the number of processors allows for a higher level of replication and a higher degree of tolerance to transient failures without incurring overhead during normal execution.

To minimize overhead during normal execution, our approach seeks to overlap the computation with communication associated with data replication. For those applications in which the cost of data replication is smaller than the cost of computation in the super-step, replication of computation state can be done without any overhead during normal execution. We refer to such applications as *computation dominant applications*. For such applications, the scheme is scalable with high efficiency even for tightly synchronized computations defined as those for which the duration of the computation steps, t_s, is small compared to the mean length of non-available periods, t_n. Applications for which the cost of data replication is larger than the computation have an overhead associated with data replication, and therefore they are referred to as *data replication dominant applications*. A more detailed discussion of the performance of the adaptive replication scheme along with the analysis can be found in (Nibhanupudi, 1998).

39.5 DESIGN AND IMPLEMENTATION OF THE ARS LIBRARY

39.5.1 Design of Adaptive Bulk-Synchronous Parallel Library

The adaptive replication scheme is developed using the Oxford BSP library (Miller, 1993; Miller and Reed, 1993). The library has been extended to provide dynamic process management and virtual synchronization as described in (Nibhanupudi and Szymanski, 1996). Using the extended library, processes can now be terminated at any time or migrated to new host machines; new processes can be created to join the parallel computation. Processes can now perform synchronization for one another, which allows for dynamic work sharing.

The run-time system to support adaptive parallelism in the Bulk-Synchronous Parallel model has been described in detail in (Nibhanupudi and Szymanski, 1998). The run-time support uses two levels of abstraction: *replication layer* and *user layer*. The replication layer implements the functionality of the adaptive replication scheme by providing primitives for replicating the computation state, detecting the failure of a process, replicating the computations of a failed process and restoring the state of a replicated process. This layer is not visible to the user; its functionality can only be accessed through the user layer, which provides the application programming interface. The user layer includes primitives for specifying the replication data and the computation state of the user process and for performing the memory management required for replication of the computation state. The adaptive extensions in the user layer include constructs to specify computation and communication super-steps. The replication and recovery mechanism is embedded into these constructs; the process of data replication, as well as the detection of failures and recovery, are transparent to the user. Figure 39.2 shows the constructs provided by the user layer.

39.5.2 Prototype Implementation

For the prototype we assume that super-steps that make use of replication contain computation only. This is not overly restrictive because, in the BSP model, data communicated in a super-step are guaranteed to be received at the destination process only by

```
/* Constructs to specify a computation superstep */

bsp_comp_sstep(stepid);
bsp_comp_sstep_end(stepid);

/* Constructs to specify replication data and allocate storage */

bsp_replication_data(void* data, long nbytes, void* store, char* tag, int subscript);
bsp_setup_replication_environment();

/* Constructs to specify computation state */

struct BspSystemState;
bsp_init_system_state(BspSystemState* bss);
bsp_reset_system_state(BspSystemState* bss);
bsp_set_system_state(BspSystemState* bss, char* tag, int subscript);
bsp_specific_system_state(BspSystemState* bss);
bsp_common_system_state(BspSystemState* bss);
void RecoveryFunction();
```

Figure 39.2   Adaptive BSP library: user layer.

the end of the super-step and therefore can only be used in the next super-step. Hence a super-step involving both computation and communication can always be split into a computation super-step and a communication super-step. This assumption greatly simplifies the design of the protocol for the recovery of failed processes. Furthermore, the protocol assumes a reliable network, so a message that is sent by a process will always be received at the destination. The prototype uses a replication level of one.

Failure detection is a tricky issue in distributed system design, as there is no way to distinguish between a failed process and a process that is simply slow. In a heterogeneous network, the computations on individual workstations often proceed at different speeds owing to differences in processor speed, characteristics of work load on the individual machines, *etc.* To compensate for the differences in processing speed, a *grace period* can be used to allow a slow predecessor to complete its computations before concluding that the predecessor has failed. However, using a grace period also delays replicating for the predecessor when required. Based on experimental results, our implementation uses no grace period. A process starts replicating for its predecessor if it has not received a message of successful completion from the predecessor by the time it finishes its own computations. However, to avoid unnecessary migrations, we abort the new replicated process and allow the predecessor to continue if the predecessor completes its computations before the replicated process or before the super-step is complete.[1] This results in a nice property of the adaptive replication scheme—any processor that is twice as slow as its successor is automatically dropped from the parallel computation and a new available host is chosen in its place. This allows the application to choose faster machines for execution from the available machines. As part of the synchronization, all participating processes receive a list of surviving processes

---

[1] The super-step is complete when synchronization has been initiated on behalf of all the participating processes.

from the master. A participating process that replicated for a predecessor aborts the new replicated process when it finds that the predecessor has successfully performed synchronization.

We are testing our adaptive replication scheme using simulated transient processors with exponential available and non-available periods. A *timer process* maintains the state of the host machine. Transitions of the host machine from an available state to a non-available state and *vice versa* are transmitted to the process via signals. The process is suspended immediately if it is performing a computationally-intensive task such as a computation super-step. Otherwise, the host is marked as unavailable and the process is suspended before entering a computationally-intensive task. The prototype is implemented on Sun Sparcstations using the Solaris operating system (SunOS 5.5). It makes use of the checkpoint-based migration scheme of Condor (Bricker *et al.*, 1992) for process migration.

It should be noted that our adaptive replication scheme protocol can be applied to other message-passing libraries such as MPI (MPI Forum, 1994). The only requirement is that the application be written in the BSP style, using a sequence of computation and communication super-steps.

39.6 APPLICATION OF ARS LIBRARY TO PARALLEL COMPUTATION

We applied the ARS library to two different applications that illustrate the performance of the scheme for a *computation dominant* application (maximum independent set) and a *data replication dominant* application (plasma simulation).

39.6.1 Maximum Independent set

A set of vertices in a graph is said to be an *independent set* if no two vertices in the set are adjacent (Deo, 1974). A *maximal independent set* is an independent set which is not a subset of any other independent set. A graph, in general, has many maximal independent sets. In the maximum independent set problem, we want to find a maximal independent set with the largest number of vertices. To find a maximal independent set in a graph G, we start with a vertex v of G in the set. We add more vertices to this set, selecting at each stage a vertex that is not adjacent to any of the vertices already in the set. This procedure will ultimately produce a maximal independent set. To find a maximal independent set with the largest number of vertices, we find all the maximal independent sets using a recursive depth-first search with backtracking (Goldberg and Hollinger, 1997). To conserve memory, no explicit representation of the graph is maintained. Instead, the connectivity information is used to search through a virtual graph. To reduce the search space, heuristics are used to prune the search space. Each processor searches a subgraph and the processors exchange information on the maximal independent set found on each processor. Since the adjacency matrix is replicated on each processor, the computation state that needs to be communicated to a successor to deal with transient failures is nil. That is, the computation state of a failed process can be recreated based on the knowledge of its identity alone. This application can therefore be categorized as a computation dominant application.

39.6.2   Plasma Simulation

The plasma Particle In Cell simulation model (Norton *et al.*, 1995) integrates, in time, the trajectories of millions of charged particles in their self-consistent electro-magnetic fields. In the replicated grid version of the plasma simulation (Nibhanupudi *et al.*, 1995), the particles are evenly distributed among the processors for sharing work load; the simulation space (field grid) is replicated on each of the processors to avoid frequent communication between processors. The computations modify the positions and velocities of the particles, forces at the grid points and the charge distribution on the grid. Hence, the computation state data that need to be replicated include the positions and velocities of the particles, the forces at the grid points and the grid charge. However, at the beginning of each super-step, all processors have the same global charge distribution of the grid and, hence, the charge data need not be replicated on a remote host. Instead, each process can save this data locally, which it can use to restore a failed predecessor. Checkpointing data locally when possible reduces the amount of data communicated for data replication. Due to the overhead associated with the communication of computation state, this application can be categorized as a replication dominant application (also see the discussion in the next section).

39.6.3   Results

Table 39.1(a) shows the execution times of maximum independent set problem on transient processors using the adaptive replication scheme with $t_a = 40$ minutes and $t_n = 20$ minutes respectively. These values for t_a and t_n are within the range of values reported in earlier works (Mutka and Livny, 1987). The measurements were taken on a network of Sun Sparc 5 workstations connected by a 10 Mbps Ethernet. The number of processors available is much larger than the degree of parallelism used in the simulations and, therefore, migration to an available processor was always possible. The execution times of the runs on transient processors using the adaptive replication scheme were compared with the execution time on dedicated processors and with the execution time on transient processors without using the scheme. Runs on transient processors without the scheme simply suspend the execution of the parallel computation when the host processor is busy. The execution time on a single processor is also shown for reference. As can be seen from these timings, the runs on transient processors using the adaptive replication scheme compare favorably with the runs on dedicated processors. Figure 39.3(a) shows a plot of these timings.

Our measurements indicate that a significant amount of computation was performed using idle workstations. As mentioned in Section 39.4.1, parallel runs using the adaptive replication scheme use the user's own host for one of the processes and use the idle machines in the network for the remaining processes. When using p processors, $\frac{p-1}{p}$ of the total computation is performed by the idle machines. In the reported runs, a significant proportion of work—for example, 84% when using 6 processors and about 92% when using 12 processors—was done using the idle processors. For the runs on dedicated processors, parallel efficiency is given by $\frac{T_s}{pT_p}$, where T_s is the sequential execution time, T_p is the parallel execution time and p is the number of processors. For the runs on non-dedicated processors, p is replaced by the effective number of pro-

Table 39.1 Execution times of (a) maximum independent set problem and (b) plasma simulation on dedicated processors, on transient processors without adaptive replication and on transient processors with adaptive replication. For the runs on transient processors with adaptive replication, the number of migrations during the lifetime of the parallel computation (#moves) is listed in parentheses. All times shown are wall-clock times in seconds.

(a) Maximum Independent Set

Single Proc	Degree of	Dedicated Processors	Transient Processors	Transient Processors with Adaptive BSP		
Mean Time	Para-llelism	Mean Time	Mean Time	Mean (#Trials)	Min (#Moves)	Max (#Moves)
10300	3	3650	12400	4350 (12)	3950 (2)	4790 (8)
	6	1840	12900	2340 (11)	1990 (1)	2900 (10)
	12	980	26650	1620 (9)	1000 (0)	2700 (9)

(b) Plasma Simulation

Single Proc	Degree of	Dedicated Processors	Transient Processors	Transient Processors with Adaptive BSP		
Mean Time	Para-llelism	Mean Time	Mean Time	Mean (#Trials)	Min (#Moves)	Max (#Moves)
not	4	840	3400	3500 (4)	3340 (3)	3774 (6)
possi-	8	750	20300	3100 (4)	2583 (5)	3503 (11)
ble	12	620	26500	2700 (3)	2150 (5)	3080 (20)

cessors, $p_{eff} = 1 + (p - 1)\frac{t_a}{t_a+t_n}$, since each processor is available only for a fraction of time equal to $\frac{t_a}{t_a+t_n}$. For the values of t_a and t_n used for these runs, $p_{eff} = \frac{2p+1}{3}$. For the dedicated runs, parallel efficiency ranges from nearly 100% (for 3 processors) to 88% (for 12 processors). For the non-dedicated runs using adaptive replication, these values range from nearly 100% (for 3 processors) to 76% at 12 processors. The corresponding values for non-dedicated runs without adaptive replication are 36% and 5%. Thus the adaptive runs are nearly as efficient as the dedicated runs and much more efficient than the transient processor runs.

Table 39.1(b) shows the results of applying the adaptive replication scheme to a plasma simulation with $N = 3,500,000$ particles. As mentioned in Section 39.6.2, the computation state data that needs to be replicated include the positions and velocities of particles in the local partition and the forces at the grid points in the local partition. The replicated data include 4 floating point numbers for each particle. As a result, for runs with 4 processors, the size of data replicated for particles is about 14 MB. On a 10 Mbps network, replicating the computation state of 3 processors takes up to 40 seconds while the computation step, t_s is half as long. In addition, the network is shared with other users, so heavy network traffic may increase the time needed for replication. Figure 39.3(b) shows a plot of the execution times on transient processors with and without adaptive replication for degrees of parallelism of 4, 8 and 12. These measurements were obtained using $t_a = 30$ minutes and $t_n = 20$ minutes

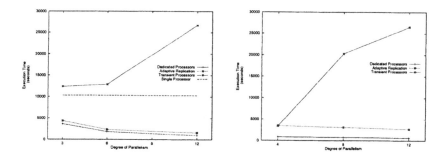

Figure 39.3   Plot showing execution times of (a) maximum independent set and (b) plasma simulation on dedicated processors, on transient processors using adaptive replication and on transient processors without adaptive replication. Execution time on a single processor is shown for comparison purposes.

respectively. Due to the overhead associated with communicating the computation state in each step, simulation runs on transient processors using the adaptive replication scheme take longer to execute compared to the runs on dedicated processors. The execution time on transient processors with adaptive replication is also longer than the sequential execution time, as estimated from the execution times on dedicated processors. However, even in this case, adaptive replication scheme is relevant for the following reasons. The execution time on transient processors with adaptive replication is still much smaller than the execution time without adaptive replication. Furthermore, the simulation used for our measurements was too large to fit on a single workstation and, hence, single processor runs were not even possible. For simulations that are too large to fit on a single workstation, parallel runs are mandatory. When dedicated machines are not available for parallel computation, adaptive replication scheme ensures that parallel runs using idle workstations complete in a reasonable time.

Any approach intended to tolerate transient failures will necessarily incur some overhead to checkpoint the computation state of the processes. Overhead incurred by replicating the computation state, as done in the adaptive replication scheme (which can be considered a form of diskless checkpointing), is no larger than the overhead caused by checkpointing to disk. The network used to obtain the measurements is a 10 Mbps Ethernet, which is quickly becoming obsolete. With a faster network such as an ATM network or a 100 Mbps Ethernet, the overhead due to data replication should be much smaller.

39.7   CONCLUSIONS

We have proposed an approach to adaptive parallelism based on the Bulk-Synchronous Parallel model to enable parallel computations to tolerate frequent transient failures and thereby adapt to the changing computing environment in a network of non-dedicated workstations. Our approach offers a general framework for adaptive parallelism and is algorithm independent. We described a protocol for replicating the computation state and replicating computations. We extended the Oxford BSP library

(Miller, 1993) with dynamic process management and virtual synchronization, and implemented the protocol on top of the extended library. The adaptive parallel extensions to the library include primitives for the specification of replication data, memory management for replication data and specification of computation state. We integrated the adaptive parallel extensions into the Oxford BSP library. The ARS library performs data replication and recovery of failed computations transparently to the user. We have demonstrated the adaptive capabilities of the library by applying it to two applications, a graph search problem and plasma simulation. Our results demonstrate that the ARS library can be used to execute parallel computations efficiently using idle machines in a network of workstations.

Acknowledgments

This work was partially supported by NSF Grant CCR-9527151. The content does not necessarily reflect the position or policy of the U.S. Government.

References

Barocas, V. and Tranquillo, R. (1994). Biphasic theory and in vitro assays of cell-fibril mechanical interactions in tissue-equivalent collagen gels. In Mow, V. *et al.*, editors, *Cell Mechanics and Cellular Engineering*, pages 185–209, New York. Springer–Verlag.

Bisseling, R. and McColl, W. (1993). Scientific computing on bulk synchronous parallel architectures. Technical Report 836, Department of Mathematics, University of Utrecht.

Bricker, A., Litzkow, M., and Livny, M. (1992). Condor technical summary. Technical Report CS-TR-92-1069, Computer Science Department, University of Wisconsin, Madison.

Cabillic, G. and Puaut, I. (1997). Stardust: An environment for parallel programming on networks of heterogeneous workstations. *J. Parallel and Distributed Computing*, 40(1):65–80.

Calinescu, R. (1995). Conservative discrete-event simulations on bulk synchronous parallel architectures. Technical Report TR-16-95, Oxford University Computing Laboratory.

Cap, C. and Strumpen, V. (1993). Efficient parallel computing in distributed workstation environments. *Parallel Computing*, 19(11):1221–1234.

Carriero, N., Freeman, E., Gelernter, D., and Kaminsky, D. (1995). Adaptive parallelism and Piranha. *Computer*, 28(1):40–49.

Deo, N. (1974). *Graph Theory with Applications to Engineering and Computer Science*. Prentice-Hall, Inc., Englewood Cliffs, N.J.

Gerbessiotis, A. and Siniolakis, C. (1996). Selection on the bulk synchronous parallel model with applications to priority queues. In *International Conference on Parallel and Distributed Processing Techniques and Applications (PDPTA'96)*, Sunnyvale, California, USA. On CD-ROM.

Goldberg, M. and Hollinger, D. (1997). Database learning: A method for empirical algorithm design. In *Workshop on Algorithm Engineering*, pages 231–239.

Gropp, W., Lusk, E., and Skjellum, A. (1994). *Using MPI:Portable Parallel Programming With the Message-Passing Interface*. MIT Press.

Kleinrock, L. and Korfhage, W. (1993). Collecting unused processing capacity: An analysis of transient distributed systems. *IEEE Transactions on Parallel and Distributed Systems*, 4(5):535–546.

Leon, J., Fischer, A., and Steenkiste, P. (1993). Fail-safe PVM: A portable package for distributed programming with transparent recovery. Technical report, School of Computer Science, Carnegie Mellon University, Pittsburgh, PA 15213.

Litzkow, M., Livny, M., and Mutka, M. (1988). Condor - a hunter of idle workstations. In *8th International Conference on Distributed Computing Systems*, pages 104–111, San Jose, California.

Miller, R. (1993). A library for bulk-synchronous parallel programming. In *British Computer Society Parallel Processing Special Group: General Purpose Parallel Computing*.

Miller, R. and Reed, J. (1993). The Oxford BSP library users' guide, version 1.0. Technical report, Oxford Parallel.

MPI Forum (1994). *MPI: A Message Passing Interface Standard*. Technical report, Message Passing Interface Forum.

Mutka, M. and Livny, M. (1987). Profiling workstations' available capacity for remote execution. In *12th Symposium on Computer Performance*, pages 2–9, Brussels, Belgium.

Nibhanupudi, M. (1998). *Adaptive Parallel Computations on Networks of Workstations*. PhD thesis, Computer Science Department, Rensselaer Polytechnic Institute.

Nibhanupudi, M., Norton, C., and Szymanski, B. (1995). Plasma simulation on networks of workstations using the bulk-synchronous parallel model. In *International Conference on Parallel and Distributed Processing Techniques and Applications (PDPTA'95)*, pages 13–20, Athens, Georgia.

Nibhanupudi, M. and Szymanski, B. (1996). Adaptive parallelism in the bulk-synchronous parallel model. In *2nd International Euro-Par Conference*, pages 311–318.

Nibhanupudi, M. and Szymanski, B. (1998). Runtime support for virtual BSP computer. To appear in the Workshop on Runtime Systems for Parallel Programming, RTSPP'98.

Norton, C., Szymanski, B., and Decyk, V. (1995). Object oriented parallel computation for plasma PIC simulation. *Communications of the ACM*, 38(10):88–100.

OpenGroup (1997). *Network Computer Profile*. Technical report, The Open Group.

Skillicorn, D., Hill, J., and McColl, W. (1997). Questions and answers about BSP. *Scientific Programming*, 6(3):249–274.

Snir, M., Otto, S., Huss-Lederman, S., and Walker, D. (1996). *MPI: The Complete Reference*. Scientific and Engg Computation Series. MIT Press.

Stellner, G. (1996). CoCheck: Checkpointing and process migration for MPI. In *International Parallel Processing Symposium*, pages 625–632.

Sunderam, V. (1990). PVM: A framework for parallel distributed computing. *Concurrency: Practice and Experience*, 2(4):315–339.

Valiant, L. (1990). A bridging model for parallel computation. *Communications of the ACM*, 33(8):103–111.

40 A NEW FAMILY OF RING-BISECTION NETWORKS AND ROUTING ALGORITHMS

Yuzhong Sun, Paul Y.S. Cheung and Xiaola Lin

Department of Electric and Electronic Engineering,
University of Hong Kong, Hong Kong

cheung@eee.hku.hk

Abstract: In this paper, we introduce a family of scalable interconnection network topologies, named *Ring-Bisection Networks* (RBNs), which are recursively constructed by adding ring edges to a cube. RBNs possess many desirable topological properties in building scalable parallel machines, such as fixed degree, small diameter, large bisection width, and symmetry. A general deadlock-free routing algorithm for RBNs is also presented and analyzed.

Keywords: scalable computer systems, ring-bisection networks, recursive cube of rings.

40.1 INTRODUCTION

A computer system is *scalable* if it can scale up its resources to accommodate ever-increasing performance and functionality demand (Hwang and Xu, 1998). In a distributed-memory parallel computer system, the design of the interconnection network is critical to the performance and scalability of the system. It is desirable for an interconnection network to have a fixed degree, small diameter, wide bisection width, and symmetry. In most existing interconnection networks, these requirements are often in conflict with one another. For example, although an $N \times N$ mesh and torus have fixed degree, their diameters are $2N$ and N respectively (hence relatively large). The node degree of an n-cube (hypercube) increases logarithmically with the size of the network though the diameter of a hypercube is small. Recently, some new network topologies have been proposed. The product of two classical topologies is a prospective method for constructing new interconnection networks (Kemal and Fernandez, 1995). The cross product of interconnection networks outperforms traditional topologies such as the mesh and hypercube in diameter, degree, and matching size (Day and

Abdel-Elah, 1997). Linear recursive networks are networks that are produced by a linear recurrence (Hsu *et al.*, 1997). However, these networks do not have the fixed degree property which is highly desirable in building scalable systems. Cube-of-rings networks may have a fixed degree and relatively small diameter but, as will be shown later, the network size that may be chosen is very limited (Cortina and Xu, 1998).

In this paper, we propose a new family of interconnection networks, initially named *Ring Bisection Network* (RBN). (Topologically, this type of network may be more appropriately described as *recursive cube of rings*). An RBN is constructed by recursive expansions on a given generation seed (GS). A GS for an RBN consists of a number of rings interconnected in a cube-like fashion. It can be created according to certain criteria. RBNs possess many desirable topological properties for building scalable parallel systems, such as fixed degree, small diameter, large bisection width and symmetry. Rings and hypercubes are special forms of the RBNs.

The paper is organized as follows. We first describe the proposed RBN topology and its construction method in Section 40.2. In Section 40.3, we examine the topological properties of RBNs. A deadlock-free message routing algorithm for RBNs is presented and analyzed in Section 40.4, followed by a conclusion in Section 40.5.

40.2   RING-BISECTION NETWORKS (RBN)

A general RBN consists of a number of rings interconnected by some links, called *cube links*. The nodes within a ring are connected by links called *ring links*. An RBN is denoted by RBN(k, r, j), where k is the dimension of the cube, r is the number of nodes on a ring, and j is the number of the expansions from the generation seed. The address of a node in an RBN is specified as $[a_{m-1}a_{m-2} \ldots a_0, b]$, where $m = k + j$; a_i is a binary bit, $0 \leq i \leq m - 1$; and $0 \leq b < r$. A node with address $[u_{m-1}a_{m-2} \ldots a_0, b]$ has k cube neighbors with addresses $[a_{m-1}a_{m-2} \ldots \bar{a}_{(b \times j + x) mod(k+j)} \ldots a_0, b]$, and two ring neighbors with addresses $[a_{m-1}a_{m-2} \ldots a_0, (b + 1) \bmod r]$ and $[a_{m-1}a_{m-2} \ldots a_0, (b - 1) \bmod r]$.

Given two parameters, an integer k and the number of nodes in the network N, the generation seed GS(k, r) for RBN(k, r, j) is created as follows:

1. The GS(k, r) consists of 2^k r-rings;

2. A node with address $[a_{m-1}a_{m-2} \ldots a_0, b]$ has k cube neighbors with addresses $[a_{m-1} \ldots \bar{a}_k, \ldots a_0, b], \ldots, [a_{m-1} \ldots a_k, \ldots \bar{a}_0, b]$. It has two ring nodes with addresses $[a_{m-1} \ldots a_0, (b + 1) \bmod r]$ and $[a_{m-1} \ldots a_0, (b - 1) \bmod r]$.

In terms of the given k and N, the parameters r and j can be determined (Sun *et al.*, 1998). Then j expansions are conducted to obtain the desired network RBN(k, r, j). Figure 40.1 depicts a generation seed GS(2,2). Figure 40.2 shows an RBN(2,2,1) obtained by one expansion from generation seed GS(2,2), and Figure 40.3 shows an RBN(2,2,2) obtained from one more expansion from RBN(2,2,1). The algorithm for constructing an RBN is described in Figure 40.4. It is important to note that, for the general cases, the actual number of nodes N in a complete RBN network may be different from the desired network size N'. For example, to build a network with 20,000 nodes ($N' = 20,000$), the size of the closest complete RBN network is 20,480, as shown in Table 40.1 (see Section 40.3.1).

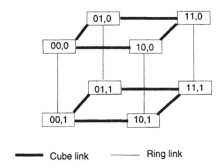

Cube link ——— Ring link

Figure 40.1 Generation seed GS(2,2) for RBN.

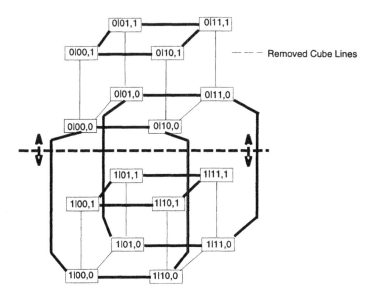

--- Removed Cube Lines

Figure 40.2 RBN(2,2,1) after one expansion from GS(2,2).

In the construction algorithm given in Figure 40.4, for a node v, the address of the node, $d_v = [a_m a_{m-1} \ldots a_0, b] = [A, b]$ (a_i is a binary bit, $0 \le i \le m$), then $o^1(d_v)$ = $[1 a_m a_{m-1} \ldots a_0, b]$ and $o^0(d_v) = [0 a_m a_{m-1} \ldots a_0, b]$ are the addresses of the next expansion of node v. In terms of the address, the expansion is done by concatenating a 0-bit or a 1-bit before the address. First, the GS is taken as the left graph G_L and its nodes are duplicated as the right graph G_R. The node set of a new graph G is the union of the node sets of the two graphs G_L and G_R. However, in the new graph G, the cube links at each node are rearranged. For example, a node v with address $[a_{k-1} \ldots a_0, b]$ in the new graph has cube neighbors connected to the node v by a

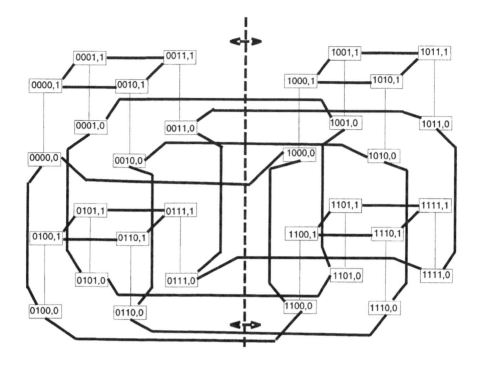

Figure 40.3   The topology of RBN(2,2,2).

cube link. These neighbors have the addresses $[a_{k-1} \cdots \overline{a}_{(i \times b + m) \, mod \, (i+k)} \cdots a_0, b]$, where $1 \leq m \leq k$ and i is the current index of expansion number.

40.3   TOPOLOGICAL PROPERTIES OF RBN

RBNs possess many desirable topological properties in building scalable parallel systems, such as fixed degree, better size match, logarithmic diameter, symmetry, and planarity. The detailed proofs of all of the properties can be found in (Sun *et al.*, 1998).

40.3.1   *General Topological Properties*

Property 1: In an RBN(k, r, j), we have the following properties:

P1.1: The number of nodes, N, is $r \times 2^{k+j}$.

P1.2: All of the nodes in the RBN(k, r, j) have the same degree, and the degree d of the network is $k + 2$ for $r > 2$, and $k + r - 1$ for $1 \leq r \leq 2$.

P1.3: The number of edges of the network, E, is given as follows:

Input: design parameters, k and N', N' is the desired number of nodes of the network to be constructed;

Output: an network RBN(k, r, j), the number of nodes of RBN(k, r, j) is N.

Procedure:

1. Select parameters, k, r, j, such that N is closest to N' and construct GS(k, r).

2. /* initialize G_L and G_R */
   Set $G_L(V, E) = $ GS(k, r) and $G_R(V, E) = $ GS(k, r).

3. For $i = 1$ to j, do the following:

   ■ /* expand the network by concatenating a 0-bit to the most significant bit of the node addresses of G_L */
   For any node v in G_L, expands its address $d_v = o^0(d_v)$;

   ■ /* expand the network by concatenating a 1-bit to the most significant bit of the node addresses of G_R */
   For any node v in G_R, expands its address $d_v = o^1(d_v)$;

   ■ /* generate the next higher-order RBN, G, by the union of G_L and G_R */
   Set $G = G_L \cup G_R$;

   ■ /* complete the generation of the next higher order RBN by the appropriate link insertion and deletion */
   Insert and delete the appropriate links in G, such that each node of G_L or G_R with address $[a_m a_{m-1} \ldots a_0, b]$ has its cube neighbors, $m = k + j - 1$, $[a_m \ldots \bar{a}_{(i \times b+1) mod(i+k)} \cdots a_0, b], \ldots, [a_m \ldots \bar{a}_{(i \times b+x) mod(i+k)} \ldots a_0, b], \ldots, [a_m \ldots \bar{a}_{(i \times b+k) mod(i+k)} \cdots a_0, b]$, where $1 \leq x \leq k$. The ring neighboring relation of each node remains unchanged.

   ■ Set $G_L = G$ and $G_R = G$.

4. Set RBN$(k, r, j) = G$.

Figure 40.4   Algorithm to construct an RBN.

$$E = \begin{cases} r \times 2^{k+j} \times 1 + k/2 & r > 2 \\ r \times 2^{k+j} \times 1/2 + k/2 & r = 2 \\ r \times 2^{k+j} \times k/2 & r = 1 \end{cases}$$

P1.4: The bisection width of the network, B, is $Num(k, r, j) \times 2^{k+j-1}$ or $Num(k, r, j)/r \times N$, where N is the number of nodes of the network, and $Num(k, r, j)$ is defined as follows. For any integers x and y, $0 \leq x \leq r - 1$ and $1 \leq y \leq k$, $Num(k, r, j)$ denotes the number of the x values satisfying $(j \times x + y) \bmod (j + k) = 1$.

Proof: The topological properties P1.1, P1.2 and P1.3 follow directly from the GS architecture and the recursive construction algorithm of RBNs. Therefore it is only necessary to give the proof of P1.4. Given a node with address $[a_{m-1} \ldots a_0, b]$, $m = k + j$, all its k cube neighbor addresses are different only in one bit position, $\bar{a}_{(j \times b + i) mod(j+k)}$, $1 \leq i \leq k$. For a given RBN(k, r, j) network, the inverse bit position of a node is determined only by the value b, $b \in \{0, 1, \ldots r - 1\}$. According to P1.1, for a value of b, there exist 2^{k+j} nodes with the same inverse bitwise position. An RBN(k, r, j) network can be divided into two RBN(k, r, j − 1) networks. The link contributing to bisection bandwidth of the RBN(k, r, j) network should have one endpoint of the link on one RBN(k, r, j − 1) network and the other endpoint on the other RBN(k, r, j − 1) network. According to the construction algorithm, the two nodes of such a link must have the same value b in their addresses. All 2^{k+j} nodes with such value b in their addresses can be divided into two groups, one in each of the lower order networks (RBN(k, r, j−1)). Therefore, the value b determines the number of links crossing the two RBN(k, r, j − 1) networks. According to the construction algorithm of RBN, the two nodes of the one link that contributes to the bisection bandwidth differ in the highest bit position of the first parts in their addresses. That is, the possible value of b should satisfy $(j \times b + y) \bmod (j + k) = 1$.   ■

The proofs of the following corollary and property can be found in (Sun et al., 1998) and are not included here for the interest of brevity.

Corollary: For an RBN(k, r, j) network, the bisection bandwidth B is:

$$2^{k+j-1} \leq B \leq 2^{k+j-1} \times \lfloor j \times (r - 1)/(j + k) \rfloor.$$

Property 2: The following topologies are special forms of RBNs:

P2.1: Ring networks are RBN(0, r, j) for any r and j.

P2.2: An n-dimensional hypercube is RBN(2, 1, j).

The actual size N of a network with a given topology is generally different from an arbitrary size N' of the desired network. The ratio N/N' is called the *match ratio*. In Table 40.1, for $N' = 20,000$, we compare the match ratio and node degree of RBNs with some popular topologies such as hypercube, cube-connected cycles (CCC), and cube of rings (COR). Clearly, RBNs have a better match ratio, because its match ratio is much closer to 1 than that of any other topology.

40.3.2   Planar Property

RBNs also possess a planar property such that an RBN(k, r, j) can be taken as the combination of two different types of planes, *cube-plane* (CP) and *ring-plane* (RP) to be defined below. This property can be used to develop efficient routing algorithms.

Definition 1: A *cube plane* CP is a subgraph of an RBN(k, r, j) such that the CP is connected and all links of the CP are cube links.

Definition 2: A *ring plane* RP is a subgraph of an RBN(k, r, j) such that the RP is connected and all links of the RP are ring links.

Table 40.1   Comparison of networks for size matching for N'=20,000.

Topology	Number of Nodes (N)	Match Ratio (N/N')	Node Degree
14-cube	16,384	0.82	14
15-cube	32,768	1.64	15
CCC	10,240	0.51	3
CCC	22,528	1.13	3
COR	12,288	0.61	6
COR	98,304	4.92	7
RBN	18,422	0.92	4
RBN	20,480	1.02	4

Property 3: For an RBN(k, r, j), let $G = G(V, E)$ denote the corresponding undirected graph of the network. The graph G consists of $C_k^2 \times 2^{k+j-2}$ CPs with 4 nodes. All CPs are connected by RPs.

Proof: According to the neighboring definition of an RBN(k, r, j), the ring neighbor relations of each node with the other nodes remain in the whole expansion construction of the RBN because each node with the address $[a_{k+j-1} \ldots a_0, b]$ should have two ring neighbors with the addresses $[a_{k+j-1}, \ldots a_0, (b + 1) \bmod r]$ and $[a_{k+j-1}, \ldots a_0, (b - 1) \bmod r]$ which are independent of the parameter j. Therefore, each node must be located in the same ring plane. In other words, in the whole construction process, all the RPs in the RBN remain unchanged. The ring neighborhood relations between nodes are reserved in the expansions.

On the one hand, the cube relations of one node with other nodes may often change after each expansion according to the construction algorithm. Now, considering the intermediate derived network RBN(k, r, q), where $1 \leq q \leq j$, let p_0 with the address $[a_{q+k-1} \ldots a_0, b]$ be an arbitrary node in the network RBN(k, r, q). We try to find a ring in a CP starting with node p_0. According to the construction algorithm, the k cube neighbors of p_0 should be $[a_{q+k-1} \cdots \overline{a}_{(q \times b) \bmod (q+k)} \cdots a_0, \ b], \ \ldots, \ [a_{q+k-1} \cdots \overline{a}_{(q \times b+i) \bmod (q+k)} \cdots a_0, b]$, $\ldots, [a_{q+k-1} \cdots \overline{a}_{(q \times b+k-1) \bmod (q+k)} \cdots a_0, b]$. We repeat the same procedure for each next node until no new node can be found. Then, we find a cycle from the node p_0 to the same node p_0. The cycle forms a CP with length 4. On the other hand, any two neighboring nodes should have only one bit different in the first parts of their addresses. Each node should simultaneously be located on the C_k^2 CPs. The number of CPs is $C_k^2 \times r \times 2^{k+j-2}$. Thus, we conclude that all CPs are connected by RPs. ■

40.3.3   RBN as a Cayley Graph

A network is symmetric if the network topology is the same looking from any node in the network. A symmetric interconnection network may simplify the design of the routers and interfaces, and thus reduce the cost of the networks. Cayley graphs have been proven to be symmetric graphs (Akers and Krishnamurthy, 1989). We show that RBNs are Cayley graphs, and therefore they are symmetric.

Property 4: An RBN(k, r, j) is a Cayley graph and is symmetric.

Proof: First, we construct a finite group G and define an associative operator * on it, $[A, b]*[X, y] = [A \times C^{k+j} + X, b+y]$. Let Q_2^n denote the set of all Boolean n-tuples under bitwise addition modulo 2, and Q_r denote the integer set $\{0, 1, \ldots, r-1\}$ under addition modulo r. The two sets Q_2^n and Q_r are proven to be Cayley graphs (Cortina and Xu, 1998). Then, we construct a finite group through ordered pairwise $Q_2^n \times Q_r$, where $n = k + j + 1$. For the associate operator *, bitwise addition modulo 2 is used for the first entry, and addition modulo r is used for the second entry, where C is an Boolean matrix:

$$\begin{bmatrix} 0 & 0 & 0 & \ldots & 0 & 1 \\ 1 & 0 & 0 & \ldots & 0 & 0 \\ 0 & 1 & 0 & \ldots & 0 & 0 \\ \ldots & & & & & \\ 0 & 0 & 0 & \ldots & 1 & 0 \end{bmatrix}$$

Second, we construct a Cayley set H using the same method in (Cortina and Xu, 1998), which consists of nodes $[0 \ldots 0, 1]$, $[0 \ldots 0, r-1]$, $[0 \ldots 1, 0]$, $[0 \ldots 010, 1]$, $\ldots, [\underbrace{0 \ldots 0}_{k-1} 10 \ldots, 0]$. H is a subset of G, which does not include the identity element $[0 \ldots 0, 0]$. Note that the inverse element of each element of H also belongs to H. Therefore, H is a Cayley set.

Finally, we can construct the RBN(k, r, j) in the group G based on H and the associative operator. For any node $[A, b]$ of the RBN(k, r, j) and each element from H, we derive one edge to each neighbor of $[A, b]$ in the RBN(k, r, j) as follows:

- $([A, b], [0 \ldots 0, 1]*[A, b]) = ([A, b], [A, b+1])$

- $([A, b], [0 \ldots r-1, 1]*[A, b]) = ([A, b], [A, b-1])$

- $([A, b], [0 \ldots 1, 1]*[A, b]) = ([A, b], [a_{k+j-1} \ldots \bar{a}_{b \times j+1}, \ldots a_0, b])$

- $([A, b], [0 \ldots 10, 1]*[A, b]) = ([A, b], [a_{k+j-1} \ldots \bar{a}_{b \times j+2}, \ldots a_0, b])$

- $([A, b], [0 \ldots 1 \ldots \underbrace{0 \ldots 0}_{k}, 1]*[A, b]) = ([A, b], [a_{k+j-1} \ldots \bar{a}_{b \times j+k}, \ldots a_0, b])$

Therefore, RBN(k, r, j) is a Cayley graph. Then, according to (Akers and Krishnamurthy, 1989), we conclude that the RBN(k, r, j) is symmetric. ■

Property 5: The diameter D of an RBN(k, r, j) is less than or equal to:

$$k + j + \lceil r/2 \rceil \times \lceil (k+j)/k \rceil.$$

Proof: Consider the longest path from node $[0 \ldots 0, 0]$ (node S) to node $[1 \ldots 1, r-1]$ (node D). According to the definition of RBN(k, r, j), S has k cube neighbors. We may choose the cube neighbor that is located in the same cube plane as S and is closest to D as the first next node. According to the construction algorithm in Figure 40.4, any two nodes differ in at most k bits. The address of a node, $[A, b]$, consists of two parts A and b, as defined in Section 40.2. A is the first part of the address and b is the second part. Without using the second part of the address of a node, we can construct a virtual cube path from S to D, $p_0 p_1 \ldots p_{m-1}$, where $m = k + j$. The second part of a node's address is used to connect determined virtual cube nodes on the path $p_0 p_1 \ldots p_{m-1}$. We assume that all links in RBNs are bidirectional. The maximum number of required RPs is $\lceil r/2 \rceil$. In each RP, the maximum number of additional links is the longest path from S to D is the sum of the two parts, $k + j + \lceil \frac{r}{2} \rceil \times \lceil \frac{k+j}{k} \rceil$.   ∎

40.4 MESSAGE ROUTING IN RBN

The basic idea of the routing algorithm is similar to the well-known e-cube routing algorithm for binary cubes (Dally and Seize, 1987). It is proved that each node of an RBN(k, r, j) is on certain CPs. In the case of $k = 2$, each node is located only on a single CP. In one CP, the addresses $[A, b]$'s of nodes show a regular change of bit patterns such that the same k bit positions in A of each node differ and the other bits remain the same. An appropriate neighboring cube plane can be chosen in the way similar to the e-cube routing. The exit node of such a chosen CP is the node closest to the destination. We call such a routing algorithm in Figure 40.5 a *hop-planar* routing algorithm. The hop-planar routing algorithm can always find a shortest path from any source node to any destination node in RBNs.

To prevent the occurrences of deadlock, two virtual channels are set up on a physical link (Dally and Seize, 1987). One of the two virtual channels, denoted by $vc1$, is used when a message traverses a link in ascending order from one node to another. The other virtual channel, denoted by $vc2$, is used when the message traverses a link in descending order, regardless of cube links or ring links. Let (a, c) or $([A_a, b_a]$, $[A_c, b_c])$ denote a link. For a cube link, A_a differs from A_c while $b_a = b_c$. For a ring link, $|b_a - b_c| = 1$ while $A_a = A_c$. It can be shown that the message routing algorithm shown in Figure 40.5 is deadlock free. The proof can be found in (Sun *et al.*, 1998).

Theorem: For an RBN(k, r, j), the message routing algorithm in Figure 40.5 can always find a shortest path from the source node to the destination node, and it is deadlock-free.

40.5 CONCLUSIONS

We have proposed a class of new topologies for interconnection network, named ring-bisection network (or recursive cube of rings), which are recursively constructed by adding ring edges to a cube. RBNs possess many desirable topological properties in building scalable parallel machines, such as fixed degree, small diameter, planarity, wide bisection width and symmetry. A deadlock-free routing algorithm for the proposed RBN networks is also developed.

Acknowledgments

Partial funding provided by Hong Kong RGC f163-C and HKU CRCG grant 337/062/0012.

Input: current node c with address $[A_c, b_c]$, source node S with address $[A_s, b_s]$ and destination node D with address $[A_d, b_d]$.

Output: an outgoing virtual channel (c, v) used to forward the message towards D (v is a neighboring node of c).

Procedure:

1. If $A_c = A_d$

   - If $b_c = b_d$, then c is D, forward message to the local processor, and stop.

   - If $b_c \neq b_d$, then select neighboring node v such that $|b_v - b_d| \leq |b_c - b_d|$.

   - If $b_c < b_v \leq b_d$ (or $b_c < b_v \geq b_d$), then select $vc1$ (or $vc2$) as the outgoing channel ($vc1$ and $vc2$ are virtual channels from c to v), and stop.

2. If $A_c \neq A_d$, then use e-cube routing with A_c as the local node address and A_d as the destination address to find A_v. If $[A_v, b_v]$ is a neighboring node of c, then select any outgoing virtual channel from c to v, and stop.

3. Select neighboring node v such that $|b_v - b_d| \leq |b_c - b_d|$.

4. If $b_c < b_v \leq b_d$ (or $b_c < b_v \geq b_d$), then select $vc1$ (or $vc2$) as the outgoing channel ($vc1$ and $vc2$ are virtual channels from c to v), and stop.

Figure 40.5   Routing algorithm for RBN.

References

Akers, S. and Krishnamurthy, B. (1989). A group-network theoretical model for symmetric interconnection networks. *IEEE Trans. on Computers*, 38(4):555–566.

Cortina, T. and Xu, X. (1998). The cube-of-rings interconnection networks. *International Journal of Foundations of Computer Science*. To appear.

Dally, W. and Seize, C. (1987). Deadlock-free message routing in multiprocessor interconnection networks. *IEEE Trans. on Computers*, 36:547–553.

Day, K. and Abdel-Elah, A. (1997). The cross product of interconnection networks. *IEEE Trans. on Parallel and Distributed Systems*, 8(2):109–118.

Hsu, W., Chung, M., and Das, A. (1997). Linear recursive networks and their applications in distributed systems. *IEEE Trans. on Parallel and Distributed Systems*, 8(7):673–680.

Hwang, K. and Xu, Z. (1998). *Scalable Parallel Computing: Technology, Architecture, Programming*. McGraw Hill.

Kemal, E. and Fernandez, A. (1995). Products of networks with logarithmic diameter and fixed degree. *IEEE Trans. on Parallel and Distributed Systems*, 6(9):963–975.

Sun, Y., Cheung, P., and Lin, X. (1998). A new family of ring-bisection networks and routing algorithms. Technical Report HKU-EEE-HPCR-100, Department of Electric and Electronic Engineering, University of Hong Kong.

Index

CPSIA information can be obtained at www.ICGtesting.com
Printed in the USA
LVOW011748210413

330169LV00003B/61/P

9 781461 375678